Memoirs of the American Mathematical Society

Number 217

W. B. Johnson, B. Maurey,
G. Schechtman, and L. Tzafriri

Symmetric structures in Banach spaces

Published by the

AMERICAN MATHEMATICAL SOCIETY

Providence, Rhode Island, USA

May 1979 · Volume 19 · Number 217 (end of volume)

ABSTRACT

In this paper detailed investigations of spaces with a symmetric basis of finite length and rearrangement invariant function spaces are presented. The emphasis is on questions arising naturally from the theory of L_p-spaces.

AMS(MOS) subject classifications (1970). 46B99, 46E30, 46E99, 47B55.

Supported in part by the National Science Foundation, NSF-MCS 76-06565, and the U. S.-Israeli Binational Science Foundation.

Library of Congress Cataloging in Publication Data

Main entry under title:

Symmetric structures in Banach spaces.

(Memoirs of the American Mathematical Society ; no. 217 ISSN 0065-9290)
Bibliography: p.
1. Banach spaces. 2. Symmetric spaces.
I. Johnson, William B., 1944- II. Series:
American Mathematical Society. Memoirs ; no. 217.
QA3.A57 no. 217 [QA322.2] 510'.8s [515'.73] 79-10225
ISBN 0-8218-2217-9

Reprinted with corrections, 1980

TABLE OF CONTENTS

ACKNOWLEDGEMENTS

Most of the research for this monograph was conducted while the authors were members of the Institute for Advanced Studies of The Hebrew University of Jerusalem. The authors want to thank the administrative staff of the Institute for providing excellent working conditions. They also thank the clerical staffs of the Institute, The Hebrew University, and The Ohio State University for the typing of the manuscript. Special thanks are due Pam Walsh for her fine typing of the final version of the monograph.

PREFACE

The most important and most interesting class of Banach spaces is undoubtedly that of L_p-spaces; $1 \leq p \leq \infty$. These spaces, which appear in many problems arising naturally in analysis, have been studied thoroughly for a long time. As a result of this intensive research, the discrete case of ℓ_p-spaces is relatively well understood by now while that of L_p-spaces over an atomless measure space, though in a quite satisfactory stage, still raises many questions of importance.

The observation that the unit vector basis of ℓ_p: $1 \leq p < \infty$ is equivalent to any of its permutations (i.e., ℓ_p, as a normed space of sequences, is invariant under permutations) led to the study of a more general class, namely that of spaces with a symmetric basis. Besides the ℓ_p-spaces; $1 \leq p < \infty$, the simplest examples of spaces with a symmetric basis are the Orlicz sequence spaces ℓ_F, where $F(t)$ is a non-decreasing convex function on $[0,\infty)$ with $F(0) = 0$ and $\lim_{t \to \infty} F(t) = \infty$ (the function $F(t)$ plays about the same role for ℓ_p as t^p does in the case of ℓ_p). In recent years, a great deal of research went into the study of spaces with a symmetric basis and, in particular, into that of Orlicz sequence spaces and, by now, there is quite clear picture of their structure. A survey of this topic can be found e.g. in [49].

The present paper has only relatively little bearing on the subject of infinite dimensional spaces with a symmetric basis but, instead, it attempts to use the methods and the knowledge accumulated in this field as a starting point in the investigation of some other classes of spaces with a symmetric structure. One of our main targets is the class of L_p-spaces over an atomless measure space or, more generally, that of rearrangement invariant (r.i.)

Received by the editor March 27, 1978.

function spaces which we consider to be the "continuous" correspondent of spaces with a symmetric basis. We are mostly interested in r.i. function spaces on $[0,1]$ and $[0,\infty)$ since, essentially speaking, these are the only distinct countably generated atomless measure spaces.

It is quite surprising that, while a great deal of effort was spent to study the discrete case of spaces with a symmetric basis, only very little has been achieved with the continuous case or r.i. function spaces. These spaces were mostly considered in connection with questions arising in interpolation theory but, from the point of view of geometry of Banach spaces, few results were obtained except in some particular cases as e.g. that of Orlicz function spaces, where probabilistic methods have been used successfully.

There are many instances in which there is no major difference between the study of $L_p(0,1)$ or $L_p(0,\infty)$-spaces and that of more general r.i. function spaces on $[0,1]$ or on $[0,\infty)$. This is the case when results are first proved for L_p-spaces and then extended, by interpolation methods, to suitable r.i. function spaces or when suitable conditions imposed on a r.i. function space X allow to adapt for X a proof which was originally designed to work only in L_p-spaces. This kind of generalization is, of course, only of limited interest though, in many cases, it explains better the true reasons why a certain proof works under given assumptions.

As the investigation of r.i. function spaces undertaken by us progressed, many new problems, which had not been anticipated in advance, arose and, in some instances, were even solved. As a result of this study we believe that we now have a better understanding of the general class of r.i. function spaces and, in particular, of that of L_p and Orlicz function spaces.

Our results indicate that there are important differences, even beyond what could have been expected in advance, between the behavior of spaces with a symmetric basis, on one hand, and that of r.i. function spaces, on the other hand. While some similarities between infinite dimensional spaces with a symmetric basis and r.i. function spaces on $[0,\infty)$ do indeed exist it appears that, in many directions, the class of r.i. function spaces on $[0,1]$ is more related to that of finite dimensional spaces with a symmetric basis. The main difference between the finite symmetric bases and the infinite ones seems to lie in the fact that in the latter class it is much easier to construct isomorphisms (for instance, such isomorphisms may be constructed by having an infinite symmetric basis be equivalent to a block basis of another such basis while, in the finite case, such a possibility simply does not exist for reasons of dimension).

In general, the methods used in the paper to study r.i. function spaces are finite dimensional. In most cases, the study of a given r.i. function space X, say, on $[0,1]$ is reduced to that of spaces with a symmetric basis of finite length by considering, for each n, the linear span of the characteristic functions $\{\chi_{[(i-1)2^{-n}, i2^{-n})}\}_{i=1}^{2^n}$. The difficulties with such an approach appear, of course, in the moment we want to put together the information gathered for various values of n. Obviously, this method requires an advanced study of spaces with a symmetric basis of finite length and, indeed, a non-trivial part of the paper is devoted to this topic which is also of independent interest. Since the investigation of finite dimensional spaces with a symmetric basis is quantitative in its nature it poses questions which rarely would have been considered in the infinite dimensional context. As we shall see in the sequel, such questions have sometimes completely different and unexpected solutions.

In addition to the finite dimensional methods described above, there are several instances where interpolation or factorization arguments of different kinds are extensively used. Also, some purely probabilistic notions (as e.g. Poisson processes) play an important role in many proofs.

Evidently, the present paper does not exhaust completely the subjects considered in it. In as much as possible, we have tried to give complete solutions to all problems encountered throughout our work but, in some cases, we were unable to come up with a totally satisfactory answer. The most important open problems connected with this study are scattered through the paper and, especially, in the Introduction.

The attempt to state and prove the results in the most general setting leads often to complicated proofs. Since some readers would be only interested in some simpler cases which have easier proofs the paper is organized in such a manner that, sometimes, the same result appears in two or even three different settings, each having another degree of generality. For instance, those results, which are new even for L_p-spaces, are first presented in a separate section with proofs specific to the L_p-case and only later on in the paper they appear in their utmost possible degree of generality, this time, obviously, with more difficult proofs.

0 INTRODUCTION

The main object of this section is to present the contents of the paper and its organization. In addition, we describe here many notions, methods and results which are used often throughout the work. The notions and results of local interest are discussed in the text only in the place where their use is required. The reader interested in a detailed account of the material used in the background of the paper is referred to [49] and [50]. The notations used here are quite standard and in most cases follow the books mentioned above.

Since the well-known inequality of Khintchine

$$A_p\Big(\sum_{i=1}^{\infty} |a_i|^2\Big)^{1/2} \le \Big(\int_0^1 \Big|\sum_{i=1}^{\infty} a_i r_i(u)\Big|^p du\Big)^{1/p} \le B_p\Big(\sum_{i=1}^{\infty} |a_i|^2\Big)^{1/2} ,$$

where $1 \le p < \infty$ and $\{r_i\}_{i=1}^{\infty}$ denotes the sequence of Rademacher functions, is used frequently throughout the paper, we keep the symbols A_p and B_p exclusively for the constants appearing in Khintchine's inequality.

The next section of the article; i.e., Section 1, is devoted to a quite thorough analysis of those subspaces of $L_p(0,1)$; $p > 2$ which have some kind of symmetric structure. The starting point is the well-known and trivial fact that each of the spaces ℓ_p ; $1 \le p < \infty$ has, up to equivalence, a unique symmetric basis. D. R. Lewis raised the question whether this fact has a local analogue; i.e., whether each of the spaces ℓ_p^n ; $1 \le p < \infty$; $n = 1,2,\ldots,$ has a unique symmetric basis. In order to state this problem as well as its solution in a precise manner, we first recall that the symmetry constant $K(\{x_i\})$ of a finite or infinite basis $\{x_i\}$ is the smallest number K such that the inequality

$$K^{-1}\Big\|\sum_i a_i x_i\Big\| \le \Big\|\sum_i \epsilon_i a_{\pi(i)} x_i\Big\| \le K\Big\|\sum_i a_i x_i\Big\|$$

holds for all permutations π of the integers, all choices of signs $\{\epsilon_i = \pm 1\}$, and all scalars $\{a_i\}$. If there is no such K we put $K(\{x_i\}) = \infty$. A basis whose symmetry constant is $\leq K$ is called, in short, K-symmetric. In those cases when we deal with very complicated expressions we use the equivalence sign $\sim$ instead of a double inequality as above (for example, instead of the inequality above we simply write

$$\|\sum_i \epsilon_i a_{\pi(i)} x_i\| \overset{K}{\sim} \|\sum_i a_i x_i\|) \quad .$$

Now, for $1 \leq p < \infty$, set $\mathfrak{F}_p = \{\ell_p^n\}_{n=1}^\infty$. We prove in the sequel that the members of $\mathfrak{F}_p$ have, up to equivalence, a unique symmetric basis. More precisely, for each $K \geq 1$, there exists a number $C_p(K)$ so that if $X \in \mathfrak{F}_p$; i.e., $X = \ell_p^n$ for some n, and $\{x_i\}_{i=1}^n$ is a K-symmetric normalized basis of X then $\{x_i\}_{i=1}^n$ is $C_p(K)$-equivalent to the unit vector basis $\{e_i\}_{i=1}^n$ of ℓ_p^n (two bases $\{x_i\}$ and $\{y_i\}$ are said to be C-equivalent if

$$C^{-1}\|\sum_i a_i x_i\| \leq \|\sum_i a_i y_i\| \leq C\|\sum_i a_i x_i\| \quad ,$$

for every choice of scalars $\{a_i\}$). In fact, the uniqueness theorem for ℓ_p^n holds in a uniform manner for $1 \leq p < \infty$ in the sense that the function $C(K) = \sup\{C_p(K)\ ;\ 1 \leq p < \infty\}$ is finite for each $1 \leq K < \infty$ (cf. 1.5 below). However, the proof of the uniformity requires additional work.

The problem of uniqueness for symmetric bases of finite length can be considered in a more general setting than that of ℓ_p^n-spaces.

Let F *be a family of finite-dimensional Banach spaces, each of which has a normalized* 1*-symmetric basis. We say that each member of* F *has, up to equivalence, a unique symmetric basis provided that there exists a function* $\psi: [1,\infty) \to [1,\infty)$ *such that if* $X \in F$ *and* $\{x_i\}$ *is a normalized K-symmetric basis of* X *then* $\{x_i\}$ *is* $\psi(K)$ *-equivalent to the normalized* 1*-symmetric basis of* X .

Observe that there is no abuse in not specifying the 1-symmetric basis of X since, by a theorem of Pelczynski and Rolewicz [69], any two normalized 1-symmetric bases of an arbitrary Banach space are 1-equivalent. In 1.4 below we prove that if $\mathcal{S}_p$ denotes the class of all subspaces of $L_p(0,1)$; $p > 2$, which have a 1-symmetric basis, then each member of $\mathcal{S}_p$ has a unique symmetric basis. It is important to point out that there is no loss of generality in treating only 1-symmetric basic sequences in $L_p(0,1)$ since, by 1.2, every finite K-symmetric basic sequence in $L_p(0,1)$ is K-equivalent to a 1-symmetric basic sequence in that space. We have not been able to decide whether there is uniformity in 1.4. This question is very important since uniformity in 1.4 as $p \to \infty$ would give a positive answer to the following problem.

Problem 0.1. Does each member of the family of all finite-dimensional Banach space with a 1-*symmetric basis have a unique symmetric basis, up tp equivalence?*

We will come back to this problem later when we discuss Section 3, where a partial solution to 0.1 is given.

The crucial point in the proofs of the above results on the uniqueness of symmetric bases lies in the fact that symmetric basic sequences in $L_p(0,1)$; $p > 2$ can be completely classified. As is well known, Kadec and Pelczynski [38] proved that, for every weakly null normalized infinite sequence $\{x_n\}_{n=1}^{\infty}$ in $L_p(0,1)$; $p > 2$ and for every $\varepsilon > 0$, there exists a subsequence $\{x_{n_i}\}_{i=1}^{\infty}$ of $\{x_n\}_{n=1}^{\infty}$ which is either $1+\varepsilon$-equivalent to the unit vector basis of ℓ_p or is equivalent to the unit vector basis of ℓ_2. The result of Kadec and Pelczynski implies that every normalized *infinite* symmetric basis sequence $\{x_n\}_{n=1}^{\infty}$ in $L_p(0,1)$; $p > 2$ is either $K(\{x_n\}_{n=1}^{\infty})$-equivalent to the unit vector basis of ℓ_p or is equivalent to the unit vector basis of ℓ_2 but, in the latter case the constant of equivalence to the basis of ℓ_2 can be large even when $K(\{x_n\}_{n=1}^{\infty})$ is close to one. Thus, in order to study finite symmetric basic sequences in $L_p(0,1)$; $p > 2$ from a quantitative point of view, a more precise classification is needed. To understand what could be expected, we first discuss some special situations. Clearly, the simplest and most natural examples of 1-symmetric normalized sequences in $L_p(0,1)$ are the unit vector basis $\{e_n\}_{n=1}^\infty$ of ℓ_p and $\{\delta_n\}_{n=1}^\infty$ of ℓ_2. Other examples can be obtained by considering sequences of the form

$$(e_n + w\delta_n)/(1 + w^p)^{1/p} \ ; \ n = 1,2,\ldots$$

in the subspace $(\ell_p \oplus \ell_2)_p$ of $L_p(0,1)$. According to the terminology of H. P. Rosenthal [70], such a basic sequence will be called a *symmetric* X_p-*basis* (whether it is finite or infinite). The main result of Section 1, namely the classification Theorem 1.1, asserts that there exists a function $D(p,K)$; $p \geq 2$, $K \geq 1$, such that every K-symmetric normalized basic sequence $\{x_i\}$ (of finite or infinite length) in $L_p(0,1)$ is $D(p,K)$-equivalent to some symmetric X_p-basis of the same length. We point out in passing that, by using a result of H. P. Rosenthal [70], it follows that any subspace of $L_p(0,1)$; $p > 2$ spanned by a K-symmetric basic sequence is $D(p,K)$-isomorphic to a $D(p,K)$-complemented subspace of $L_p(0,1)$.

The classification Theorem 1.1 has also several consequences which do not involve however the notion of symmetric basis. In order to present the first such application, we have to discuss the concept of rearrangement invariant (r.i.) spaces of functions. Since we deal mostly with separable r.i. function spaces there is no loss of generality in considering only the canonical cases of r.i. function spaces on $[0,1]$ and $[0,\infty)$. A r.i. function space X on the interval $I = [0,1]$ is a Banach space of equivalent classes of measurable functions on I such that

(i) X is a Banach lattice with respect to the pointwise order.

(ii) For every automorphism τ of I (i.e., an invertible transformation τ from I onto itself so that, for any measurable subset E of I, $\mu(\tau^{-1}E) = \mu(E)$) and every $f \in X$, also $f(\tau) \in X$ and $\|f(\tau)\| = \|f\|$.

(iii) $L_\infty(I) \subset X \subset L_1(I)$, with norm one embeddings.

(iv) $L_\infty(I)$ is dense in X .

In the presence of conditions (i), (ii), and (iv), condition (iii) simply means that the function $f \equiv 1$ has norm one in X. Condition (iii) is meant to exclude trivial cases as, e.g., when X consists only of constant functions. Often r.i. function spaces are defined without imposing condition (iv), in which case one has to distinguish between minimal r.i. function spaces; i.e., those satisfying also (iv), and maximal r.i. function spaces; i.e., those which do not have any r.i. enlargement (the interest in minimal and maximal r.i. function spaces lies in the fact that, under these two assumptions, the usual interpolation theorems work). Since the separability of X implies that (iv) is necessarily satisfied, we have included this condition as an axiom.

A r.i. function space on the interval $I = [0,\infty)$ is a Banach space of measurable functions on I which satisfies (i) and (ii) (with the convention that we consider automorphisms τ from I *into* I) and, instead of (iii) and (iv), the following two conditions

(iii') $L_1(I) \cap L_\infty(I) \subset X \subset L_1(I) + L_\infty(I)$, with norm one embeddings.

(iv') The simple functions with bounded support are dense in X.

The space $L_1(I) + L_\infty(I)$ is defined as the set of all measurable functions f on $[0,\infty)$ endowed with the norm

$$\|f\| = \inf\{\|g\|_{L_1(I)} + \|h\|_{L_\infty(I)} \; ; \; f = g + h\} .$$

$L_1(I) + L_\infty(I)$ is clearly a r.i. function space on $[0,\infty)$. Again, it is easily verified that condition (iii') is equivalent to the fact that $\chi_{[0,1]}$ has norm one in X. In general, the dual X^* of a r.i. function space X need not be a space of functions but, for many purposes, one can use instead the subspace X' of X^* consisting of the integrals; i.e., of those $x^* \in X^*$ for which there exists a measurable function φ so that

$$x^*(f) = \int_I f\varphi d\mu ,$$

for every $f \in X$. The conditions imposed on X guarantee that X' is a norm determining subspace of X^* which satisfies (i), (ii) and (iii), respectively, (iii'). Therefore, the closure of the simple functions with bounded support in X' is a genuine r.i. function space, according to the above definition.

A sizeable part of this paper is devoted to r.i. function spaces. In Section 1, however, we prove only some results on L_p-spaces which follow directly from the classification Theorem 1.1. For instance, in 1.6, we show that if a r.i. function space X on an interval I, where I is either $[0,1]$ or $[0,\infty)$, embeds isomorphically into $L_p(0,1)$, for some $p > 2$, then, up to an equivalent norm, X is equal to $L_p(I)$, $L_2(I)$ or to $L_p(I) \cap L_2(I)$, the last possibility being of interest only in the case when $I = [0,\infty)$. From 1.6 it is deduced immediately that, for any $1 \leq p < \infty$, $L_p(0,1)$ has a unique representation as a r.i. function space on $[0,1]$. This statement should be understood in the sense that any r.i. function space X on $[0,1]$ which is isomorphic to $L_p(0,1)$, is already equal to $L_p(0,1)$, up to an equivalent renorming. This result is known to be true for $p = 1$ and $p = \infty$ (cf. [44], [48]). Surprisingly, the situation for $L_p(0,\infty)$, is quite different: this space has two distinct representations as a r.i. function space on $[0,\infty)$ if $1 < p \neq 2 < \infty$. For $p > 2$, the second representation is $L_p(0,\infty) \cap L_2(0,\infty)$ while, for $1 < p < 2$, it is $L_p(0,\infty) + L_2(0,\infty)$. The fact that $L_p(0,\infty) \cap L_2(0,\infty)$; $p > 2$ is isomorphic to $L_p(0,\infty)$ follows from Pelczynski's decomposition method and the result of H. P. Rosenthal [70] that X_p-spaces are isomorphic to complemented subspaces of L_p. Alternatively, this fact is proved in a more general setting in Section 8.

As another application of 1.1 we can prove (cf. 1.8 below) that a normalized unconditional basic sequence $\{x_n\}_{n=1}^{\infty}$ in $L_p(0,1)$; $p > 2$ is either equivalent to the unit vector basis of ℓ_p, or ℓ_2 is block finitely representable in it; i.e., for every $\varepsilon > 0$ and every m, there are m disjoint blocks $\{u_j\}_{j=1}^{m}$ of $\{x_n\}_{n=1}^{\infty}$ (or, equivalently, m consecutive blocks of some permutation of $\{x_n\}_{n=1}^{\infty}$) so that $\{u_j\}_{j=1}^{m}$ is $1+\varepsilon$-equivalent to the unit vector basis of ℓ_2. This result, which was discovered earlier by L. Dor and T. Starbird [20], actually can be proved in a more general context by a similar method and, therefore, its proof is postponed to Section 2. The result 1.8 is used later to deduce the following two assertions: (1) For every $1 < p < \infty$ there exists a number N_p (where N_p is about equal to $2^{-1/4}\max\{p^{1/4},(p/(p-1))^{1/4}\}$) so that every monotone unconditional basic sequence in $L_p(0,1)$; $1 < p < \infty$, whose span is N_p-complemented in $L_p(0,1)$, is already equivalent to the unit vector basis of ℓ_p (1.10 below). (2) A normalized unconditional basis of ℓ_p ; $1 < p < \infty$ is equivalent to the unit vector basis unless it has a large ($\geq 2^{1/2}\max\{p^{1/2},(p/(p-1))^{1/2}\}$) unconditional constant (1.12 below).

Section 1 is concluded with a quantitive variant of the aforementioned result of Kadec and Pelczynski asserting that any weakly null sequence $\{x_n\}_{n=1}^\infty$ in $L_p(0,1)$; $p > 2$ contains a subsequence which is K-equivalent to some X_p-basis, where K is bounded from above by a number *independent* of $\{x_n\}_{n=1}^\infty$. The proof of this theorem (1.14 below) is based on a result of Burkholder [13] on martingale differences.

The next section is devoted to some generalizations of results proved in the first section for L_p-spaces; $p > 2$, to Banach lattices of type 2. Part of these results have applications also in other sections of this paper. We begin by presenting some background material. A Banach space Y is said to be of type p, for some $1 \leq p \leq 2$, (cotype q, for some $q \geq 2$) if there exists a constant $M < \infty$, called the type p-constant (cotype q constant) so that

$$\int_0^1 \|\sum_{i=1}^n r_i(u)y_i\|^p du \leq M^p \sum_{i=1}^n \|y_i\|^p$$

$$\int_0^1 \|\sum_{i=1}^n r_i(u)y_i\|^q du \geq M^{-q} \sum_{i=1}^n \|y_i\|^q ,$$

for every sequence $\{y_i\}_{i=1}^n$ in Y. By a well-known result of J.P. Kahane [39], the norm in $L_p(0,1)$ or $L_q(0,1)$ of the expression $\|\sum_{i=1}^n r_i(u)y_i\|$ can be replaced by the norm in any other $L_r(0,1)$-space without affecting the definitions of the type and cotype. The type and the cotype are notions which can be defined for every Banach space. In the particular case of Banach lattices it is more convenient to work with the related notions of p-convexity and q-concavity. Generally speaking, a lattice of functions X is p-convex for some $p > 1$ if the expression $|||f||| = \| |f|^{1/p}\|^p$ is equivalent to a norm on the set of all f for which $|f|^{1/p} \in X$, and q-concave, for some $q < \infty$, if its dual is a $q/(q-1)$-convex Banach lattice. In order to define these two notions in an intrinsic way in the setting of general Banach lattices (i.e., lattices which are not given as spaces of functions) we need to be able to give a precise meaning to expressions of the form $(\sum_{i=1}^{n} |x_i|^r)^{1/r}$, where $r \geq 1$ and $\{x_i\}_{i=1}^n$ are arbitrary elements in a Banach lattice. This can be achieved trivially when X possesses a functional representation though, a-priori, it could be that the definition of $(\sum_{i=1}^{n} |x_i|^r)^{1/r}$ would depend on the particular functional representation which was used. In order to preserve the full generality and also to avoid technical difficulties arising from this approach, we prefer to use a method described by Krivine [41]. His approach shows that, for every Banach lattice X, every real function $f(t_1,\ldots,t_n)$ on $\mathbb{R}^n$ which is homogeneous of degree one and for every sequence $\{x_i\}_{i=1}^n$ in X, there is a procedure to define the vector $f(x_1,\ldots,x_n) \in X$ in a unique way so that if two such homogeneous functions $f(t_1,\ldots,t_n)$ and $g(t_1,\ldots,t_n)$ satisfy the inequality $f(t_1,\ldots,t_n) \leq g(t_1,\ldots,t_n)$ for every $(t_1,\ldots,t_n) \in \mathbb{R}^n$ then also $f(x_1,\ldots,x_n) \leq g(x_1,\ldots,x_n)$ in X. For example, from the numerical inequality $(\sum_{i=1}^{n} |t_i|^q)^{1/q} \leq (\sum_{i=1}^{n} |t_i|^p)^{1/p}$ for $p < q$ we deduce that $(\sum_{i=1}^{n} |x_i|^q)^{1/q} \leq (\sum_{i=1}^{n} |x_i|^p)^{1/p}$ holds for any sequence $\{x_i\}_{i=1}^n$ of elements of an arbitrary Banach lattice. Once these facts have been clarified, we can define the above notions as follows.

A Banach lattice X is said to be p-convex; $1 < p < \infty$ with p-convexity constant $\leq M$ (q-concave with q-concavity constant $\leq M$) provided that

$$\|(\sum_{i=1}^{n} |x_i|^p)^{1/p}\| \le M(\sum_{i=1}^{n} \|x_i\|^p)^{1/p}$$

$$(\ \|(\sum_{i=1}^{n} |x_i|^q)^{1/q}\| \ge M^{-1}(\sum_{i=1}^{n} \|x_i\|^q)^{1/q})\ ,$$

for every choice of vectors $\{x_i\}_{i=1}^{n}$ in X .

While in general the p-convexity and q-concavity are properties of the particular lattice structure under consideration, the cases $1 < p \le 2$ and $q \ge 2$ behave particularly well. For instance, it was shown by Dubinski, Pelczynski and Rosenthal [21] that if X is a 2-concave Banach lattice with constant $\le M$ then every lattice Y , which is isomorphic to a subspace of X , is also 2-concave. In 10.3 it is proved that a similar result holds if X is q-concave with $q > 2$. Also, it can be shown that if X is a lattice which is p-convex and r-concave for some $1 < p \le 2$ and $r < \infty$, then any other lattice Y , which is isomorphic to a subspace of X , is p-convex, too.

The following observation illustrates well the importance of q-concavity for $q < \infty$. Let $\{x_i\}_{i=1}^{n}$ be a K-unconditional basic sequence in a q-concave Banach lattice with q-concavity constant $\le M$, for some $1 < q < \infty$ and some $M < \infty$. By the q-concavity of X and Khintchine's inequality, we get that

$$\|\sum_{i=1}^{n} x_i\| \ge K^{-1}\int_0^1 \|\sum_{i=1}^{n} r_i(u)x_i\|du \ge K^{-1}\|\int_0^1 |\sum_{i=1}^{n} r_i(u)x_i|du\| \ge$$

$$\ge A_1K^{-1}\|(\sum_{i=1}^{n} |x_i|^2)^{1/2}\|$$

and also

$$\|\sum_{i=1}^{n} x_i\| \le K(\int_0^1 \|\sum_{i=1}^{n} r_i(u)x_i\|^q du)^{1/q} \le KM\|(\int_0^1 |\sum_{i=1}^{n} r_i(u)x_i|^q du)^{1/q}\|$$

$$\le B_qKM\|(\sum_{i=1}^{n} |x_i|^2)^{1/2}\|\ .$$

In conclusion, we have

$$(*)\qquad K^{-1}2^{-1/2}\|(\sum_{i=1}^{n}|x_i|^2)^{1/2}\| \le \|\sum_{i=1}^{n} x_i\| \le B_q KM\|(\sum_{i=1}^{n}|x_i|^2)^{1/2}\| \ .$$

Since the proof of the left-hand side inequality does not use the assumption of q-concavity imposed on X it is easily deduced by a duality argument that if $\{x_i\}_{i=1}^{\infty}$ is an unconditional basis of a *complemented* subspace of X then there exists a constant $D < \infty$ so that

$$D^{-1}\|(\sum_{i=1}^{n}|a_i x_i|^2)^{1/2}\| \le \|\sum_{i=1}^{n} a_i x_i\| \le D\|(\sum_{i=1}^{n}|a_i x_i|^2)^{1/2}\| \ ,$$

for every choice of scalars $\{a_i\}_{i=1}^{n}$. This result is used; e.g., in Section 9 for the Haar basis of a r.i. function space on $[0,1]$.

As is by now well known (cf. [50] and [59]), any σ-order complete, σ-order continuous, purely non-atomic separable Banach lattice is isometric and order isomorphic to a lattice of functions on $[0,1]$ which contains $L_\infty(0,1)$ and is contained in $L_1(0,1)$ with norm one injections. If X is such a lattice which is p-convex and q-concave for some $1 \le p < q < \infty$ with p-convexity and q-concavity constants one, then a lot more can be said: if L is a representation of X as a function lattice on $[0,1]$ as above, then $L_q(0,1) \subset L \subset$ $\subset L_p(0,1)$ with the injections having norm one. Indeed, assume that X (and thus also L) is p-convex with constant 1 . For any non-negative t we get

$$\int_0^1 (tx^p(u) + 1)^{1/p}du \le \|(tx^p + 1)^{1/p}\|_L \le (t\|x\|_L^p + 1)^{1/p}$$

with equality for $t = 0$. So for x being a simple function

$$\frac{1}{p}\int_0^1 x^p(u)du = \frac{\partial(\int_0^1 (tx^p(u)+1)^{1/p}du)}{\partial t}(0) \le \frac{\partial(t\|x\|_L^p + 1)^{1/p}}{\partial t}(0) = \frac{1}{p}\|x\|_L^p$$

so that

$$\|x\|_p \le \|x\|_L$$

and the proof can be completed by a simple density and duality argument.

We discuss now briefly another notion which is related to that of p-convexity and q-concavity but does not coincide with it. A Banach lattice X is said to satisfy an upper, respectively, lower r-estimate; $1 < r < \infty$ (for mutually disjoint elements) if there exists a constant $M < \infty$ so that

$$\|\sum_{i=1}^{n} x_i\| \leq M(\sum_{i=1}^{n} \|x_i\|^r)^{1/r} ,$$

respectively,

$$\|\sum_{i=1}^{n} x_i\| \geq M^{-1}(\sum_{i=1}^{n} \|x_i\|^r)^{1/r} ,$$

for every choice of pairwise disjoint vectors $\{x_i\}_{i=1}^n$ in X. It is clear that a p-convex lattice satisfies an upper p-estimate but the converse need not be true. However, the existence of an upper r-estimate for a lattice X implies that X is p-convex for every $1 < p < r$ (cf. [54]).

We return now to the presentation of the contents of Section 2. As in the first section, the main tool of Section 2 is a classification theorem for symmetric basic sequences in Banach lattices of type 2 or, equivalently, in Banach lattices which are 2-convex and q-concave for some $q > 2$. This result (Theorem 2.1) states that every finite symmetric basic sequence $\{x_i\}_{i=1}^n$ in a lattice as above behaves like a sort of symmetric X_p-space in which, however, the expression $(\sum_{i=1}^{n} |a_i|^p)^{1/p}$ is replaced by an average in the sense of $\ell_q^{n!}$ of terms having the form $\|\max_{1\leq i\leq n} |a_{\pi(i)}x_i|\|$, where π is an arbitrary permutation of the integers $1,2,\ldots,n$. For technical reasons, q is taken to be an even integer. It is not clear whether one can simplify 2.1 as to assert, for example, that the norm of an arbitrary linear combination $\sum_{i=1}^{n} a_i x_i$ is given by the formula

$$\max\{\|\sum_{i=1}^{n} a_i z_i\| , \frac{\|\sum_{i=1}^{n} x_i\|}{\sqrt{n}} (\sum_{i=1}^{n} |a_i|^2)^{1/2}\} ,$$

where $\{z_i\}_{i=1}^n$ is a suitable sequence of disjoint elements in the lattice. The proof of 2.1 is based on an extrapolation argument (cf. [19] and [41]), which is presented in Lemma 2.2.

Despite its complicated statement, Theorem 2.1 is useful for many applications. For instance, by using it, we are able to prove a generalization of 1.8 to lattices of type 2. It is shown in 2.3 that a Banach lattice X, which is isomorphic to a subspace of a lattice Y of type 2 with an upper r-estimate for some $r > 2$, either itself satisfies such an upper r-estimate or ℓ_2 is lattice finitely representable in X; i.e., for every $\varepsilon > 0$ and every integer m, there exists a sequence $\{x_i\}_{i=1}^m$ of pairwise disjoint vectors in X which is $1+\varepsilon$-equivalent to the unit vector basis of ℓ_2^m. Theorem 2.3 can be considerably strengthened when X is a r.i. function space on $[0,1]$. In this case, if Y is of type 2 and r-convex for $r > 2$ then either X itself is r-convex too or, up to an equivalent norm, X equals $L_2(0,1)$ (2.6 below). This result will be used in Section 5 in order to prove that some r.i. function spaces on $[0,1]$ have a unique r.i. structure on this interval.

In Section 3 we study some questions related to unconditional and, especially, symmetric bases of finite length which, in some sense, are on one side of Hilbert space. The precise meaning of this restriction varies from case to case and, in each instance, is stated precisely.

Section 3 begins with a simplified proof of a surprising result due to C. Schütt [75] (Theorem 3.1 below). This theorem asserts that, in a finite dimensional Banach space X with a monotone normalized unconditional basis $\{x_i\}_{i=1}^n$ which is dominated by the unit vector basis of ℓ_2^n (i.e., $\|\sum_{i=1}^n a_i x_i\| \leq M(\sum_{i=1}^n |a_i|^2)^{1/2}$ for some constant $M < \infty$ and for every choice of scalars $\{a_i\}_{i=1}^n$), the expression $\lambda(n) = \|\sum_{i=1}^n x_i\|$ is, up to a constant depending only on M, an invariant of the space X. More precisely,

$$\lambda(n)/M\sqrt{2} \le \gamma_\infty(X) \le d(X,\ell_\infty^n) \le \lambda(n) \quad ,$$

where $\gamma_\infty(X)$ is the factorization constant of X through $L_\infty(0,1)$, namely,

$$\gamma_\infty(X) = \inf\{\|T\|\cdot\|S\|;\ T\colon X \to L_\infty(0,1);\ S\colon L_\infty(0,1) \to X;\ ST = \text{identity on } X\}.$$

It is easily checked that, for a finite dimensional or separable Banach space X , $\gamma_\infty(X)$ is the infimum of the norms of all possible projections from $L_\infty(0,1)$ onto some isometric image of X in $L_\infty(0,1)$. In view of this fact, $\gamma_\infty(X)$ is called the projection constant of X in $L_\infty(0,1)$. By replacing $L_\infty(0,1)$ with $L_p(0,1)$; $1 \le p < \infty$, one can similarly define the factorization constant of X through $L_p(0,1)$; namely $\gamma_p(X)$.

Besides giving here a relatively simple proof of Schütt's result we also show (3.1 (ii) below) that if $\{x_i^*\}_{i=1}^n$ is the sequence of biorthogonal functionals associated to the monotone normalized unconditional basis $\{x_i\}_{i=1}^n$ of X (which is assumed to be dominated by the unit vector basis of ℓ_2^n), then the expression $\mu(n) = \|\sum_{i=1}^n x_i^*\|$ is an invariant of X , or X^* , too. The exact statement asserts that

$$\mu(n) \le \pi_1(X) \le M\sqrt{2}\mu(n) \quad ,$$

where $\pi_1(X)$ denotes the 1-absolutely summing norm of the identity in X (i.e., the smallest constant $C < \infty$ for which

$$\sum_{i=1}^m \|z_i\| \le C \sup_{\epsilon_i = \pm 1} \|\sum_{i=1}^m \epsilon_i z_i\| \quad ,$$

for every choice of $\{z_i\}_{i=1}^m$ in X). The restriction imposed on $\{x_i\}_{i=1}^n$ to be dominated by the usual basis of ℓ_2^n is probably needed but so far the following question is open.

Problem 0.2. Does there exist a sequence of finite dimensional Banach spaces $\{X_n\}_{n=1}^\infty$ *with* $\dim X_n = n$ *so that each* X_n *has two normalized K-unconditional bases* $\{x_i^{(n)}\}_{i=1}^n$ *and* $\{y_i^{(n)}\}_{i=1}^n$ *(with* K *being indepen-*

dent of n*) and*

$$\lim_{n\to\infty} \|\sum_{i=1}^{n} x_i^{(n)}\| / \|\sum_{i=1}^{n} y_i^{(n)}\| = 0 \ ?$$

In connection with 0.2 we mention that it is relatively easy to construct a sequence of Banach spaces $\{X_n\}_{n=1}$ with $\dim X_n = n$ such that the X_n's have two asymptotically non-equivalent normalized K-unconditional bases (a general method for constructing such examples can be derived from 4.7 below).

Theorem 3.1 is used to deduce in 3.4 that if X is an n-dimensional Banach space with two normalized K-unconditional bases $\{x_i\}_{i=1}^n$ and $\{y_i\}_{i=1}^n$ then the ratio $\|\sum_{i=1}^{n} x_i\|/\|\sum_{i=1}^{n} y_i\|$ is bounded and bounded away from zero by constants depending only on K and on the type 2 constant of X. In other words, if X is of type 2 then the expression $\|\sum_{i=1}^{n} x_i\|$ is essentially the same for all normalized K-unconditional bases $\{x_i\}_{i=1}^n$ of X. A similar result is, of course, true for $\|\sum_{i=1}^{n} x_i^*\|$.

Garling and Gordon [30] showed that if an n-dimensional Banach space has a K-symmetric normalized basis then $\gamma_\infty(X)\pi_1(X) \le nK^2$. But, by the results mentioned above, in a Banach space X having a normalized monotone unconditional basis $\{x_i\}_{i=1}^n$, which is M-dominated by the unit vector basis of ℓ_2^n, the product $\|\sum_{i=1}^{n} x_i\| \cdot \|\sum_{i=1}^{n} x_i^*\|$, is, up to a constant depending only on M, equal to $\gamma_\infty(X)\pi_1(X)$. It follows that the ratio

$$\rho_n = \|\sum_{i=1}^{n} x_i\| \cdot \|\sum_{i=1}^{n} x_i^*\|/n$$

measures the degree of symmetricity of X in the sense that if X has a K symmetric basis then K should exceed, up to a constant, the number $\sqrt{\rho_n}$ (3.6 below).

The next question considered in this section is related to Problem 0.1. Although a complete solution is not given here, we present a result of a positive nature which extends some of the theorems proved in Section 1. Our Theorem 3.8 below asserts that for any fixed $q < 2$ and $M < \infty$, each member of the family $\mathcal{C}_{q,M}$, of all finite dimensional Banach spaces having a normalized 1-symmetric basis, which is q-concave with q-concavity constant $\leq M$, has, up to equivalence, a unique symmetric basis. The role of q-concavity for $q < 2$ in 3.8 is to eliminate the situation when the vectors of a second normalized symmetric basis $\{y_i\}_{i=1}^n$ of a space $X \in \mathcal{C}_{q,M}$ with a normalized q-concave 1-symmetric basis $\{x_i\}_{i=1}^n$, are "flat" when represented as linear combinations of the x_i's. This happens in ℓ_2^n, if, for example, we take the Walsh system over the unit vector basis of ℓ_2^n. Our proof for 3.8 actually shows that, under the $q < 2$ concavity assumption, a certain fixed percentage of the y_i's have "peaks" whose altitude is bounded from below by a fixed number.

It is interesting to compare the uniqueness theorem 3.8 with the possibility of constructing, for every $1 < q < 2$, an *infinite dimensional* Orlicz sequence space ℓ_{M_q} which has uncountably many mutually non-equivalent symmetric bases, all of which are q-concave with the same q-concavity constant (cf [46]).

The study of finite dimensional Banach spaces with an unconditional or symmetric basis is continued in Section 4, which begins with a localization of M. Zippin's [82] characterization of perfectly homogeneous bases. We recall that a normalized basis $\{x_n\}_{n=1}^\infty$ is said to be perfectly homogeneous if it is equivalent to each of its normalized block bases. Actually, the fact that perfectly homogeneous bases coincide with the unit vector bases of c_0 and ℓ_p ; $1 \leq p < \infty$ was proved in [82] under the weaker assumption that the basis is equivalent to each of its normalized block bases with constant coefficients. The perfect homogeneity of a basis $\{x_n\}_{n=1}^\infty$ yields that the basis is subsymmetric. This fact, which is useful in the infinite case, does not seem to be suitable for bases of finite length, and therefore, we prefer to define the perfect homogeneity of a finite basis in a manner which will ensure the symmetricity of the basis.

The degree of perfect homogeneity of a normalized basis $\{x_i\}_{i=1}^n$ of finite length is measured by a constant, denoted $PH(\{x_i\}_{i=1}^n)$. This constant is defined as the smallest number K so that every sequence of disjoint blocks having the form $\{\sum_{i\in\sigma_j} \pm x_i/\|\sum_{i\in\sigma_j} x_i\|\}_{j=1}^k$, with $\{\sigma_j\}_{j=1}^k$ being mutually disjoint subsets of $\{1,2,\ldots,\}$, is K-equivalent to $\{x_i\}_{i=1}^k$. Since the symmetry constant $K(\{x_i\}_{i=1}^n)$ is clearly $\le PH(\{x_i\}_{i=1}^n)$ we may as well assume that $\{x_i\}_{i=1}^n$ is a-priori 1-symmetric.

The local variant (4.2 below) of Zippin's result states that any normalized 1-symmetric basis $\{x_i\}_{i=1}^n$ is $160PH(\{x_i\}_{i=1}^n)^6$-equivalent to the unit vector basis of ℓ_p^n, where $p = \log n/\log \|\sum_{i=1}^n x_i\|$. The numbers 160 and 6 appearing above are not the best constants and, obviously, of no importance. What this result asserts is that the unit vector bases of ℓ_p^n; $1 \le p \le \infty$, $n = 1,2,\ldots$, are the only symmetric bases having a "small" PH-constant.

One of the main applications of perfectly homogeneous bases of infinite length is found in [51], where it is shown that ℓ_1, ℓ_2 and c_o are the only spaces with a unique unconditional basis. The proof of this assertion requires the use of the well-known decomposition method of Pelczynski. In 4.5 we present a finite dimensional decomposition method, the proof of which is a bit more difficult than that of the finite dimensional case. Roughly speaking, 4.5 says that, for any block basis with constant coefficients $\{y_i\}_{i=1}^k$ of a 1-symmetric normalized basis $\{x_i\}_{i=1}^n$, $[x_i]_{i=1}^n$ is M-isomorphic to $[y_i]_{i=1}^k \oplus [x_i]_{i=1}^{n-k}$, where M is a universal constant. This local version of the decomposition method is then used in 4.8 (and 4.9) to prove that Euclidean spaces are the only "smooth" (= far away from ℓ_1^n and ℓ_∞^n) finite dimensional spaces which have a unique unconditional basis. The precise meaning of the smoothness condition is given in 4.9.

It should be noted that there is no exact local analogue to the aforementioned result that ℓ_1, ℓ_2 and c_o are the only infinite dimensional spaces with a unique unconditional basis since 1.12 already implies that every normalized basis of ℓ_p^n ; $p > 2$ with unconditional constant less than, approximately, $\sqrt{p}$, is C-equivalent to the unit vector basis of ℓ_p^n , for some C independent of n However, there may be a joint characterization of ℓ_1^n, ℓ_2^n and ℓ_∞^n ; $n = 1,2,.$ as the only spaces of finite dimension satisfying some quantitative form of uniqueness of the unconditional basis.

Section 5 is devoted mainly to the generalization of some of the results from Section 3 to the "continuous" case; i.e., the case of r.i. function spaces on $[0,1]$ The section begins with an embedding result (5.1 below) asserting that if a q-concave; $q < 2$ r.i. function space X on $[0,1]$ is isomorphic to a subspace of a r.i. function space Y on $[0,1]$ or on $[0,\infty)$, which is of some cotype $s < \infty$, then either the norm in X dominates that in Y (i.e., $\|f\|_Y \leq V\|f\|_X$, for some $V < \infty$ and every $f \in X$) or Y contains a sequence of mutually disjoint elements equivalent to the Haar basis of X . From 5.1 it is easy to deduce that if X and Y are two isomorphic q-concave; $q < 2$ r.i. function spaces on $[0,1]$ then, up to an equivalent norm, X is equal to Y . This however does not suffice to ensure the uniqueness of the r.i. structure on $[0,1]$ of every q-concave; $q < 2$ r.i. function space X on $[0,1]$ since we cannot assume a-priori that all other possible r.i. representations of X are q-concave, too. This is the case, however, if X is B-convex (i.e., if X does not contain uniformly isomorphic copies of ℓ_1^n for all n) since, under this assumption, X^* is a proper r-convex r.i. function space on $[0,1]$, where $1/r + 1/q = 1$, and therefore, by 2.6, any other representation of X^* as a r.i. function space on $[0,1]$ is also r-convex or, equivalently, any representation of X as a r.i. function space on $[0,1]$ is q-concave.

In order to prove in general that every q-concave; $q < 2$ r.i. function space X on $[0,1]$ has unique r.i. structure on $[0,1]$ (5.5 below) some more work involving, among other methods, interpolation and change of density, is needed. Actually, by using these arguments, it is possible to prove the more general Theorem 5.6 which asserts that when X is a r.i. function space on [0,1] or on [0,∞) and the subspace of functions in X which are supported on [0,1] is different from $L_1(0,1)$, then every r.i. function space Y on [0,1], which embeds isomorphically into X as a complemented subspace is, up to equivalent renorming, equal to $X_{|[0,1]}$ or to $L_2(0,1)$.

The fact that 5.1 and 5.6 are proved under the same restrictions as is the finite version 3.8 (in contrast to the existence of counterexamples among q-concave; $q < 2$ infinite dimensional spaces with a symmetric basis [56]) suggests that many results on symmetric bases of finite length might have suitable analogues for r.i. function spaces on [0,1]. It is also important to point out that, contrary to the finite case (see Problem 0.1), in the case of r.i. spaces on [0,1] we have been able to construct an example (10.1 below) which shows that the q-concavity assumption with $q < 2$ is essential for 5.5. The construction of this example is based on a factorization method from [18] which has already been used by W.J. Davis [17] to embed isomorphically a uniformly convex space with an unconditional basis into a uniformly convex space with a symmetric basis as a complemented subspace. In the present case, we want to embed the universal space U of Pelczynski into a r.i. function space on [0,1] as a complemented subspace. We recall that U is a space with a normalized unconditional basis $\{u_n\}_{n=1}^{\infty}$ having the property that every normalized unconditional basis in an arbitrary Banach space is equivalent to a subsequence of $\{u_n\}_{n=1}^{\infty}$. A simple way of constructing this space (cf. [74]) is achieved by letting $\{u_n(t)\}_{n=1}^{\infty}$ be a sequence of norm one functions which is dense in the unit sphere of C(0,1) and by setting

$$\|\sum_{n=1}^{\infty} a_n u_n\| = \sup_{\varepsilon_n = \pm 1} \|\sum_{n=1}^{\infty} \varepsilon_n a_n u_n(t)\|_{C(0,1)} .$$

Applying the factorization method mentioned above for the basis $\{u_n\}_{n=1}^{\infty}$ of U and any two L_p-spaces, say $L_r(0,1)$ and $L_s(0,1)$ with $1 < r < s < \infty$, we construct a r.i. function space $S_{r,s}$ on [0,1] which contains a sequence of disjointly supported functions $\{f_n\}_{n=1}^{\infty}$ equivalent to $\{u_n\}_{n=1}^{\infty}$. The functions f_n ; $n = 1,2,\ldots$ can be constructed so that they are constant on their support. Since $S_{r,s}$ is an interpolation space between $L_r(0,1)$ and $L_s(0,1)$, the usual Haar system in it is an unconditional basis and, thus $S_{r,s}$ is isomorphic to a complemented subspace of U. From this fact and the decomposition method [63] it would follow that $S_{r,s}$ is actually isomorphic to U provided that we know that $[f_n]_{n=1}^{\infty}$ is complemented in $S_{r,s}$ (note that it could happen that the conditional expectation operator on the σ-field generated by the supports of the f_n's has range in $S_{r,s}^{**}$). However, we can overcome this difficulty by working with a simultaneously convexified and concavified version U_o of U which ensures that the above space $S_{r,s}$ will be reflexive.

Actually, by using different values of r and s in the above construction, we get that U_o has uncountably many mutually non-equivalent representations as a r.i. function space on $[0,1]$. A more careful construction of U_o will even ensure that there are 2-concave r.i. function space on $[0,1]$ which do not have a unique representation. If, instead of $L_r(0,1)$ and $L_s(0,1)$, we use $L_r(0,\infty)$ and $L_s(0,\infty)$ it follows that the above space U_o has also uncountably many mutually non-equivalent representations as a r.i. function space on $[0,\infty)$. This fact solves a problem raised by B. S. Mitjagin [60] who asked whether $L_p(0,1)$; $1 \le p \le \infty$ are the only spaces having a representation as a r.i. function space both on $[0,1]$ and $[0,\infty)$. In fact, the class of r.i. function spaces with dual representation on $[0,1]$ and $[0,\infty)$ is very large, as will be shown in Section 8.

The peculiar construction of the Example 10.1 makes it a-priori improbable that such a space would appear in analysis in a natural way. Therefore, it is quite surprising that such spaces can be constructed as to embed isomorphically in such classical spaces as the L_p-spaces. In 10.4 we show indeed that, for every $1 < p < 2$, there exists a subspace V_p of $L_p(0,1)$ having uncountably many mutually non-equivalent representations as a r.i. function space on $[0,1]$

which, in addition, has an unconditional basis $\{v_n\}_{n=1}^{\infty}$ with the property that every unconditional basic sequence in $L_p(0,1)$ is equivalent to a subsequence of $\{v_n\}_{n=1}^{\infty}$. In particular, V_p contains $L_p(0,1)$ as a complemented subspace since the Haar basis of $L_p(0,1)$ is unconditional for $1 < p < \infty$.

Theorem 5.5. is not the best result on the uniqueness of r.i. function spaces on [0,1] presented in the paper. This topic is also studied in Section 6 where an attempt is made not to restrict the discussion only to q-concave; $q < 2$ r.i. function spaces on [0,1]. The main result proved in this section, namely 6.1, classifies all the possible isomorphic embeddings of one r.i. function space X on [0,1] into another r.i. function space Y on [0,1] or $[0,\infty)$, which is of some cotype $s < \infty$. The basic difference between 6.1 and 5.1 is that the assumption of q-concavity for $q < 2$ imposed on X in 5.1 is replaced in 6.1 by the requirement that the Haar basis in X be unconditional. The unconditionality of the Haar basis in X does not follow from the q-concavity of X so, strictly speaking, 6.1 is not a generalization of 5.1. On the other hand, the assumption made in 6.1 is a very reasonable one and it holds, for example, if X is super-reflexive or, more precisely, if and only if the Boyd indices α_X and β_X of X satisfy $0 < \beta_X$ and $\alpha_X < 1$. These indices, which are defined in Section 8, can be characterized by the fact, due to Boyd [9], [10], that $[1/\alpha_X, 1/\beta_X]$ is the largest interval so that, whenever $1 \leq p < 1/\alpha_X, 1/\beta_X < q$ and T is a linear operator of weak type (p,p) and (q,q), then T is also bounded as an operator on X. This interval can be also characterized as the smallest interval containing all the numbers $1 \leq r \leq \infty$ such that, for every $\epsilon > 0$ and every integer n, there exists a sequence $\{f_i\}_{i=1}^n$ of pairwise disjoint functions with the same distribution in X which is $1+\epsilon$-equivalent to the unit vector basis of ℓ_r^n.

Theorem 6.1 shows that, for X and Y as above, if X is isomorphic to a subspace of Y then either the norm in X dominates that in Y or the Haar basis of X is equivalent to a sequence of mutually disjoint functions in Y , or X is $L_2(0,1)$, up to an equivalent norm. Recall that the first two conclusions of 6.1 appear also in 5.1 while the third is obviously meaningless when X is q-concave for $q < 2$. The proof of 6.1 is quite long and complicated but, in our opinion, some of the arguments used there may be of intrinsic interest (already in Section 9 we use some of the methods developed for 6.1). The main tool of the proof consists of, or, perhaps, begins with, the construction of a sequence of Y-valued measures associated to every isomorphism from X into Y . A careful analysis of the behavior of the limit of the above sequence of vector valued measures makes it possible to distinguish between the different possible conclusions of 6.1. The reader might find some interest also in the definition of new systems of functions, called gaussian Haar systems, which are, in fact, block bases of a certain form of the usual Haar basis equivalent to it. It is however quite difficult to go here into more details without lengthening undesirably the introduction.

Theorem 6.1 has many interesting applications most of which deal with Orlicz functions and are presented in Section 7. In Section 6, however, we deduce from 6.1 and its proof that a superreflexive r.i. function space X on [0,1] has a unique representation as a r.i. function space on [0,1] provided that the Haar basis in X is unconditional but not equivalent to any sequence of pairwise disjoint functions in X (6.13 below). This result shows again the distinct behavior of those r.i. function spaces on [0,1] in which the Haar basis is equivalent to a sequence of mutually disjoint functions, as happens in the spaces introduced in 10.1 and 10.4.

One class of r.i. function spaces, for which it is very convenient to apply the results proved in Section 6, is that of Orlicz function spaces. For simplicity, we restrict the discussion to Orlicz functions (i.e., non decreasing convex functions F on $[0,\infty)$ such that $F(0) = 0$ and $\lim_{t\to\infty} F(t) = \infty$) which satisfy the Δ_2-condition at 0 and at ∞ , in the case when the underlying measure space is $[0,\infty)$, or only at 0 when the underlying interval is [0,1] . We also assume that $F(1) = 1$ in order to ensure that the norm of $\chi_{[0,1]}$ in $L_F(0,1)$ or $L_F(0,\infty)$ is equal to one.

We study the class of Orlicz function spaces in Section 7. This class has already been the object of intensive research (cf. e.g., [12], [16], [47]) but, since almost always probabilistic methods were used, most of the results obtained apply only to 2-concave Orlicz function spaces (which are subspaces of $L_1(0,1)$). The aim of Section 7 is to give, in addition to some results of a general nature, a quite complete description of isomorphic embeddings and finite representability in a 2-convex Orlicz function space. The main reason why the results of Section 6 are particularly effective in the study of Orlicz function spaces lies in the fact proved in [47] that the Haar basis of no r.i. function space X on $[0,1]$, except perhaps $L_2(0,1)$, is equivalent to a sequence of mutually disjoint functions in an Orlicz function space on $[0,1]$ or on $[0,\infty)$. Some consequences of this observation and 6.1 are stated in 7.1. The most important one asserts that every reflexive Orlicz function space $L_F(0,1)$ has a unique representation as a r.i. function space on $[0,1]$; in particular, we get that if a reflexive $L_F(0,1)$ is isomorphic to a space $L_G(0,1)$ then already F is equivalent to G at ∞ ; i.e., the ratio $F(t)/G(t)$ stays bounded and bounded away from zero as $t \to \infty$. Actually, a stronger result is true: if a r.i. function space X on $[0,1]$ is isomorphic to a complemented subspace of a reflexive $L_F(0,\infty)$ then, up to an equivalent norm, either $X = L_2(0,1)$ or $X = L_F(0,1)$. The reflexivity assumption is needed above in order to facilitate the use of 6.1 for L_F and its dual; we do not know whether the uniqueness of the r.i. structure on $[0,1]$ remains valid for non-reflexive Orlicz function spaces on $[0,1]$.

The rest of Section 7 is devoted to a study of 2-convex Orlicz function spaces (i.e., when F is such that $F(\sqrt{t})$ is equivalent to a convex function). Theorems 7.5 and 7.6 describe completely the class of r.i. function spaces X on $[0,1]$, respectively, on $[0,\infty)$ which are finitely crudely representable in a 2-convex Orlicz function space on $[0,1]$ or $[0,\infty)$. (We recall that a space Y is said to be finitely crudely representable in a space Z if there exists a constant $D < \infty$ such that, for every finite dimensional subspace Y_o of Y, there is a subspace Z_o of Z so that $d(Y_o,Z_o) \leq D$). These r.i. function spaces X must be, according to 7.5 or 7.6, Orlicz function spaces L_G and G should belong, in each case, to a certain closed convex family of Orlicz functions generated in some way by F.

The proofs of 7.5 and 7.6 are not very complicated but somewhat of a technical nature. One fact (7.2 below) of a general character used in these proofs, which might be of interest in other instances, asserts that, in any r.i. function space on $[0,1]$ or $[0,\infty)$, the norm of the sum of a sequence $\{f_i\}_{i=1}^n$, of positive functions with given individual distributions, is minimal when the f_i's ; $1 \leq i \leq n$ are taken as to be disjointly supported. This fact is used in 7.3 in order to prove that if $\{x_i\}_{i=1}^n$ is a K-unconditional basic sequence of functions with given individual distributions in a r.i. function space on $[0,1]$ or $[0,\infty)$ which is of type 2 then the norm of any expression of the form $\|\sum_{i=1}^n a_i x_i\|$ is, up to a constant depending only on K, minimal when the x_i's ; $1 \leq i \leq n$ are disjointly supported.

Section 7 ends with some results on those r.i. function spaces Y on $[0,1]$ which embed isomorphically into a p-convex; $p > 2$ Orlicz function space L_F on $[0,1]$ or $[0,\infty)$. Of particular interest is the fact, proved in 7.7, that any such Y must be itself an Orlicz function space $L_G(0,1)$ which is also lattice isomorphic to a sublattice of L_F unless it is equal to $L_2(0,1)$.

The results in Section 8 are inspired by the problem raised by B.S. Mitjagin [60] (already mentioned above) whether $L_p(0,1)$; $1 \leq p \leq \infty$ are the only spaces which can be represented as r.i. function spaces on $[0,1]$ as well as on $[0,\infty)$. We have seen that example 10.1 gives a negative answer to this question but, since this space has quite an artificial construction, we have continued the investigation in order to determine how wide is the class of those spaces which have a double representation as r.i. function spaces on both $[0,1]$ and $[0,\infty)$. Surprisingly enough, this class contains all the r.i. function spaces on $[0,1]$ whose Boyd indices are non-trivial. More precisely, it is shown in 8.6 that every r.i. function space X on $[0,1]$ with $0 < \beta_X$ and $\alpha_X < 1$ is isomorphic to a r.i. function space Y_X on $[0,\infty)$. The space Y_X is the completion of the simple functions f on $[0,\infty)$ which have a bounded support under the norm

$$\|f\|_{Y_X} = \max\{\|f^*\chi_{[0,1]}\|_X \, , \, (\sum_{n=0}^{\infty} (\int_n^{n+1} f^*(t)dt)^2)^{1/2}\} ,$$

where f^* denotes the decreasing rearrangement of f. That $\|\cdot\|_{Y_X}$ is a norm follows from the fact that the left hand side expression in the maximum is equal to

$$\sup\{\|f\chi_E\|_X \; ; \; \mu(E) = 1\}$$

while the right hand side one is equal to

$$\sup\{(\sum_{n=0}^{\infty} (\int_{E_n} f(t)dt)^2)^{1/2}; \; E_n \cap E_m = \emptyset \text{ for } n \neq m \text{ and } \mu(E_n) = 1; n = 1,2,\dots\}.$$

To understand better the significance of this norm we point out that it is actually equivalent to the expression

$$\max\{\|f^*\chi_{[0,1]}\|_X \, , \, \|f^*\chi_{[1,\infty]}\|_{L_2(0,\infty)}\} ,$$

which, in general, is not a norm. In other words, the restriction of Y_X to $[0,1]$ coincides with the original space X while, at ∞, Y_X behaves as $L_2(0,\infty)$. As is easily verified, the space Y_X is equal to $L_p(0,\infty) \cap L_2(0,\infty)$, respectively $L_p(0,\infty) + L_2(0,\infty)$, when $X = L_p(0,1)$ for $p > 2$, respectively, $1 < p < 2$. Therefore, 8.6 completes, in particular, the part of 1.7 left unproved in Section 1.

The proof of 8.6 is of a probabilistic nature and it involves the notion of symmetrized Poisson process. The starting point is, of course, the usual

Poisson process, $\{N_t\}_{0 \leq t < \infty}$ which is a stationary (i.e., the distribution function of $N_t - N_s$ depends on the difference $t - s$ but not on the particular values of $0 \leq s \leq t$) process with independent increments (i.e., for any $0 \leq t_0 \leq t_1 \leq \ldots \leq t_k$, $\{N_{t_j} - N_{t_{j-1}}\}_{j=1}^k$ is a sequence of independent random variables) over some probability space $(\Omega, \mathcal{A}, P)$ taking only non-negative integer values such that $N_0 = 0$ and

$$P(\{N_t = n\}) = e^{-\theta t}(\theta t)^n/n! \ ; \ n = 0,1,2,\ldots,$$

where θ is a parameter which, in this work, is taken to be equal $\frac{1}{2}$. Though the existence of the Poisson process is proved in many books on probability theory, we present in Section 8, for the convenience of the reader, a concrete construction of this process requiring only a limited knowledge of probability. By taking the difference of two independent copies of the Poisson process we obtain a new process $\{Z_t\}_{0 \leq t < \infty}$ which is called the symmetrized Poisson process. With the aid of $\{Z_t\}_{0 \leq t < \infty}$, we define an operator T from the step functions on $[0,\infty)$ into $L_1(\Omega, \mathcal{A}, P)$ by putting

$$T\chi_{[s,t)} = Z_t - Z_s \ ; \ 0 \leq s \leq t < \infty ,$$

and extending linearly. The crucial point in the proof of 8.6 is to show that if X is a r.i. function space on $[0,1]$ with non-trivial Boyd indices then T extends uniquely to an isomorphism from the corresponding space Y_X onto a complemented subspace of X . Since Y_X clearly contains a complemented copy of X the proof of 8.6 is then concluded by using the decomposition method.

Of particular interest is the case when X is an Orlicz function space $L_F(0,1)$. It follows easily from the definition of the norm in Y_{L_F} that this space is also an Orlicz function space $L_G(0,\infty)$, where, for $G(t)$, we can take any Orlicz function equivalent to t^2 at 0 and to $F(t)$ at ∞ . The space $L_G(0,\infty)$, corresponding in the above manner to $L_F(0,1)$ is related to some results of Dacunha-Castelle and Schreiber [16]. This work indicates that there is a strong possibility that an Orlicz function space $L_H(0,1)$ is isomorphic to a subspace of a 2-concave Orlicz space $L_F(0,1)$ if and only if H is equivalent to a function of the form

$$\int_0^\infty (F(st) \wedge s^2t^2)/(F(s)\wedge s^2)\nu(ds) ,$$

where ν is a probability measure on $(0,\infty)$. But, by the result 7.7, any $H \in \mathfrak{F}$ has the property that $L_H(0,1)$ embeds isomorphically as a sublattice of Y_X, where $X = L_F(0,1)$; i.e., of $L_G(0,1)$, associated above to $L_F(0,1)$. Therefore, we raise formally the following open question.

Problem 0.3. Suppose that an Orlicz space $L_H(0,1)$ *is isomorphic to a subspace of a 2-concave Orlicz space* $L_F(0,1)$. *Does then* $L_H(0,1)$ *embed isomorphically as a sublattice of the space* Y_{L_F}, *corresponding to* $L_F(0,1)$ *?*

We want to make now some remarks on the degree of generality of 8.6. The condition imposed there that X be a r.i. function space on [0,1] with non-trivial Boyd indices is not, in fact, a necessary condition. For instance, the space $L_F(0,1)$, where $F(t) = e^t - 1$, is isomorphic to Y_{L_F} in spite of the fact that $\beta_{L_F} = 0$. On the other hand, some restrictions have to be imposed in 8.6 since, as shown in 8.17, the space $L_F(0,1)$, where $F(t) = t(\log t)^{1/4}$ for $t \geq 1$, is isomorphic to no r.i. function space on $[0,\infty)$. This question can be viewed from a different angle, too. For instance, it is shown in 8.14 that there exist rather nice r.i. function spaces on $[0,\infty)$ which are isomorphic to no r.i. function space on [0,1].

The space Y_X, associated to X by 8.6, has in some sense a certain uniqueness feature: it is proved in 8.12 that if a super-reflexive r.i. function space X on [0,1] is isomorphic to a r.i. function space Y on $[0,\infty)$ which differs from Y_X (even up to an equivalent norm) then the sequences $\{\chi_{[(i-1)n^{-1},in^{-1})}\}_{i=1}^\infty$; $n = 1,2,\ldots$ in Y must be uniformily equivalent to sequences of pairwise disjoint functions in X. That the latter possibility in 8.12 can actually hold is shown by an example presented in detail at the end of Section 8.

The approach described above can be also used to recapture some known results from the theory of L_p-spaces. For example, it is proved in 8.7 that any sequence of characteristic functions of mutually disjoint integrable subsets $\{A_n\}_{n=1}^{\infty}$ of $[0,\infty)$ with $\sum_{\mu(A_n)<\epsilon} \mu(A_n) = +\infty$ for every $\epsilon > 0$ spans in an arbitrary r.i. function space Y on $[0,\infty)$ a subspace U_Y which, up to isomorphism, does not depend on the particular sequence $\{A_n\}_{n=1}^{\infty}$ used. This subspace U_Y can be seen as a natural generalization of the space X_p of H. P. Rosenthal [70] since, for $p > 2$, the space U_Y corresponding to $Y = L_p(0,\infty) \cap L_2(0,\infty)$ is, as is easily verified, identical to X_p. In general, if X is a r.i. function space on $[0,1]$ with non-trivial Boyd indices and Y_X is the r.i. function space on $[0,\infty)$ associated to X by 8.6 then U_{Y_X} is isomorphic to the span in X of any sequence of independent symmetrized Poisson variables $\{Z_n\}_{n=1}^{\infty}$ which satisfy

$$\lim_{n\to\infty} \int_\Omega Z_n^2 dP = 0 \quad \text{and} \quad \sum_{n=1}^{\infty} \int_\Omega Z_n^2 dP = +\infty .$$

Another matter of a similar nature is that of embedding isometrically the space $L_q(0,1)$ into $L_p(0,1)$ when $1 \le p < q < 2$. As is well known, (cf. [11]) this embedding is achieved by using a family of independent q-stable random variables. In the present paper we prove directly (8.8 and 8.9) that if the function $t^{-1/q}$; $1 < q < 2$ belongs to some r.i. function space Y on $[0,\infty)$ then Y contains a sublattice isometric to $L_q(0,\infty)$. In particular, since $t^{-1/q} \in L_p(0,\infty) + L_2(0,\infty)$ for $1 < p < q$, we get that $L_q(0,\infty)$ embeds isometrically as a sublattice of $L_p(0,\infty) + L_2(0,\infty)$ which, in turn, is isomorphic to $L_p(0,\infty)$. In fact, there is no essential difference between our approach using the Poisson process and the classical one which uses, as mentioned above, the q-stable random variables. The reason lies in the fact that the map T associated above to the symmetrized Poisson process maps the function $t^{-1/q}$ into a q-stable random variable.

Section 9 is devoted mainly to the study of embeddings of a r.i. function space into itself. The first result in this direction was obtained by Enflo and Starbird [22], who proved that a subspace of $L_1(0,1)$ which is isomorphic to $L_1(0,1)$ contains a further subspace which is isomorphic to $L_1(0,1)$ and complemented in $L_1(0,1)$. We prove here a similar result for a large class of r.i. function spaces. More precisely, Theorem 9.1 below states that if X is a r.i. function space on $[0,1]$ which is s-concave for some $s < \infty$, the index α_X of X satisfies $\alpha_X < 1$ and the Haar system in X is not equivalent to a sequence of disjoint functions in X, then any subspace of X which is isomorphic to X contains a further subspace isomorphic to X and complemented in X. In particular, we get that the conclusion holds for $X = L_p(0,1)$, $1 < p < \infty$ and even for any reflexive Orlicz function space on $[0,1]$.

Example 10.4, which was already mentioned, shows also that one cannot drop the assumption that the Haar system in X is not equivalent to a sequence of disjoint functions in X. We suspect that by the same method of proof, with a little more effort one can weaken the assumption "X is s-concave for some $s < \infty$" to "$\beta_X > 0$". However, it seems that one cannot get more than this by the same method of proof since we use crucially the fact that the Haar system in X is unconditional which, in turn, is equivalent to $0 < \beta_X \leq \alpha_X < 1$ (cf. [50]). We thus propose

Problem 0.4. Let X be a r.i. function space on $[0,1]$ *such that the Haar system in X is not equivalent to a sequence of disjoint functions in X (this is automatically satisfied in the interesting remaining case; i.e., when the Haar system is conditional). Does every subspace of X which is isomorphic to X contain another subspace isomorphic to X and complemented in X? Is it true under the assumption that X is s-concave for some* $s < \infty$ *?*

In the sequel we introduce a class of operators from X to X called operators with property A by means of an analytic definition (see definition 9.9 below) and we prove that under the assumptions of the theorem $T: X \to X$ has property A if and only if TX contains a (complemented in X) subspace isomorphic to X. Property A comes to replace the notion of E-operator in [22]; however, the definition is quite complicated and some better definition may be found (so we left the notion "E-operator" open for it).

The proof of Theorem 9.1 in the general case is quite complicated and it depends heavily on Section 6 both practically and in spirit, so the reader is advised to have a good understanding of Section 6 (which we consider the deepest part of this paper) before trying to read intensively Section 9. We however included (for the stubborn reader) a proof of Theorem 9.1 in the case $X = L_p(0,1)$ $1 < p < \infty$ which, except for the notion of gaussian Haar system, is independent from Section 6. The proof for $2 < p < \infty$ illustrates the proof in the general case so we put it inside the section, while the proof for $1 < p < 2$ digresses from the line of ideas in the section, so it appears in an appendix to Section 9.

Section 10 is devoted to examples. Besides the examples 10.1 and 10.4 concerning spaces with many representations as a r.i. function space, Section 10 contains also an example of a space X with a symmetric basis such that the collection of p's for which ℓ_p is lattice finitely representable in X does not form an interval (10.6 below).

1 SYMMETRIC STRUCTURES IN SUBSPACES OF $L_p(0,1)$; $p > 2$.

The main result of this section is the following theorem on the classification of subspaces of $L_p(0,1)$; $p > 2$ with a symmetric basis.

Theorem 1.1: *For every* $p > 2$ *and every* $K \geq 1$ *there exists a number* $D = D(p,K) < \infty$ *so that if* $\{x_i\}_{i=1}^n$ *is a finite normalized basic sequence in* $L_p(0,1)$; $p > 2$ *whose symmetry constant* $K(\{x_i\}_{i=1}^n)$ *is* $\leq K$ *then*

$$D^{-1}\left\|\sum_{i=1}^n a_i x_i\right\| \leq \max\left\{\left(\sum_{i=1}^n |a_i|^p\right)^{1/p}, \frac{\left\|\sum_{i=1}^n x_i\right\|}{\sqrt{n}}\left(\sum_{i=1}^n |a_i|^2\right)^{1/2}\right\} \leq D\left\|\sum_{i=1}^n a_i x_i\right\|,$$

for every choice of scalars $\{a_i\}_{i=1}^n$.

This yields that every finite or infinite normalized K-symmetric basic sequence in $L_p(0,1)$; $p > 2$ is $D(p,K)$-equivalent to some symmetric X_p-basis. In the proof of 1.1 we use the notion of exchangeable random variables. A finite sequence $\{f_i\}_{i=1}^n$ of random variables (functions) over a probability space is called *symmetrically exchangeable* if, for every permutation π of the integers $\{1,2,\ldots,n\}$ and for every choice of signs $\epsilon_i = \pm 1$, the probability distribution in R^n of $\{f_i\}_{i=1}^n$ is the same as that of $\{\epsilon_i f_{\pi(i)}\}_{i=1}^n$. The classical definition of exchangeability, as considered in probability theory, involves only invariance under permutations.

The following lemma is known (see e.g. [12]).

Lemma 1.2: *A K-symmetric normalized basic sequence* $\{x_i\}_{i=1}^n$ *in* $L_p(0,1)$; $1 \leq p < \infty$ *is K- equivalent to a normalized sequence* $\{f_i\}_{i=1}^n$ *of symmetrically exchangeable random variables in* $L_p(0,1)$.

Proof: Let H_n be the family of all distinct pairs $(\pi, \{\epsilon_j\}_{j=1}^n)$, where π is a permutation of the integers $\{1,2,\ldots,n\}$ and $\{\epsilon_j\}_{j=1}^n$ a sequence of signs. Let $\{I_h\}_{h\in H_n}$ be a partition of $[0,1]$ into mutually disjoint intervals of length $1/(n!2^n)$ and, for $h \in H_n$, let ζ_h be the increasing linear map from the interval I_h onto $[0,1]$. For $1 \le i \le n$ and $h = (\pi,\{\epsilon_j\}_{j=1}^n) \in H_n$ put

$$f_i(t) = \epsilon_i\, x_{\pi(i)}(\zeta_h(t)) \; ; \; t \in I_h .$$

Then, the functions $\{f_i\}_{i=1}^n$ clearly form a normalized sequence of symmetrically exchangeable randon variables in $L_p(0,1)$ which is K-equivalent to $\{x_i\}_{i=1}^n$.

□

Remark: Observe that the sequence $\{f_i\}_{i=1}^n$ can be constructed so that $(\sum_{i=1}^n |f_i|^2)^{1/2}$ is a constant. Indeed, assume without loss of generality that $\bigcup_{i=1}^n \operatorname{supp} x_i = [0,1]$ and change the density to make $(\sum_{i=1}^n |x_i|^2)^{1/2}$ a constant; then $(\sum_{i=1}^n |f_i|^2)^{1/2}$ is the same constant.

Proof of 1.1. Fix $p > 2$ and $K \ge 1$. Let $\{x_i\}_{i=1}^n$ be a normalized basic sequence in $L_p(0,1)$ whose symmetry constant $K(\{x_i\}_{i=1}^n)$ is $\le K$. By 1.2 and the remark, $\{x_i\}_{i=1}^n$ is K-equivalent to a normalized sequence $\{f_i\}_{i=1}^n$ of symmetrically exchangeable random variables such that $(\sum_{i=1}^n |f_i|^2)^{1/2}$ is the constant $C = \|(\sum_{i=1}^n |x_i|^2)^{1/2}\|$. Since $\{f_i\}_{i=1}^n$ is, in particular, a monotone unconditional basic sequence and the space $L_p(0,1)$ has cotype p with constant 1 we get that $\|\sum_{i=1}^n a_i f_i\| \ge (\sum_{i=1}^n |a_i|^p)^{1/p}$

Observe that $\{f_i\}_{i=1}^n$ is orthogonal in $L_2(0,1)$ and each f_i has L_2-norm $C/\sqrt{n}$; therefore,

$$K\|\sum_{i=1}^n a_i x_i\| \ge \|\sum_{i=1}^n a_i f_i\| \ge \|\sum_{i=1}^n a_i f_i\|_2 = (\sum_{i=1}^n |a_i|^2)^{1/2} C/\sqrt{n} .$$

Of course,

$$\|\sum_{i=1}^{n} x_i\| \le K\|\sum_{i=1}^{n} f_i\| \le K\, B_p\|(\sum_{i=1}^{n} |f_i|^2)^{1/2}\| = K\, B_p\, C\ .$$

Hence,

$$\|\sum_{i=1}^{n} a_i x_i\| \ge B_p^{-1}K^{-2}\max\{(\sum_{i=1}^{n} |a_i|^p)^{1/p}\ ,\ \frac{\|\sum_{i=1}^{n} x_i\|}{\sqrt{n}}\ (\sum_{i=1}^{n} |a_i|^2)^{1/2}\}\ .$$

In order to prove the right hand side inequality we first notice that, by (*) (see the Introduction), we have that

$$\|\sum_{i=1}^{n} a_i f_i\| \le B_p(\int_0^1 (\sum_{i=1}^{n} |a_i f_i(t)|^2)^{p/2}dt)^{1/p}\ .$$

Thus,

$$\|\sum_{i=1}^{n} a_i f_i\|^p \le B_p^p \sum_{j=1}^{n} |a_j|^2 \int_0^1 (\sum_{i=1}^{n} |a_i f_i(t)|^2)^{(p-2)/2}|f_j(t)|^2dt \le$$

$$\le 2^{(p-2)/2}\, B_p^p[\sum_{j=1}^{n} |a_j|^2 \int_0^1 (\sum_{i\ne j} |a_i f_i(t)|^2)^{(p-2)/2}|f_j(t)|^2dt + \sum_{j=1}^{n} |a_j|^p]\ ;$$

i.e.,

$$\|\sum_{i=1}^{n} a_i f_i\|^p \le 2^{p/2}\, B_p^p\max\{\sum_{j=1}^{n}|a_j|^p, \sum_{j=1}^{n}|a_j|^2 \int_0^1(\sum_{i\ne j}|a_i f_i(t)|^2)^{(p-2)/2}|f_j(t)|^2dt\}.$$

The argument above is based on some ideas from H.P. Rosenthal [70]. Suppose now that $a_i = 0$ for $i \ge [n/2] + 1$. Then, by the exchangeability of $\{f_i\}_{i=1}^n$, we get that

$$\int_0^1 (\sum_{i\ne j} |a_i f_i(t)|^2)^{(p-2)/2}\ |f_j(t)|^2dt = \int_0^1(\sum_{i\ne j}|a_i f_i(t)|^2)^{(p-2)/2}|f_k(t)|^2dt$$

for $k \ge [n/2] + 1$ and $j < [n/2] + 1$. Thus, by averaging over $[n/2] + 1 \le$ $\le k \le n$ we have

$$\int_0^1 \Big(\sum_{i\neq j} |a_i f_i(t)|^2\Big)^{(p-2)/2} |f_j(t)|^2 dt \leq$$

$$\leq 2n^{-1}\int_0^1 \Big(\sum_{i\neq j} |a_i f_i(t)|^2\Big)^{(p-2)/2}\Big(\sum_{k=[n/2]+1} |f_k(t)|^2\Big)dt \leq$$

$$\leq 2n^{-1}\Big\|\Big(\sum_{i=1}^n |a_i f_i|^2\Big)^{1/2}\Big\|_{p-2}^{p-2} C^2 \leq 2n^{-1}C^2\Big\|\sum_{i=1}^n a_i f_i\Big\|_p^{p-2} .$$

Consequently, for scalars as above (i.e., $a_i = 0$ for $i \geq [n/2] + 1$), we get that

$$\Big\|\sum_{i=1}^n a_i f_i\Big\| \leq 2^{\frac{p+2}{4}} \cdot B_2^{p/2} \max\Big\{\Big(\sum_{j=1}^n |a_j|^p\Big)^{1/p} , \frac{C}{\sqrt{n}}\Big(\sum_{j=1}^n |a_j|^2\Big)^{1/2}\Big\} .$$

A similar inequality holds for arbitrary scalars $\{a_i\}_{i=1}^n$ and for $\{x_i\}_{i=1}^n$ replacing $\{f_i\}_{i=1}^n$ provided the constant preceding the maximum is multiplied by $2K^2$. This completes the proof, because $C = \|(\sum_{i=1}^n |x_i|^2)^{1/2}\| \leq K\|\sum_{i=1}^n x_i\|$.

□

A simple consequence of 1.1 is that two finite K-symmetric basic sequences $\{x_j\}_{j=1}^n$ and $\{y_j\}_{j=1}^n$ in $L_p(0,1)$; $p > 2$ are D^2-equivalent if $w_1 = \|\sum_{j=1}^n x_j\|/\sqrt{n}$ is equal to $w_2 = \|\sum_{j=1}^n y_j\|/\sqrt{n}$. The full significance of the expression $w = \|\sum_{j=1}^n x_j\|/\sqrt{n}$ is put in evidence by the following result which is actually true in a more general context (see Section 2).

Proposition 1.3: *Let* $\{x_i\}_{i=1}^n$ *be a finite K-symmetric normalized basis of a subspace* X *of* $L_p(0,1)$; $1 \leq p < \infty$. *Then, the distance coefficient* $d(X,\ell_2^n)$ *satisfies*

$$(i) \quad K^{-2}\frac{\|\sum_{i=1}^n x_i\|}{\sqrt{n}} \leq d(X,\ell_2^n) \leq \frac{K^3}{A_p^2}\frac{\|\sum_{i=1}^n x_i\|}{\sqrt{n}} \qquad \text{if} \quad 1 \leq p \leq 2$$

$$\text{or}(ii) \quad K^{-2}\frac{\sqrt{n}}{\|\sum_{i=1}^n x_i\|} \leq d(X,\ell_2^n) \leq K^3 B_p^2 \frac{\sqrt{n}}{\|\sum_{i=1}^n x_i\|} \qquad \text{if} \quad p > 2 .$$

Proof: (ii) For $p > 2$ we know, by the first part of the proof of 1.1, that $\|\sum_{i=1}^n a_i x_i\| \geq B_p^{-1} K^{-2} n^{-1/2} \|\sum_{i=1}^n x_i\| (\sum_{i=1}^n |a_i|^2)^{1/2}$. On the other hand, by (*), $\|\sum_{i=1}^n a_i x_i\| \leq K B_p (\sum_{i=1}^n |a_i|^2)^{1/2}$. Hence, by using the formal identity mapping between the basis $\{x_i\}_{i=1}^n$ and the unit vector basis of ℓ_2^n, we get that $d(X, \ell_2^n) \leq K^3 B_p^2 n^{1/2} / \|\sum_{i=1}^n x_i\|$. Now it is known that this formal identity gives the distance up to a constant. To see this, let T be an invertible operator from X into $L_2(0,1)$ so that $\|T\| \cdot \|T^{-1}\| = d(X, \ell_2^n)$. Let H_n, $\{I_h\}_{h \in H_n}$ and $\{\zeta_h\}_{h \in H_n}$ have the same meaning as in the proof of 1.2, and, for $h = (\pi, \{\epsilon_j\}_{j=1}^n) \in H_n$ and $1 \leq i \leq n$, put

$$g_i(t) = \epsilon_i (T x_{\pi(i)}) (\zeta_h(t)) \; ; \; t \in I_h .$$

Then, as is easily verified, the functions $\{g_i\}_{i=1}^n$ form a sequence of symmetrically exchangeable random variables in $L_2(0,1)$ and

$$K^{-1} \|T^{-1}\|^{-1} \|\sum_{i=1}^n a_i x_i\| \leq \|\sum_{i=1}^n a_i g_i\| \leq K \cdot \|T\| \; \|\sum_{i=1}^n a_i x_i\| .$$

Since $\{g_i\}_{i=1}^n$ is an orthogonal system in a Hilbert space it follows that

$$K \|T\| \; \|\sum_{i=1}^n x_i\| \geq \|\sum_{i=1}^n g_i\| = \|g_1\| n^{1/2} \geq K^{-1} \|T^{-1}\|^{-1} n^{1/2} \quad ;$$

i.e., $d(X, \ell_2^n) \geq K^{-2} n^{1/2} / \|\sum_{i=1}^n x_i\|$.

The proof of (i) is similar. □

Proposition 1.3 shows that, for any subspace X of $L_p(0,1)$; $1 \leq p < \infty$ with a symmetric basis $\{x_i\}_{i=1}^n$, the expression $\|\sum_{i=1}^n x_i\|$ is an invariant of the space X rather than of the particular symmetric basis $\{x_i\}_{i=1}^n$. This fact together with 1.1 imply the following result.

Corollary 1.4: *Let* G_p *be the family of all subspaces of* $L_p(0,1)$; $p > 2$ *which have a* 1*-symmetric basis. Then each member of* G_p *has a unique symmetric basis.*

Notice that, by 1.2, 1.4 remains valid even if we only require that all members of G_p have a K-symmetric basis for some fixed $1 < K < \infty$ (rather than $K = 1$).

Corollary 1.4 shows, in particular, that each member of the family $\mathfrak{F}_p = \{\ell_p^n\}_{n=1}^\infty$; $1 < p < \infty$ has a unique symmetric basis. This fact is also true in the cases $p = 1$ and $p = \infty$ for which there is even a uniqueness theorem for all unconditional bases of ℓ_1^n, respectively ℓ_∞^n ; $n = 1,2,\ldots$ (cf. [44]). The following result shows that the uniqueness of the symmetric basis of each member of $\mathfrak{F}_p$ is actually uniform on the whole range $1 \leq p \leq \infty$.

Theorem 1.5: *Each member of the family*

$$\mathfrak{F} = \{\ell_p^n ; 1 \leq p \leq \infty , n = 1,2,\ldots\}$$

has, up to equivalence, a unique symmetric basis.

Proof: Fix $1 < r < 2$ and let $r' > 2$ be so that $1/r + 1/r' = 1$. The constants appearing in the proof of 1.4 show that each member of the class $\{\ell_p^n ; r \leq p \leq r' , n = 1,2,\ldots\}$ has, up to equivalence, a unique symmetric basis (in other words, for any $K \geq 1$ there exists a number $0 < \theta(K,r) < \infty$, independent of $r \leq p \leq r'$, so that any K-symmetric normalized basis of ℓ_p^n, with $r \leq p \leq r'$ and $n = 1,2,\ldots$, is $\theta(K,r)$-equivalent to the unit vector basis of ℓ_p^n).

Let now $\{x_i\}_{i=1}^n$ be a normalized K-symmetric basis of ℓ_p^n for some $1 \leq p < r$ and some $n = 1,2,\ldots$. By 1.2, $\{x_i\}_{i=1}^n$ is K-equivalent to a normalized sequence $\{f_i\}_{i=1}^n$ of symmetrically exchangeable random variables in $L_p(0,1)$. The exchangeability of this sequence implies that

$$\|\sum_{i=1}^{n} f_i\|^p = \int_0^1 \|\sum_{i=1}^{n} r_i(u)f_i\|^p du \le \int_0^1 (\sum_{i=1}^{n} |f_i(t)|^2)^{p/2} dt =$$

$$= \int_0^1 (\sum_{i=1}^{n} |f_i(t)|^{2-p} \cdot |f_i(t)|^p)^{p/2} dt \le \int_0^1 \max_{1\le j\le n} |f_j(t)|^{(2-p)p/2} (\sum_{i=1}^{n} |f_i(t)|^p)^{p/2} dt.$$

Thus, by Hölder's inequality with the indices $2/(2-p)$ and $2/p$, we get that

$$\|\sum_{i=1}^{n} f_i\|^p \le [\int_0^1 \max_{1\le j\le n} |f_j(t)|^p dt]^{(2-p)/2} \cdot [\int_0^1 \sum_{i=1}^{n} |f_i(t)|^p dt]^{p/2} =$$

$$= n^{p/2} \cdot [\int_0^1 \max_{1\le j\le n} |f_j(t)|^p dt]^{(2-p)/2} .$$

Oh the other hand, by 1.3(i), it follows that

$$\|\sum_{i=1}^{n} f_i\| \ge K^{-1} \|\sum_{i=1}^{n} x_i\| \ge A_p^2 K^{-3} n^{1/2} d(\ell_p^n, \ell_2^n) = A_p^2 K^{-3} n^{1/p} .$$

Consequently, $\int_0^1 \max_{1\le j\le n} |f_j(t)|^p dt \ge (A_p K^{-2})^{4p/(2-p)} \cdot n$ and, since $4p/(2-p)$ is an increasing function of p on the interval $[1,r]$, we have that

$$\int_0^1 \max_{1\le j\le n} |f_j(t)|^p dt \ge (A_p K^{-2})^{4r/(2-r)} \cdot n .$$

Since $\{f_i\}_{i=1}^n$ are symmetrically exchangeable random variables it is possible to split the interval $[0,1]$ into n mutually disjoint subsets $\{E_i\}_{i=1}^n$, each having measure equal to $1/n$, so that

$$\int_{E_i} |f_i(t)|^p dt \ge (A_p K^{-2})^{4r/(2-r)} .$$

(the set E_i is chosen so that $\max_{1\le j\le n} |f_j(t)| = |f_i(t)|$ for $t \in E_i$).

It follows that

$$(\sum_{i=1}^{n} |a_i|^p)^{1/p} \geq \|\sum_{i=1}^{n} a_i f_i\| \geq A_p(\int_0^1 (\sum_{i=1}^{n} |a_i f_i(t)|^2)^{p/2} dt)^{1/p} =$$

$$= A_p(\sum_{j=1}^{n} \int_{E_j} (\sum_{i=1}^{n} |a_i f_i(t)|^2)^{p/2} dt)^{1/p} \geq A_p(A_p K^{-2})^{4r/(2-r)}(\sum_{i=1}^{n} |a_j|^p)^{1/p} .$$

This proves that $\{x_i\}_{i=1}^n$ is $(K^2 A_p^{-1})^{(3r+2)/(2-r)}$-equivalent to the unit vector basis of ℓ_p^n so that we have uniformity on the interval $[1,r]$ because $A_1 \leq A_p$ for $1 \leq p \leq 2$. By duality, we have uniformity also on the interval $[r',\infty]$.

□

Remarks. 1. The arguments used to prove 1.5 are based on a method of L. Dor [19]. 2. We do not know whether a uniformity theorem is also valid for the family of all subspaces of $L_p(0,1)$; $2 \leq p < \infty$ which have a 1-symmetric basis.

We pass now to some applications to rearrangement invariant (r.i.) function spaces on $[0,1]$ and on $[0,\infty)$.

Theorem 1.6: *(i) Let* X *be a r.i. function space on* $[0,1]$ *which is isomorphic to a subspace of* $L_p(0,1)$; $p>2$. *Then, up to an equivalent renorming,* X *coincides either with* $L_p(0,1)$ *or with* $L_2(0,1)$.

(ii) Let X *be a r.i. function space on* $[0,\infty)$ *which is isomorphic to a subspace of* $L_p(0,1)$; $p>2$. *Then, up to an equivalent renorming,* X *coincides with one of the spaces* $L_p(0,\infty)$, $L_2(0,\infty)$ *or* $L_p(0,\infty) \cap L_2(0,\infty)$; *i.e., the space of all functions on* $[0,\infty)$ *such that*

$$\|f\| = \max(\|f\|_p , \|f\|_2) < \infty .$$

Proof: Case (i): Let T be an isomorphism from X into $L_p(0,1)$; $p > 2$. For every integer n and every $1 \leq i \leq 2^n$, let $x_{n,i}$ be the characteristic function of the interval $[(i-1)2^{-n}, i2^{-n})$. Then, for every n , the sequence $\{Tx_{n,i}/\|x_{n,1}\|_X\}_{i=1}^{2^n}$ is a $\|T\|\cdot\|T^{-1}\|$-symmetric basic sequence in $L_p(0,1)$ which, by 1.1, implies the existence of a constant $D < \infty$, independent of n , so that, for every n and every choice of $\{a_i\}_{i=1}^{2^n}$, we have that

$$D^{-1}\|\sum_{i=1}^{2^n} a_i x_{n,i}\|_X \le \|x_{n,1}\|_X \max\{(\sum_{i=1}^{2^n} |a_i|^p)^{1/p}, (\sum_{i=1}^{2^n} |a_i|^2)^{1/2}/\|x_{n,1}\|_X\sqrt{2^n}\}$$

$$\le D \|\sum_{i=1}^{2^n} a_i x_{n,i}\|_X$$

In other words, if f is a simple function over the field generated by the intervals $\{[(i-1)2^{-n}, i2^{-n})\}_{i=1}^{2^n}$ then,

$$D^{-1}\|f\|_X \le \max\{2^{n/p}\|x_{n,1}\|_X \|f\|_p , \|f\|_2\} \le D\|f\|_X .$$

In particular, for $f \equiv 1$, we get that $2^{n/p}\|x_{n,1}\|_X \le D$ for all n . Let $\alpha = \liminf_{n\to\infty} 2^{n/p}\|x_{n,1}\|_X$ (so $0 \le \alpha \le D$) . Then, for any step function f over the dyadic intervals, we have

$$(*) \qquad D^{-1}\|f\|_X \le \max\{\alpha\|f\|_p, \|f\|_2\} \le D\|f\|_X .$$

Hence, if $\alpha = 0$ then X is, up to an equivalent norm, equal to $L_2(0,1)$ while if $\alpha > 0$ then X is equal to $L_p(0,1)$. This finishes the proof of case (i).

Case (ii): Denote by X_s the subspace of X consisting of all functions which vanish outside $[0,s]$. Since X_s is isometric and order isomorphic to a r.i. function space on $[0,1]$ by an operator U such that $\|Uf\|_p = a_s\|f\|_p$ and $\|Uf\|_2 = b_s\|f\|_2$ for every $f \in X_s$ and for some a_s and b_s which depend only on s , we get from (*) above that

$$D^{-1}\|f\|_X \leq \max\{\alpha\, a_s\|f\|_p,\ b_s\|f\|_2\} \leq D\,\|f\|_X\ ,$$

for every $f \in X_s$. Passing to a limit as $s \to \infty$, we conclude that there exist constants a and b such that

$$D^{-1}\|f\|_X \leq \max\{a\,\|f\|_p,\ b\,\|f\|_2\} \leq D\,\|f\|_X$$

and case (ii) follows by inspecting the alternatives $b = 0$, $a = 0$, and $a, b > 0$.

□

The above proof shows that if a r.i. function space on $[0,1]$ embeds isomorphically into $L_p(0,1)$ for $2 < p < \infty$, then the norm in X is, up to a constant which depends only on the isomorphism constant of the embedding of X into $L_p(0,1)$, and on p, given by

$$\|f\|_X = \max\{\alpha\,\|f\|_p,\ \|f\|_2\}$$

for some $0 \leq \alpha \leq 1$. A similar remark holds in the case of a r.i. function space on $[0,\infty)$.

Corollary 1.7: *(i) For every* $1 \leq p \leq \infty$ *the space* $L_p(0,1)$ *has a unique representation as a r.i. function space on* $[0,1]$*; i.e., if a r.i. function space* X *on* $[0,1]$ *is isomorphic to* $L_p(0,1)$*, then* X *is, up to an equivalent renorming, equal to the space* $L_p(0,1)$.

(ii) The space $L_p(0,\infty)$, $1<p<\infty$, $p \neq 2$, *has exactly two representations as a r.i. function space on* $[0,\infty)$: $L_p(0,\infty)$ *and* $L_p(0,\infty) \cap L_2(0,\infty)$ *if* $p > 2$ *and* $L_p(0,\infty)$ *and* $L_p(0,\infty) + L_2(0,\infty)$ *if* $1<p<2$. ($L_p(0,\infty) + L_2(0,\infty)$ *is the space of all functions on* $[0,\infty)$ *such that*

$\|f\| = \inf\{\|g\|_p + \|h\|_2\ ;\ g \in L_p(0,\infty),\ h \in L_2(0,\infty)\ ;\ g + h = f\} < \infty$).

In order to prove (ii) in the case $2 < p < \infty$ it is enough, in view of 1.6, to prove that $X = L_p(0,\infty) \cap L_2(0,\infty)$ is isomorphic to $L_p(0,\infty)$. Since $X_1 = \{f \in X\ ;\ f(t) = 0 \text{ for } t > 1\}$ is isomorphic to $L_p(0,1)$, it is sufficient to prove that X is isomorphic to a complemented subspace of $L_p(0,1)$. Notice that for each n and m, the sequence $\{x_{n,i}/\|x_{n,1}\|\}_{i=1}^m$ in X is equivalent, with constants independent of n and m, to the first m unit vectors of a (symmetric) X_p space. Thus, by [70], this space is λ-isomorphic to a λ-complemented subspace of $L_p(0,1)$, where λ is independent of n and m. It follows now by a usual compactness (or ultraproduct) argument that X is isomorphic to a complemented subspace of $L_p(0,1)$. The case $1 < p < 2$ follows by duality.

□

Remark: A slightly different proof that $L_p(0,\infty) \cap L_2(0,\infty)$ is isomorphic to L_p $(p > 2)$ will be given in Section 8 together with more general results of the same kind.

We wish to discuss now some questions concerning infinite dimensional spaces with an unconditional basis or with a Banach lattice structure which can be embedded isomorphically into $L_p(0,1)$ for some $p > 2$. Each known example of such a space is either, itself, an $L_p(\mu)$-space up to an equivalent norm or has the property that ℓ_2 is *lattice finitely representable* in it; i.e., for every $\epsilon > 0$ and every integer k, there exists a sequence $\{x_j\}_{j=1}^k$ of k disjoint elements of X which is $1+\epsilon$-equivalent to the unit vector basis of ℓ_2^k. The following result shows that this situation has actually a general character. A previous proof (for the discrete case) was given by L. Dor and T. Starbird [20].

Theorem 1.8: A Banach lattice X, *which is linearly isomorphic to a subspace of* $L_p(0,1)$ *for some* $p > 2$, *is either linearly isomorphic and order equivalent to an* $L_p(\mu)$ *-space for a suitable measure* μ *or* ℓ_2 *is lattice finitely representable in* X.

The proof of 1.8, which is an immediate consequence of 2.3, will be given following the proof of 2.3.

It is easy to check that 1.8 admits a local version which, roughly speaking, says that a normalized unconditional basic sequence $\{x_i\}_{i=1}^n$ in $L_p(0,1)$; $p > 2$ is either equivalent to the unit vector basis of ℓ_p^n , with a constant of equivalence independent of the particular sequence $\{x_i\}_{i=1}^n$, or it has $k(n)$ disjoint blocks which are 2-equivalent to the unit vector basis of $\ell_2^{k(n)}$, where $k(n) \to \infty$ as $n \to \infty$.

We wish to discuss now some questions concerning Banach spaces X which are isomorphic to a complemented subspace of $L_p(0,1)$ for some $1 < p < \infty$. For this study it is convenient to use the factorization constant in $L_p(0,1)$; $1 < p < \infty$, namely

$$\gamma_p(X) = \inf\{\|T\|\cdot\|S\| \; ; \; T: X \to L_p(0,1),\ S: L_p(0,1) \to X,\ ST = \text{identity on } X\}.$$

For instance, by using the embedding of ℓ_2 onto the span of the Rademacher functions in $L_p(0,1)$, it is easily checked that $\gamma_p(\ell_2) \leq \max\{B_p, B_{p'}\}$, where $1/p + 1/p' = 1$. It is well-known that, for $q > 2$, the Khintchine constant B_q satisfies $B_q \leq \sqrt{q}$. Hence, $\gamma_p(\ell_2) \leq \max\{\sqrt{p}\ , \sqrt{p'}\}$ and it is known (cf. [32]) that $\gamma_p(\ell_2)$ behaves like $\sqrt{p}$ when $p \to \infty$. In order to state our next result we also need the following definition which, for $K = 1$, coincides with that of a Banach lattice. Let $K \geq 1$; a K-lattice X is a Banach space with a consistent structure of a vector lattice so that $\|x\| \leq K\|y\|$, whenever $x, y \in X$ and $|x| \leq |y|$. This notion is clearly a generalization of that of a space with an unconditional basis having unconditional constant $\leq K$.

Theorem 1.9: *Let* X *be a* K*-lattice for some* $K \geq 1$. *If* X *is linearly isomorphic to a complemented subspace of* $L_p(0,1)$ *for some* $1<p<\infty$ *and* $K\gamma_p(X)^2 < \gamma_p(\ell_2)/\sqrt{2}$ *then* X *is linearly isomorphic and order equivalent to an* $L_p(\mu)$ *-space for a suitable measure* μ .

Proof: Let X_0 be the (usual) Banach lattice obtained from X by renorming it with the equivalent norm

$$\|x\|_0 = \sup\{\|y\|;\ |y| \le |x|\};\ x \in X\ .$$

Observe that there is no loss of generality in assuming that X is linearly isomorphic to a complemented subspace of $L_p(0,1)$ for some $1 < p < 2$. If X, or equivalently X_0, is not linearly isomorphic and order equivalent to an $L_p(\mu)$-space then, by 1.8, ℓ_2 is lattice finitely representable in X_0^*. Since by [65] (see also [55]), Hilbert spaces are uniformly complemented in $L_{p'}(0,1)$ (where $1/p + 1/p' = 1$) it follows that there exists a constant $C < \infty$ so that, for every $\epsilon > 0$ and every integer n, there is a sequence $\{y_i\}_{i=1}^n$ of disjoint elements of X_0^* which is $1+\epsilon$-equivalent to the unit vector basis of ℓ_2^n and such that span $\{y_i\}_{i=1}^n$ is the range of a projection Q of norm $\le C$ in X_0^*. For every such projection Q there exists a sequence $\{z_i\}_{i=1}^n$ of elements of X_0 such that

$$Qy = \sum_{i=1}^n y(z_i)y_i\ ;\ y \in X_0^*$$

By using a diagonal principle (cf. [79] or [49] p. 20) we can assume without loss of generality that, for each $1 \le i \le n$, the support of z_i is contained in that of y_i (we use the possibility of representing X_0 and X_0^* as lattices of measurable functions on the *same* probability space). This observation shows that X itself contains uniformly isomorphic copies of ℓ_2^n; $n = 1,2,\ldots$, each of which is spanned by disjoint vectors (these copies are also uniformly complemented in X but we shall not make use of this fact). Moreover, by a theorem of Krivine [42], X contains, for each n, an almost isometric copy of ℓ_2^n, which is spanned by disjoint vectors. Fix $\epsilon > 0$ and n, and let $\{x_j\}_{j=1}^n$ be a normalized sequence of disjoint elements of X which is $1+\epsilon$-equivalent to the unit vector basis of ℓ_2^n. Let $\{x_j^*\}_{j=1}^n$ be a sequence of elements of X^* so that $x_j^* x_j = 1$ and $\|x_j^*\| = 1$ for all $1 \le j \le n$.

Let χ_{E_j} denote the characteristic function of the support of x_j and put

$$Px = \sum_{j=1}^{n} x_j^*(x\chi_{E_j})x_j \ ; \ x \in X .$$

Then P is a projection from X onto span $\{x_j\}_{j=1}^n$ and

$$\|Px\| \le (1+\varepsilon)\Big(\sum_{j=1}^{n} |x_j^*(x\chi_{E_j})|^2\Big)^{1/2} \le (1+\varepsilon)\Big(\sum_{j=1}^{n} \|x\chi_{E_j}\|^2\Big)^{1/2} .$$

Since $\gamma_p(X) < \infty$ there exists operators $T\colon X \to L_p(0,1)$ and $S\colon L_p(0,1) \to X$ such that ST = identity on X and $\|S\| \cdot \|T\| < (1+\varepsilon)\,\gamma_p(X)$. Hence,

$$\|Px\| \le (1+\varepsilon)\,\|S\|\Big(\sum_{j=1}^{n} \|T(x\chi_{E_j})\|^2\Big)^{1/2} \le$$

$$\le (1+\varepsilon)\,A_p^{-1}\|S\|\Big(\int_0^1 \Big\|\sum_{j=1}^{n} r_j(u)T(x\chi_{E_j})\Big\|_p^p du\Big)^{1/p} \le (1+\varepsilon)A_p^{-1}\|S\|\cdot\|T\|\Big(\int_0^1 \Big\|\sum_{j=1}^{n} r_j(u)x\chi_{E_j}\Big\|^p du\Big)1/p$$

Since $\{x\chi_{E_j}\}_{j=1}^n$ is a sequence with unconditional constant $\le K$ and $A_p \ge A_1 = 1/\sqrt{2}$ we get that

$$\|Px\| \le (1+\varepsilon)^2\, K\, \gamma_p(x)\, \sqrt{2}\, \|x\| \ ;$$

i.e., that $\|P\| \le (1+\varepsilon)^2\, K\, \gamma_p(X)\, \sqrt{2}$.

Now, by mapping ℓ_2^n into $L_p(0,1)$ via the vectors $\{x_j\}_{j=1}^n$ and the operator T on the one hand, and by mapping $L_p(0,1)$ onto ℓ_2^n via S,P and $\{x_j\}_{j=1}^n$ on the other hand,it follows that

$$\gamma_p(\ell_2^n) \le (1+\varepsilon)^2\|T\|\cdot\|S\|\cdot\|P\| \le (1+\varepsilon)^5 K\, \gamma_p(X)^2\sqrt{2} \ ;$$

i.e., that $\gamma_p(\ell_2) \le K\, \gamma_p(X)^2\, \sqrt{2}$ since both ε and n are arbitrarily chosen. This, however, contradicts our hypothesis.

□

There are some interesting consequences of 1.9 which we would like to single out.

Corollary 1.10: *Let* P *be a bounded linear projection in* $L_p(0,1)$; $1<p<\infty$ *such that its range* X *is a K-lattice (not necessarily a sublattice of* $L_p(0,1)$ *). If* $K\|P\|^2 < \gamma_p(\ell_2)/\sqrt{2}$ *then* X *itself is linearly isomorphic and order equivalent to an* $L_p(\mu)$*-space for a suitable measure* μ .

Corollary 1.11: *Let* X *be a Banach lattice which is linearly isomorphic to an* $L_p(\nu)$ *-space;* $1<p<\infty$ *and* $d(X,L_p(\nu))^2<\gamma_p(\ell_2)/\sqrt{2}$. *Then* X *is linearly isomorphic and order equivalent to an* $L_p(\mu)$ *-space for a suitable measure* μ .

Corollary 1.12: *Every normalized unconditional basis of* ℓ_p ; $1<p\neq 2<\infty$ *which is not equivalent to the unit vector basis of* ℓ_p *has unconditional constant* $K \geq \gamma_p(\ell_2)/\sqrt{2}$.

Remark : It is well known (e.g., cf[63]) that ℓ_p ; $1 < p \neq 2 < \infty$ does not have a unique unconditional basis. However, 1.12 shows that if we restrict our attention to normalized unconditional bases of ℓ_p whose unconditional constant is less than $\gamma_p(\ell_2)/\sqrt{2} \approx \max\{\sqrt{p/2}\ , \sqrt{p/2(p-1)}\ \}$ then, up to equivalence, the unit vector basis of ℓ_p is indeed the unique unconditional basis of ℓ_p .

Note that Theorem 1.9 as well as Corollaries 1.10, 1.11 and 1.12 are interesting especially when $p >> 2$ and they are vacuous when p is close to 2 . However, by a suitable compactness argument, we prove that certain versions of 1.9 and 1.12 are still valid even for values of p which are close to 2 .

Theorem 1.13: *For each* $1<p\neq 2<\infty$ *there exists a number* $K_p > 1$ *such that every* K_p *-lattice, which is linearly isometric to an* $L_p(\mu)$ *-space, is already linearly isomorphic and order equivalent to some* $L_p(\nu)$ *-space, for a suitable measure* ν . *In particular, every normalized unconditional basis of* ℓ_p ; $1<p\neq 2<\infty$, *which is not equivalent to the unit vector basis of*

ℓ_p, *has unconditional constant* $\geq K_p$.

Proof: Suppose that, for some $1 < p \neq 2 < \infty$, there exists no $K_p > 1$ as in the statement of the theorem. Then, for every integer n, there is a $1 + \frac{1}{n}$-lattice X_n which is linearly isometric to an L_p-space but which is not simultaneously order equivalent and linearly isomorphic to such a space. Let $\mathcal{F}$ be a free ultrafilter on the integers and let $U = \prod_{n=1}^{\infty} X_n/\mathcal{F}$ be the ultraproduct of the family $\{X_n\}_{n=1}^{\infty}$ (for details on ultraproducts of Banach spaces see [15]). As is easily checked, the space U is a Banach lattice with the order induced by that of the X_n's and, furthermore, U is linearly isometric to an $L_p(\lambda)$-space. Hence, by the characterization of Banach lattices which are linearly isometric to an L_p-space (proved by H. E. Lacey and P. Wojtaszczyk [43] and S. J. Bernau and H. E. Lacey [7]), U is linearly isometric and order equivalent to the direct sum $L_p(\Omega_1,\Sigma_1,\lambda_1) \oplus_p L_p(\Omega_2,\Sigma_2,\lambda_2;E_p^2)$, where $(\Omega_1,\Sigma_1,\lambda_1)$ and $(\Omega_2,\Sigma_2,\lambda_2)$ are suitable measure spaces, E_p^2 is the two dimensional Banach lattice obtained when ℓ_p^2 is endowed with the monotone unconditional basis $\{(e_1+e_2)/2^{1/p}, (e_1-e_2)/2^{1/p}\}$ ($\{e_1,e_2\}$ stands here for the unit vector basis of ℓ_p^2) and $L_p(\Omega_2,\Sigma_2,\lambda_2;E_p^2)$ is the space of all E_p^2-valued functions whose norms are p-integrable with respect to λ_2.

By 1.8, each of the X_n's contains a sequence of n disjoint normalized vectors which is $1 + \frac{1}{n}$ - equivalent to the unit vector basis of ℓ_2^n. Hence, by the definition of the ultraproduct of Banach lattices we get that $U = L_p(\Omega_1, \Sigma_1, \lambda_1) \oplus_p L_p(\Omega_2, \Sigma_2, \lambda_2; E_p^2)$ contains a normalized sequence of disjoint vectors $\{u_n\}_{n=1}^{\infty}$ which is 1-equivalent to the unit vector basis of ℓ_2. This is however, a contradiction since $\{u_n\}_{n=1}^{\infty}$ satisfies the condition

$$2^{-1}\Big(\sum_{n=1}^{\infty} \|u_n\|^p\Big)^{1/p} \leq \Big\|\sum_{n=1}^{\infty} u_n\Big\| \leq 2\Big(\sum_{n=1}^{\infty} \|u_n\|^p\Big)^{1/p}.$$

Indeed, this follows easily from the fact that the lattice $L_p(\Omega_2,\Sigma_2,\lambda_2;E_p^2)$ is linearly isometric (but not order equivalent) to $L_p(\Omega_2,\Sigma_2,\lambda_2;\ell_p^2)$ and, under this isometry, a sequence of disjoint elements of $L_p(\Omega_2,\Sigma_2,\lambda_2;E_p^2)$ is mapped into a sequence of functions in $L_p(\Omega_2,\Sigma_2,\lambda_2;\ell_p^2)$ which, for each $\omega \in \Omega_2$, takes on at most two (vector) values different from zero. □

We present now a quantitative version of a well-known theorem of Kadec and Pelczynski [38].

Theorem 1.14: *For every* $2 < p < \infty$ *there exists a constant* K_p *such that every normalized weakly null sequence in* $L_p(0,1)$ *has a subsequence* $\{x_n\}_{n=1}^{\infty}$ *which is* K_p *-equivalent to a symmetric* X_p *basis; i.e., there exists* $0\le w\le 1$ *such that for all scalars* $\{a_n\}_{n=1}^{\infty}$,

$$K_p^{-1}\|\sum_{n=1}^{\infty} a_n x_n\|_p \le \max\{(\sum_{n=1}^{\infty} |a_n|^p)^{1/p}, w(\sum_{n=1}^{\infty} |a_n|^2)^{1/2}\} \le K_p\|\sum_{n=1}^{\infty} a_n x_n\|_p .$$

The main step in the proof of 1.14 is a theorem from Burkholder [13]. Recall that, if $(\Omega,\mathfrak{F},P)$ is a probability space and $\mathfrak{F}_1 \subseteq \mathfrak{F}_2 \subseteq \dots \subseteq \mathfrak{F}$ is an increasing sequence of σ-fields, a sequence $\{d_n\}_{n=1}^{\infty}$ of $L_1(\Omega,\mathfrak{F},P)$ functions is called a *martingale difference* sequence with respect to $\{\mathfrak{F}_n\}_{n=1}^{\infty}$ if d_n is $\mathfrak{F}_n$ measurable and $E_n d_{n+1} = 0$ for $n = 1,2,\dots$, where E_n is the conditional expectation with respect to $\mathfrak{F}_n$.

Theorem [13], [14] : *For every* $2 < p < \infty$ *there exists a constant* C_p *such that for any sequence* $\{d_n\}_{n=1}^{\infty}$ *of martingale differences with respect to* $\{\mathfrak{F}_n\}_{n=1}^{\infty}$

$$C_p^{-1}\|\sum_{n=1}^{\infty} d_n\|_p \le \max\{\|(\sum_{n=1}^{\infty} E_n d_{n+1}^2)^{1/2}\|_p, (\sum_{n=1}^{\infty} \|d_n\|_p^p)^{1/p}\} \le C_p\|\sum_{n=1}^{\infty} d_n\|_p .$$

Proof of 1.14: Passing to a subsequence and perturbing, we can assume that the given sequence $\{y_n\}_{n=1}^{\infty}$ is a block basis of the Haar system $\{h_{n,i}\}_{i=1,n=0}^{2^n\ \ \infty}$ in $L_p(0,1)$. Since the Haar system is a sequence of martingale differences it follows that $\{y_n\}_{n=1}^{\infty}$ is a sequence of martingale differences. Moreover, the sequence $\{\mathfrak{F}_n\}_{n=1}^{\infty}$ of σ-fields associated to $\{y_n\}_{n=1}^{\infty}$ consists of finite σ-fields. It follows that for each n there exists a subsequence $\{y_{n_k}\}_{k=1}^{\infty}$ of $\{y_n\}_{n=1}^{\infty}$ such that $\{E_n y_{n_k}^2\}_{k=1}^{\infty}$ converges in $L_{p/2}$. By a simple diagonal argument there exist a subsequence $\{y_{n_k}\}_{k=1}^{\infty}$ and a sequence $\{z_n\}_{n=1}^{\infty}$ of elements of $L_{p/2}(0,1)$ such that

$$E_n\, y_{n_k}^2 \to z_n \quad \text{as} \quad k \to \infty$$

in $L_{p/2}(0,1)$ for each $n = 1,2,\dots$. Clearly, $\|z_n\|_{p/2} \leq 1$ and $\{z_n\}_{n=1}^{\infty}$ forms a martingale with respect to $\{\mathcal{F}_n\}_{n=1}^{\infty}$. Since $p/2 > 1$ we get, by the martingale convergence theorem in $L_{p/2}(0,1)$ (cf. [57]), that $\{z_n\}_{n=1}^{\infty}$ converges in $L_{p/2}(0,1)$, say to z . We can thus choose an increasing sequence of integers $\{n_k\}_{k=1}^{\infty}$ such that

$$\|E_{n_k}\, y_{n_{k+1}}^2 - z\|_{p/2} < 2^{-k} .$$

Given a sequence $\{a_k\}_{k=1}^{\infty}$ of scalars, notice that

$$\|(\sum_{k=1}^{\infty} E_{n_k}(a_{k+1}^2 y_{n_{k+1}}^2))^{1/2}\|_p \leq (\|z\|_{p/2} \sum_{k=1}^{\infty} a_{k+1}^2)^{1/2} + \max_{1\leq k<\infty} |a_{k+1}|$$

and

$$\|(\sum_{k=1}^{\infty} E_{n_k}(a_{k+1}^2 y_{n_{k+1}}^2))^{1/2}\|_p \geq (\|z\|_{p/2} \sum_{k=1}^{\infty} a_{k+1}^2)^{1/2} - \max_{1\leq k<\infty} |a_{k+1}| .$$

Since $\max_{1\leq k<\infty} |a_{k+1}| \leq (\sum_{k=1}^{\infty} |a_k|^p)^{1/p}$ we get from Burkholder's theorem (for $d_k = a_k y_{n_k}$) that

$$(3C_p)^{-1}\|\sum_{k=1}^{\infty} a_k y_{n_k}\|_p \leq \max\{\|z\|_{p/2}^{1/2} (\sum_{k=1}^{\infty} a_k^2)^{1/2},\ (\sum_{k=1}^{\infty} |a_k|^p)^{1/p} \leq$$

$$\leq 3C_p\|\sum_{k=1}^{\infty} a_k\, y_{n_k}\|_p .$$

□

We conjecture that a somewhat stronger version of 1.14 is true: *There exists a constant* K *such that for every* $2 < p < \infty$ *every normalized weakly null sequence in* $L_p(0,1)$ *has a K-symmetric subsequence.*

The assertion above is true for even values of p and with $K = 2 + \epsilon$ for any $\epsilon > 0$. Here is a sketch of a proof for this fact.

Let $2 \leq p$ be an even integer and let $\{x_i\}_{i=1}^{\infty}$ be a weakly null normalized sequence in $L_p(0,1)$. Passing to a subsequence we may assume that $\{x_i\}_{i=1}^{\infty}$ is a block basis of the Haar system in $L_p(0,1)$ and that for every integer $1 \leq s < p$ there exists $x^{(s)} \in L_{p/s}(0,1)$ such that

$$x_i^s \to x^{(s)} \quad \text{as} \quad i \to \infty \quad \text{weakly in} \quad L_{p/s}(0,1) .$$

For convenience in notation we put $x^{(p)} \equiv 1$ If $\{x_i\}_{i=1}^{\infty}$ does not contain a subsequence which is 2-equivalent to the unit vector basis of ℓ_p , then by [38] there exists a constant K_1 such that

$$\left(\sum_{i=1}^{\infty} a_i^2\right)^{1/2} \leq K_1 \left\| \sum_{i=1}^{\infty} a_i x_i \right\| ,$$

for all sequences $\{a_i\}_{i=1}^{\infty}$ of scalars, so we assume that this is really the case. Given a positive decreasing to zero sequence $\{\epsilon_i\}_{i=1}^{\infty}$ we can assume that for all integers $\{s_j\}_{j=1}^{n}$ with $s_j \geq 1$ and $\sum_{j=1}^{\infty} s_j = p$ we have

$$(+) \qquad \left|\int_0^1 x_{i_1}^{s_1} \cdots x_{i_j}^{s_j} x^{(s_{j+1})} \cdots x^{(s_n)} d\mu - \int_0^1 x_{i_1}^{s_1} \cdots x_{i_{j-1}}^{s_{j-1}} x^{(s_j)} \cdots x^{(s_n)} d\mu\right| < \\ < \epsilon_{i_j}$$

for all $i_1 < i_2 <.. < i_j$ This in turn implies that for all $\{s_j\}_{j=1}^{n}$ with $s_j \geq 1$ and $\sum_{j=1}^{n} s_j = p$, and all $i_1 < i_2 <\ldots< i_n$

$$\left|\int_0^1 x_{i_1}^{s_1} \cdots x_{i_n}^{s_n} d\mu - \int_0^1 x^{(s_1)} \cdots x^{(s_n)} d\mu\right| < p\epsilon_1 .$$

For $\{s_j\}_{j=1}^n$ as above define $\alpha_{s_1,\ldots,s_n}$ by the identity

$$\left(\sum_{i=1}^{\infty} a_i\right)^p = \sum_{s_1+\ldots+s_n=p} \alpha_{s_1,\ldots,s_n} \sum_{i_1<\ldots<i_n} a_{i_1}^{s_1}\cdots a_{i_n}^{s_n}$$

for all sequences $\{a_i\}_{i=1}^{\infty}$ of scalars where the first sum is over all n and all sequences of integers $\{s_j\}_{j=1}^n$ such that $s_j \geq 1$ $j = 1,\ldots,n$ and $\sum_{j=1}^n s_j = p$.

Then,

$$\int_0^1 \left|\sum_{i=1}^{\infty} a_i x_i\right|^p d\mu = \sum_{s_1+\ldots+s_n=p} \alpha_{s_1,\ldots,s_n} \sum_{i_1<\ldots<i_n} a_{i_n}^{s_1}\cdots a_{i_n}^{s_n} \int_0^1 x_{i_1}^{s_1}\cdots x_{i_n}^{s_n}\, d\mu .$$

We would like to evaluate the difference in absolute value between the last expression and

$$\sum_{s_1+\ldots s_n=p} \alpha_{s_1,\ldots,s_n} \sum_{i_1<\ldots<i_n} a_{i_1}^{s_1}\cdots a_{i_n}^{s_n} \int_0^1 x^{(s_1)}\cdots x^{(s_n)}\, d\mu .$$

Let Σ' denote the sum over all $s_1,\ldots,s_n$ such that $s_i \geq 2$, $i = 1,\ldots,n$, $\sum_{i=1}^n s_i = p$ and let Σ'' denote the sum over all $s_1,\ldots,s_n$ such that $s_i \geq 1$, $i = 1,\ldots,n$, $\sum_{i=1}^n s_i = p$ and at least one s_i is one. Assume also that $\left\|\sum_{i=1}^{\infty} a_i x_i\right\| = 1$. Then $|a_i| \leq K_2$, $i = 1,2,\ldots$, where K_2 is the unconditionality constant of the Haar system.

Now,

$$\Sigma' \alpha_{s_1,\dots,s_n} \sum_{i_1<\dots<i_n} |a_{i_1}^{s_1}\dots a_{i_n}^{s_n}|\, \Big|\int_0^1 x_{i_1}^{s_1}\dots x_{i_n}^{s_n} d\mu - \int_0^1 x^{(s_1)}\dots x^{(s_n)} d\mu\Big|$$

$$\le \Sigma' \alpha_{s_1,\dots,s_n} \prod_{j=1}^{n} \sum_{i=1}^{\infty} |a_i|^{s_j} \cdot p \cdot \epsilon_1$$

$$\le p\cdot\epsilon_1 \cdot \Sigma' \alpha_{s_1,\dots,s_n} \Big(\sum_{i=1}^{\infty} a_i^2\Big)^{p/2}$$

$$\le p\cdot\epsilon_1 \ \Sigma' \alpha_{s_1,\dots,s_n} \cdot K_1^p \cdot \Big\|\sum_{i=1}^{\infty} a_i x_i\Big\|^p .$$

On the other hand, if k is the last index for which $s_j = 1$ (in case $s_1,\dots,s_n$ belongs to the range of Σ'') then, since $x^{(s_k)} \equiv 0$, we get from (+)

$$\sum_{i_1<\dots<i_n} |a_{i_1}^{s_1}\dots a_{i_n}^{s_n}|\, \Big|\int_0^1 x_{i_1}^{s_1}\dots x_{i_n}^{s_n} d\mu - \int_0^1 x^{(s_1)}\dots x^{(s_n)} d\mu\Big|$$

$$= \sum_{i_1<\dots<i_n} |a_{i_1}^{s_1}\dots a_{i_n}^{s_n}|\, \Big|\int_0^1 x_{i_1}^{s_1}\dots x_{i_n}^{s_n} d\mu - \int_0^1 x_{i_1}^{s_1}\dots x_{i_{k-1}}^{s_{k-1}} x^{(s_k)}\dots x^{(s_n)} d\mu\Big|$$

$$\le \sum_{i_1<\dots<i_n} |a_{i_1}^{s_1}\dots a_{i_n}^{s_n}| \cdot p \cdot \epsilon_{i_k}$$

$$\le \sum_{i_1<\dots<i_k} |a_{i_1}^{s_1}\dots a_{i_k}^{s_k}| \prod_{j=k+1}^{n} \sum_{i=1}^{\infty} |a_i|^{s_j} \cdot p \cdot \epsilon_{i_k}$$

$$\le \sum_{i_1<\dots<i_k} \epsilon_{i_k} \cdot K_2^p \cdot p\Big(\sum_{i=1}^{\infty} a_i^2\Big)^{\sum_{j=k+1}^{n} s_j/2}$$

$$\le \sum_{i=1}^{\infty} i^k \cdot \epsilon_i \cdot K_2^p \cdot K_1^p \cdot p .$$

So

$$\Sigma'' \alpha_{s_1,\ldots,s_n} \sum_{i_1<\ldots<i_n} |a_{i_1}^{s_1}\ldots a_{i_n}^{s_n}| \left|\int_0^1 x_{i_1}^{s_1}\ldots x_{i_n}^{s_n} d\mu - \int_0^1 x^{(s_1)}\ldots x^{(s_n)} d\mu\right|$$

$$\leq \Sigma'' \alpha_{s_1,\ldots,s_n} \sum_{i=1}^{\infty} i^k \cdot \epsilon_i \cdot K_2^p \cdot K_1^p \cdot p$$

and we conclude that if the ϵ_i are chosen properly, then the difference between $1 = \|\sum_{i=1}^{\infty} a_i x_i\|$ and

$$\sum_{s_1+\ldots+s_n=p} \ \sum_{i_1<\ldots<i_n} a_{i_1}\ldots a_{i_n} \int_0^1 x^{(s_1)}\ldots x^{(s_n)} d\mu$$

is smaller than ϵ. Since the last expression is invariant under permutations we get that

$$\|\sum_{i=1}^{\infty} a_i x_i\| \leq (1+\epsilon) \|\sum_{i=1}^{\infty} a_i x_{\pi(i)}\|$$

for any permutation π of the integers and any sequence $\{a_i\}_{i=1}^{\infty}$ of scalars. From this we get also the invariance of the norm under change of signs up to a constant $2+2\epsilon$.

2 SUBSPACES OF BANACH LATTICES OF TYPE 2.

We begin this section with a characterization of finite symmetric basic sequences in a Banach lattice of type 2 which resembles and also generalizes Theorem 1.1 on the classification of symmetric basic sequences in $L_p(0,1)$; $p > 2$.

A Banach lattice is of type 2 if and only if it is 2-convex and q-concave for some $q < \infty$ (cf. [54]). In applications, however, it is more convenient to use the setting of 2-convex and q-concave Banach lattices. For technical reasons we take q to be an even integer.

Theorem 2.1: For every $K \geq 1$, $M \geq 1$ *and every integer* $m \geq 1$ *there exists a constant* $D = D(K,M,m) < \infty$ *such that if* $\{y_i\}_{i=1}^n$ *is a finite K-symmetric normalized basic sequence in a Banach lattice* Y *which is 2-convex and 2m-concave with both constants* $\leq M$ *then, for every choice of scalars* $\{a_i\}_{i=1}^n$,

$$D^{-1}\|\sum_{i=1}^n a_iy_i\| \leq \max\{(\sum_{\pi}\|\max_{1\leq i\leq n} |a_{\pi(i)}y_i|\|^{2m}/n!)^{1/2m}, \frac{\|\sum_{i=1}^n y_i\|}{\sqrt{n}}(\sum_{i=1}^n |a_i|^2)^{1/2}\} \leq$$

$$\leq D\|\sum_{i=1}^n a_iy_i\| ,$$

where $\sum_{\pi}$ *refers to summation over all permutations* π *of* $\{1,\ldots,n\}$.

We need first the following known inequalities of Holder type (cf. [41]).

Lemma 2.2: Let Z *be a Banach lattice. Then for every sequence* $\{z_i\}_{i=1}^n$ *of elements of* Z *, every sequence of positive reals* $\{\theta_i\}_{i=1}^n$ *with* $\sum_{i=1}^n \theta_i = 1$ *and every pair* $1 < p < r < \infty$ *, we have* *(i)* $\| |z_1|^{\theta_1} \dots |z_n|^{\theta_n} \| \leq \|z_1\|^{\theta_1} \dots \|z_n\|^{\theta_n}$.

(ii) $\|(\sum_{i=1}^{n} |z_i|^r)^{1/r}\| \leq \|(\sum_{i=1}^{n} |z_i|^p)^{1/p}\|^{p/r} \| \max_{1\leq i\leq n} |z_i| \|^{1-p/r}$.

Proof: It obviously suffices to prove *(i)* for $n = 2$. Let $0 < \theta < 1$ and observe that, for any pair of positive reals α and β we have $\alpha^{1-\theta}\beta^{\theta} \leq (1-\theta)\alpha + \theta\beta$. Since this is a numerical inequality between expressions which are homogeneous of degree one it follows from [41] that, for every $\lambda > 0$, we obtain

$$\| |z_1|^{1-\theta} \cdot |z_2|^{\theta} \| = \| |\lambda^{1/(1-\theta)} z_1|^{1-\theta} |\lambda^{-1/\theta} z_2|^{\theta} \| \leq (1-\theta)\lambda^{1/(1-\theta)} \|z_1\| + \theta\lambda^{-1/\theta}\|z_2\|$$

and the desired result is achieved by taking $\lambda = (\|z_2\|/\|z_1\|)^{\theta(1-\theta)}$.

In order to prove *(ii)* we put $\theta = 1 - p/r$ and notice that

$$(\sum_{i=1}^{n} |z_i|^r)^{1/r} \leq (\sum_{i=1}^{n} |z_i|^p)^{1/r} \cdot \max_{1\leq i\leq n} |z_i|^{(r-p)/r} = (\sum_{i=1}^{n} |z_i|^p)^{(1-\theta)/p} \cdot \max_{1\leq i\leq n} |z_i|^{\theta} .$$

Hence, by *(i)* , we get that

$$\|(\sum_{i=1}^{n} |z_i|^r)^{1/r}\| \leq \|(\sum_{i=1}^{n} |z_i|^p)^{1/p}\|^{1-\theta} \cdot \| \max_{1\leq i\leq n} |z_i| \|^{\theta} .$$

□

We are now prepared for the *proof of* 2.1. Let $\{y_i\}_{i=1}^n$ be a sequence as above. In the first part of the proof we must distinguish between the absolute value $|y|$ of an element $y \in Y$ and the absolute value $|\sum_{i=1}^{n} a_i y_i|_0 = \sum_{i=1}^{n} |a_i| y_i$ of an element $\sum_{i=1}^{n} a_i y_i$ when this vector is considered as an element of the K-lattice generated by the unconditional basis $\{y_i\}_{i=1}^n$.

As we have already mentioned, Y is of type 2, and therefore the K-lattice generated by the sequence $\{y_i\}_{i=1}^n$ is also of type 2; in particular, it also is 2-convex. More precisely, we have that the 2-convexity constant of $[y_j]_{j=1}^n$ is at most $C = \sqrt{2}\, M^2K^2B_{2m}$; indeed, for any sequence $\{f_i\}_{i=1}^k$ in $[y_j]_{j=1}^n$,

$$\|(\sum_i |f_i|_0^2)^{1/2}\| = \|\sum_j (\sum_i f_i^2(j))^{1/2} y_j\|$$

$$\leq KA_1^{-1}\|\sum_j (\int_0^1 |\sum_i f_i(j) r_i(u)|\,du) y_j\| \quad \text{(by the unconditionality of } \{y_j\}_{j=1}^n \text{ and Khintchine's inequality).}$$

$$\leq A_1^{-1}K^2 \int_0^1 \|\sum_{j,i} f_i(j) r_i(u) y_j\|\,du$$

$$\leq \sqrt{2}\, K^2 (\int_0^1 \|\sum_i r_i(u) f_i\|^{2m} du)^{1/2m}$$

$$\leq M\sqrt{2}\, K^2 \|(\int_0^1 |\sum_i r_i(u) f_i|^{2m} du)^{1/2m}\| \quad \text{(2m-concavity of } Y\text{)}$$

$$\leq M\sqrt{2}\, K^2 B_{2m} \|(\sum_i f_i^2)^{1/2}\| \quad \text{(Khintchine in } L_{2m}\text{)}$$

$$\leq M^2\sqrt{2}\, K^2 B_{2m} (\sum_i \|f_i^2\|)^{1/2} \quad \text{(2-convexity of } Y\text{)}.$$

Let $\{\sigma_j\}_{j=1}^n$ denote the set of all distinct cyclic permutations of the integers $\{1,2,..,n\}$. Using the 2-convexity of the K-lattice $([y_j], |\cdot|_0)$, we get that

$$(\sum_{i=1}^n |a_i|^2)^{1/2} \|\sum_{i=1}^n y_i\| = \|\sum_{i=1}^n (\sum_{j=1}^n |a_{\sigma_j(i)}|^2)^{1/2} y_i\| = \|(\sum_{j=1}^n |\sum_{i=1}^n a_{\sigma_j(i)} y_i|_0^2)^{1/2}\| \leq$$

$$\leq C(\sum_{j=1}^n \|\sum_{i=1}^n a_{\sigma_j(i)} y_i\|^2)^{1/2} \leq CK\sqrt{n}\, \|\sum_{i=1}^n a_i y_i\|$$

i.e.

$$CK\|\sum_{i=1}^n a_i y_i\| \geq \frac{\|\sum_{i=1}^n y_i\|}{\sqrt{n}} (\sum_{i=1}^n |a_i|^2)^{1/2} ,$$

for every choice of scalars $\{a_i\}_{i=1}^n$. On the other hand, for every permutation π of $\{1,\ldots,n\}$,

$$\|\sum_{i=1}^n a_i y_i\| \geq K^{-1}\int \|\sum_{i=1}^n a_{\pi(i)} r_i(t) y_i\| dt \geq$$

$$\geq K^{-1}\|\int_0^1 |\sum_{i=1}^n a_{\pi(i)} r_i(t) y_i| dt\| \geq K^{-1}\| \max_{1\leq i\leq n} |a_{\pi(i)} y_i|\|$$

from which the right-hand side inequality of 2.1 with $D = CK$ is deduced by averaging in the sense of ℓ_{2m} over all permutations π of $\{1,2,\ldots,n\}$.

The proof of the left-hand side inequality is harder. Let π be an arbitrary permutation of the integers and $\{a_i\}_{i=1}^n$ an arbitrary sequence of scalars. By the 2m-concavity of Y and (*) from the Introduction we get that

$$\|\sum_{i=1}^n a_i y_i\| \leq K\|\sum_{i=1}^n a_{\pi(i)} y_i\| \leq KMB_{2m}\|(\sum_{i=1}^n |a_{\pi(i)} y_i|^2)^{1/2}\| = KBM_{2m}\|[(\sum_{i=1}^n |a_{\pi(i)} y_i|^2)^m]^{1/2m}\| \leq$$

$$\leq KMB_{2m}\|(\sum_{\ell=1}^m \sum_{\substack{s_1+\ldots+s_\ell=m \\ s_1,\ldots,s_\ell\geq 1}} \sum_{\substack{i_1,\ldots,i_\ell \\ \text{distinct}}} |a_{\pi(i_1)} y_{i_1}|^{2s_1}\ldots|a_{\pi(i_\ell)} y_{i_\ell}|^{2s_\ell})^{1/2m}\| .$$

Hence, by averaging in the sense of ℓ_{2m} over all possible permutations π of the integers $\{1,2,\ldots,n\}$ and by separating the term corresponding to $\ell = m$ from all the other terms, we get that

$$\|\sum_{i=1}^n a_i y_i\| \leq KMB_{2m}(n!)^{-1/2m}(S_1 + S_2) ,$$

where $S_1 = (\sum_\pi \|(\sum_{\substack{i_1,\ldots,i_m \\ \text{distinct}}} |a_{\pi(i_1)} y_{i_1}|^2\ldots|a_{\pi(i_m)} y_{i_m}|^2)^{1/2m}\|^{2m})^{1/2m}$,

and

$$S_2 = \Big(\sum_\pi \Big\| \sum_{\ell=1}^{m-1} \sum_{\substack{s_1+\ldots+s_\ell=m \\ s_1,\ldots,s_\ell\geq 1}} \sum_{\substack{i_1,\ldots,i_\ell \\ \text{distinct}}} |a_{\pi(i_1)}y_{i_1}|^{2s_1}\ldots|a_{\pi(i_\ell)}y_{i_\ell}|^{2s_\ell}\Big)^{1/2m}\Big\|^{2m}\Big)^{1/2m}.$$

From now on we assume, as we clearly may, that n is an even integer such that $n/2 > m$.

We first evaluate S_1 by using the 2m-concavity of Y;

$$S_1 \leq M\Big\|\Big(\sum_\pi \Big(\sum_{\substack{i_1,\ldots,i_m \\ \text{distinct}}} |a_{\pi(i_1)}y_{i_1}|^2\ldots|a_{\pi(i_m)}y_{i_m}|^2\Big)\Big)^{1/2m}\Big\| \leq$$

$$= M\Big\|\Big(\sum_{\substack{i_1,\ldots,i_m \\ \text{distinct}}} \Big(\sum_\pi |a_{\pi(i_1)}|^2\ldots|a_{\pi(i_m)}|^2\Big)|y_{i_1}|^2\ldots|y_{i_m}|^2\Big)^{1/2m}\Big\| =$$

$$= M((n-m)! \sum_{\substack{j_1,\ldots,j_m \\ \text{distinct}}} |a_{j_1}|^2\ldots|a_{j_m}|^2)^{1/2m}\Big\|\Big(\sum_{\substack{i_1,\ldots,i_m \\ \text{distinct}}} |y_{i_1}|^2\ldots|y_{i_m}|^2\Big)^{1/2m}\Big\| \leq$$

$$\leq M((n-m)!)^{1/2m}\Big(\sum_{i=1}^n |a_i|^2\Big)^{1/2} \Big\|\Big(\sum_{i=1}^n |y_i|^2\Big)^{1/2}\Big\| \leq$$

$$\leq M(n!)^{1/2m}\sqrt{2/n}\Big(\sum_{i=1}^n |a_i|^2\Big)^{1/2} A_1^{-1}\int_0^1 \Big\|\sum_{i=1}^n r_i(u)y_i\Big\|du \leq$$

$$\leq 2MK(n!)^{1/2m}\frac{\|\sum_{i=1}^n y_i\|}{\sqrt{n}}\Big(\sum_{i=1}^n |a_i|^2\Big)^{1/2}.$$

In order to evaluate S_2, for every $1 \leq \ell \leq m-1$ and every sequence $s_1,\ldots,s_\ell \geq 1$ so that $s_1+\ldots+s_\ell = m$, we put

$$S(\ell;s_1,\ldots,s_\ell) = \Big(\sum_\pi \Big\|\Big(\sum_{\substack{i_1,\ldots,i_\ell \\ \text{distinct}}} |a_{\pi(i_1)}y_{i_1}|^{2s_1}\ldots|a_{\pi(i_\ell)}y_{i_\ell}|^{2s_\ell}\Big)^{1/2m}\Big\|^{2m}\Big)^{1/2m}.$$

Then there is a constant $R_m < \infty$, depending only on m , such that

$$S_2 \leq \sum_{\ell=1}^{m-1} \sum_{\substack{s_1+\ldots+s_\ell=m \\ s_1,\ldots,s_\ell\geq 1}} S(\ell;s_1,\ldots,s_\ell) \leq R_m \max\{S(\ell;s_1,\ldots,s_\ell);\ \begin{matrix} 1\leq \ell \leq m-1 \\ s_1+\ldots+s_\ell = m \\ s_1,\ldots,s_\ell \geq 1 \end{matrix}\} .$$

Consequently, for some $1 \leq \ell \leq m-1$ and some $s_1,\ldots,s_\ell \geq 1$ with $s_1+\ldots+s_\ell = m$, we get that

$$S_2 \leq R_m\, S(\ell;s_1,\ldots,s_\ell) \quad .$$

We shall evaluate now the expression $S(\ell;s_1,\ldots,s_\ell)$ for this specific value of ℓ and $s_1,\ldots,s_\ell$. For this purpose we consider the lattice $Z = \ell_{2m}^{n!}(Y)$; i.e., the space of all sequences $z = (z(\pi))_\pi$ of length $n!$ with $z(\pi) \in Y$ for all π in which the norm is defined by

$$\|z\|_Z = (\sum_\pi \|z(\pi)\|^{2m})^{1/2m} \quad .$$

Then, by 2.1(i) applied for $\theta_j = s_j/m$; $j = 1,\ldots,\ell$ and the vectors $\{z_j\}_{j=1}^{\ell}$ in Z defined by $z_j(\pi) = (\sum_{i=1}^{n} |a_{\pi(i)}y_i|^{2s_j})^{1/2s_j}$, it follows that

$$S(\ell;s_1,\ldots,s_\ell) \leq \|z_1^{\theta_1}\ldots z_\ell^{\theta_\ell}\|_Z \leq \|z_1\|_Z^{\theta_1}\ldots\|z_\ell\|_Z^{\theta_\ell} \quad .$$

But, since $\ell \leq m-1$, at least one of the integers $s_1,\ldots,s_\ell$, say s_1 , is ≥ 2 . Using this fact and estimating the norm in $\ell_{2s_1}^n$ by that in ℓ_4^n and the norm in $\ell_{2s_j}^n$; $1 < j \leq \ell$, by that in ℓ_2^n we get that

$$S(\ell;s_1,\ldots,s_\ell) \leq (\sum_\pi \|(\sum_{i=1}^{n} |a_{\pi(i)}y_i|^4)^{1/4}\|^{2m})^{\theta_1/2m} \times$$

$$\times (\sum_\pi \|(\sum_{i=1}^{n} |a_{\pi(i)}y_i|^2)^{1/2}\|^{2m})^{(1-\theta_1)/2m} \leq$$

$$\leq \left(\sum_{\pi} \|(\sum_{i=1}^{n} |a_{\pi(i)}y_i|^4)^{1/4}\|^{2m}\right)^{\theta_1/2m} A_1^{\theta_1 - 1} K^{1-\theta_1} (n!)^{(1-\theta_1)/2m} \|\sum_{i=1}^{n} a_i y_i\|^{1-\theta_1} .$$

In conclusion, if we take into account the estimates obtained for S_1 and S_2 it follows that there exists a constant $C_1 = C_1(K,M,m) < \infty$ such that

$$\|\sum_{i=1}^{n} a_i y_i\| \leq C_1 A_1^{1-\theta_1} \max\left\{ \frac{\|\sum_{i=1}^{n} y_i\|}{\sqrt{n}} (\sum_{i=1}^{n} |a_i|^2)^{1/2} , (n!)^{-1/2m} S(\ell; s_1, \ldots, s_\ell)\right\} .$$

Thus, if the maximum is not attained at the first term then

$$\|\sum_{i=1}^{n} a_i y_i\| \leq C_1 K^{1-\theta_1} (n!)^{-\theta_1/2m} \left(\sum_{\pi} \|(\sum_{i=1}^{n} | a_{\pi(i)} y_i|^4)^{1/4}\|^{2m}\right)^{\theta_1/2m} \|\sum_{i=1}^{n} a_i y_i\|^{1-\theta_1} .$$

Since $\theta_1 \geq 2/m$ there is a constant $C_2 = C_2(K,M,m) < \infty$ such that

$$\|\sum_{i=1}^{n} a_i y_i\| \leq C_2 \left(\sum_{\pi} \|(\sum_{i=1}^{n} |a_{\pi(i)} y_i|^4)^{1/4}\|^{2m}/n! \right)^{1/2m} .$$

Now, we use again 2.2(ii) in the lattice $Z = \ell_{2m}^{n!}(Y)$, this time with $p = 2$, $r = 4$ and $z_i(\pi) = a_{\pi(i)} y_i/(n!)^{1/2m}$; $i = 1,2,\ldots,n$. It follows that

$$\|\sum_{i=1}^{n} a_i y_i\| \leq C_2 \|(\sum_{i=1}^{n} |z_i|^4)^{1/4}\|_Z \leq C_2 \|(\sum_{i=1}^{n} |z_i|^2)^{1/2}\|_Z^{1/2} \|\max_{1\leq i\leq n} |z_i| \|_Z^{1/2} .$$

But

$$\|(\sum_{i=1}^{n} |z_i|^2)^{1/2}\|_Z = \left(\sum_{\pi} \|(\sum_{i=1}^{n} |a_{\pi(i)} y_i|^2)^{1/2}\|^{2m}/n!\right)^{1/2m} \leq K A_1^{-1} \|\sum_{i=1}^{n} a_i y_i\|$$

and this implies that

$$\|\sum_{i=1}^{n} a_i y_i\| \leq A_1^{-1} K C_2^2 \|\max_{1\leq i\leq n} |z_i|\|_Z = A_1^{-1} K C_2^2 \left(\sum_{\pi} \|\max_{1\leq i\leq n} |a_{\pi(i)} y_i|\|^{2m}/n!\right)^{1/2m} .$$

This, of course, completes the proof.

□

Remarks: 1. Notice that the 2-convexity assumption is used only to prove the inequality:

$$(\ddagger) \qquad \left\| \sum_{i=1}^{n} a_i y_i \right\| \geq D^{-1} \frac{\left\| \sum_{i=1}^{n} y_i \right\|}{\sqrt{n}} \left(\sum_{i=1}^{n} |a_i|^2 \right)^{1/2} .$$

Thus the classification formula is valid in a 2m-concave lattice for any symmetric sequence which satisfies $(\ddagger)$.

2. In the case $Y = L_p(0,1)$; $p > 2$ one can easily deduce from 2.1 an alternative but more complicated proof of the classification theorem 1.1.

3. It is not clear whether the converse to 2.1 is true, i.e., whether for every sequence $\{y_i\}_{i=1}^n$ is a Banach lattice Y of type 2 and for a suitable $q > 2$, the norm

$$\|\{a_i\}_{i=1}^n\| = \max\{ (\sum_{\pi} \| \max_{1 \leq i \leq n} |a_{\pi(i)} y_i| \|^q / n!)^{1/q} , \frac{\left\| \sum_{i=1}^{n} y_i \right\|}{\sqrt{n}} (\sum_{i=1}^{n} |a_i|^2)^{1/2} \}$$

defines a space with a symmetric basis which embeds isomorphically in Y with a constant independent of the particular sequence $\{y_i\}_{i=1}^n$. Obviously, this norm defines a subspace of $\ell_q^{n!}(Y)$, but $\ell_q^{n!}(Y)$ need not embed into Y nor be finitely representable in it. In case the converse to 2.1 is false it would be quite interesting to modify the statement as to give a two-way characterization of finite symmetric basic sequences in a Banach lattice of type 2.

As a first application of 2.1 we prove the following generalization of 1.8.

Theorem 2.3: Let X *be a Banach lattice of type 2 which has an upper r-estimate for disjoint elements, for some* $r > 2$. *Then every lattice* X , *which is isomorphic to a subspace of* Y , *either has itself an upper r-estimate for disjoint elements or* ℓ_2 *is lattice finitely representable in* X .

We present first a lemma which follows easily from [35] by a simple convexification procedure.

Lemma 2.4: Let X *be a 2-convex Banach lattice. Then either* X *has an upper* p*-estimate for disjoint elements, for some* $p > 2$ *, or* ℓ_2 *is lattice finitely representable in* X .

Proof: Using a result of Krivine [42], ℓ_2 is lattice finitely representable in X if and only if there exists a C such that for every n there are disjoint norm one vectors $\{y_i\}_{i=1}^n$ in X such that $d(\ell_2^n,[y_i]_{i=1}^n) \leq C$. Therefore, we may assume, by renorming X by an equivalent norm, that the 2-convexity constant of X is 1 . (cf. [25]).

Assume that there is a k so that for every sequence $\{y_i\}_{i=1}^k$ of disjoint vectors in X , $d(\ell_2^k,[y_i]_{i=1}^k) \geq \sqrt{10}$. Let $\{x_i\}_{i=1}^n$ be an arbitrary sequence of disjoint norm one vectors in X and define a 1-unconditional norm on R^n by

$$|||\sum_{i=1}^n a_i e_i||| = \|\sum_{i=1}^n |a_i|^{1/2} x_i\|^2 .$$

This norm has the property that for every sequence $\{z_i\}_{i=1}^k$ of vectors disjointly supported relative to the basis $\{e_i\}_{i=1}^n$, $d(\ell_1^k,[z_i]_{i=1}^k) \geq 10$. It follows now from lemma III.1 of [35] that the norm $|||\cdot|||$ satisfies a $p = p(k) > 1$ upper estimate with constant ≤ 3 . This clearly yields that X satisfies an upper 2p-estimate with constant $\leq \sqrt{3}$.

□

We also need the following lemma.

Lemma 2.5: Let Y *be an r-convex Banach lattice with r-convexity constant* $\leq K$ *for some* $1 < r < \infty$ *and some* $M < \infty$ *. Let* $\{y_i\}_{i=1}^m$ *be a fixed sequence of vectors in* Y *and define a norm on the space* R^n *of all sequences* $a = \{a_i\}_{i=1}^n$ *of scalars, by putting*

$$|||a||| = (\sum_{\pi} \| \max_{1\le i\le n} |a_i y_{\pi(i)}| \|^q)^{1/q} ,$$

where the sum is extended over all permutations π *of the integers* $\{1,2,\ldots,m\}$, *and* $q \geq r$. *Then the space* $Z_n = (R^n, |||\cdot|||)$, *endowed with the pointwise order, is an r-convex Banach lattice whose r-convexity constant is* $\leq M$.

The assertion remains valid if the r-convexity is replaced by the existence of an upper r-estimate for disjoint elements.

Proof: Since, for $q \geq r$, a direct sum in the ℓ_q-sense of r-convex Banach lattice with constant $\leq M$ is also r-convex with r-convexity constant $\leq M$ it is clearly enough to prove the claim for a norm on R^n of the form

$$\|a\|_0 = \| \max_{1\le i\le n} |a_i y_i| \| .$$

Let $a^{(k)} = \{a_i^{(k)}\}_{i=1}^n$; $k = 1,\ldots,m$ be a sequence of elements in the corresponding lattice $W_n = (R^n, \|\cdot\|_0)$. Then

$$\|(\sum_{k=1}^m |a^{(k)}|^r)^{1/r}\|_0 = \| \max_{1\le i\le n} (\sum_{k=1}^m |a_i^{(k)} y_i|^r)^{1/r}\| \leq$$

$$\leq \|(\sum_{k=1}^m \max_{1\le i\le n} |a_i^{(k)} y_i|^r)^{1/r}\| \leq M(\sum_{k=1}^m \| \max_{1\le i\le n} |a_i^{(k)} y_i| \|^r)^{1/r} = M(\sum_{k=1}^m \|a^{(k)}\|_0^r)^{1/r} .$$

The case when Y has an upper r-estimate for disjoint elements is treated similarly. □

Remark: In 2.5, note that if $n = m$, then $|||a|||$ can be written as $(\sum_{\pi} \| \max_{1\le i\le n} |a_{\pi(i)} y_i| \|^q)^{1/q}$.

Proof of 2.3: Let $\{\varphi_i\}_{i=1}^n$ be a sequence of disjoint elements in X so that $\sum_{i=1}^n \|\varphi_i\|^r = n$ and, for $1 \leq i \leq n$, let $\hat{\varphi}_i = \{\varphi_i(\pi)\}_{\pi}$ be the element of the Banach lattice $\ell_r^{n!}(X)$ (already discussed in the proof of 2.1) for which $\varphi_i(\pi) = \varphi_{\pi(i)}/(n!)^{1/r}$. The vectors $\{\varphi_i\}_{i=1}^n$ form a symmetric basic sequence in $\ell_r^{n!}(X)$ with $\|\hat{\varphi}_i\| = 1$ for all $1 \le i \le n$, as is easily verified.

Observe now that, by our assumptions, $\ell_r^{n!}(X)$ is isomorphic to a subspace of the lattice $\ell_r^{n!}(Y)$ and the isomorphism T between $\ell_r^{n!}(X)$ and this subspace of $\ell_r^{n!}(Y)$ has the same norms (for T and T^{-1}) as the isomorphism from X into Y. Moreover, $\ell_r^{n!}(Y)$ is of type 2 and, in particular, q-concave for some even integer $q \ge r$. Hence, we can apply 2.1 and conclude the existence of a constant $D = D(Y,T) < \infty$, independent of the choice of the sequence $\{\varphi_i\}_{i=1}^n$, so that

$$D^{-1}\|\sum_{i=1}^n a_i\hat{\varphi}_i\| \le \max\{(\sum_\pi \|\max_{1\le i\le n} |a_{\pi(i)}T\hat{\varphi}_i|\|^q/n!)^{1/q}, \|\sum_{i=1}^n \hat{\varphi}_i\|(\sum_{i=1}^n |a_i|^2/n)^{1/2}\}$$

$$\le D\|\sum_{i=1}^n a_i\hat{\varphi}_i\| ,$$

for every sequence of scalars $\{a_i\}_{i=1}^n$.

Suppose that n is a multiple of k, such that $n\, w_n^{2r/(r-2)} \ge k$, where $w_n = \|\sum_{i=1}^n \hat{\varphi}_i\|/n^{1/2}$. Put $m = n/k$ and $\hat{\psi}_j = (\sum_{i=(j-1)m+1}^{jm} \hat{\varphi}_i)/w_n m^{1/2}$; $j = 1,\dots,k$. If M is the constant of the upper r-estimate for disjoint elements in Y, then, by 2.5 and the remark following, we get that

$$\|\hat{\psi}_j\| \le DM\|T\| \max\{(\sum_{i=(j-1)m+1}^{jm} \|\hat{\varphi}_i\|^r)^{1/r}, w_n m^{1/2}\}/w_n m^{1/2} \le$$

$$\le DM\|T\| \max\{m^{1/r}/w_n m^{1/2}, 1\} = DM\|T\| \; ; \; 1 \le j \le k$$

since $w_n m^{1/2} \ge m^{-(r-2)/2r} \cdot m^{1/2} = m^{1/r}$.

Assume now that ℓ_2 is not disjointly finitely representable in X. Then, by 2.4, X has an upper p-estimate for disjoint elements, for some $p > 2$, and it remains to show that $p \ge r$. If $p < r$ then also the lattice $\ell_r^{n!}(X)$ has such an upper p-estimate with some constant C. Since $\{\hat{\varphi}_i\}_{i=1}^n$, and therefore $\{\hat{\psi}_j\}_{j=1}^k$ too, are disjoint elements in $\ell_r^{n!}(X)$ it follows that

$$\|\sum_{j=1}^k \hat{\psi}_j\| \leq C(\sum_{j=1}^k \|\hat{\psi}_j\|^p)^{1/p} \leq CDM\,\|T\|\,k^{1/p} .$$

On the other hand, $\|\sum_{j=1}^k \hat{\psi}_j\| = k^{1/2}$ which implies that $k \leq (CDM\|T\|)^{2p/(p-2)}$.

As an immediate consequence, we get that

$$\sup_n n\,w_n^{2r/(r-2)} = R^{2r/(r-2)} < \infty \ ;$$

i.e.,

$$\|\sum_{i=1}^n \varphi_i\| = \|\sum_{i=1}^n \hat{\varphi}_i\| = w_n n^{1/2} \leq Rn^{1/r} = R(\sum_{i=1}^n \|\varphi_i\|^r)^{1/r} .$$

This completes the proof since, except for a normalization constant which is not essential, $\{\varphi_i\}_{i=1}^n$ is an arbitrary sequence of disjoint elements in X.

□

As announced in Section 1, Theorem 2.3 can be used to give a *proof of* 1.8. Indeed, if X is a Banach lattice which is isomorphic to a subspace of $L_p(0,1)$; $p > 2$ and ℓ_2 is not disjointly finitely representable in X, then, by 2.3, X has an upper p-estimate for disjoint elements. On the other hand, X has also a lower p-estimate of the same form since $L_p(0,1)$ has cotype p This implies the existence of a constant $K < \infty$ so that every sequence $\{x_i\}_{i=1}^n$ of disjoint elements in X is K-equivalent to the unit vector basis of ℓ_p^n. Moreover, it is known (cf. [50]) that, X is linearly isometric and order equivalent to a Banach lattice $\tilde{X}$ of measurable functions over some probability space (Ω,Σ,ν). This implies, as is easily seen that $\tilde{X}$ is, up to an equivalent norm, equal to $L_p(\Omega,\Sigma,\nu)$.

The result proved in 2.3 can be considerably improved when X is not only a Banach lattice but a r.i. function space on $[0,1]$.

Theorem 2.6: Let Y *be a Banach lattice of type 2 which is r-convex, for some* $r > 2$. *Then every r.i. function space* X *on* $[0,1]$*, which is* *isomorphic to a subspace of* Y, *is either itself r-convex or it is equal to* $L_2(0,1)$, *up to an equivalent norm.*

We need first a very simple lemma.

Lemma 2.7: Let Z *be a r.i. function space on* $[0,1]$ *which is r-convex (r-concave) with constant* $\le M$ *for some* $1 < r < \infty$ *and some* $M < \infty$. *Then*

$$\|f\|_Z \ge \|f\|_r / M \quad ; \quad f \in Z$$

$$(\|f\|_Z \le M \|f\|_r \quad ; \quad f \in L_r(0,1)) .$$

Proof: Since the proof of one case can be easily deduced by duality from the proof of the other case we shall discuss here only the situation when Z is r-concave. Let $\{\sigma_j\}_{j=1}^n$ be the set of all distinct cyclic permutations of $\{1,2,\dots,n\}$ and consider a simple function of the form $z = \sum_{i=1}^n a_i \chi_{[(i-1)n^{-1}, in^{-1})}$. Then, by using the r-concavity of Z, we have that

$$\|z\|_Z^r = \sum_{j=1}^n \|\sum_{i=1}^n a_{\sigma_j(i)} \chi_{[(i-1)n^{-1}, in^{-1})}\|_Z^r / n \le$$

$$\le M^r \|(\sum_{j=1}^n |\sum_{i=1}^n a_{\sigma_j(i)} \chi_{[(i-1)n^{-1}, in^{-1})}|^r)^{1/r}\|_Z^r / n \le$$

$$\le M^r \|\sum_{i=1}^n (\sum_{j=1}^n |a_{\sigma_j(i)}|^r)^{1/r} \chi_{[(i-1)n^{-1}, in^{-1})}\|_Z^r / n = M^r \sum_{i=1}^n |a_i|^r / n$$

since $\|\chi_{[0,1]}\|_Z = 1$ (see the assumptions imposed on r.i. spaces in the Introduction).

That is, for every simple function as above, we have shown that $\|z\|_Z \le \le M\|z\|_r$. Obviously, this inequality extends to general simple functions and, therefore, to every $f \in L_r(0,1)$.

□

Proof of 2.6: The fact that Y is of type 2 implies that Y is s-concave for some $s < \infty$ and, of course, there is no loss of generality in assuming that s is an even integer. Hence, we can apply 2.1 with $s = 2m$.

Let $M \geq 1$ be so that both the r-convexity and the s-concavity constants of Y are $\leq M$. Let T be an isomorphism from X into Y and let $K \geq 1$ be a constant such that $\|T^{-1}\|\,\|T\| \leq K$. For every integer n and for $1 \leq i \leq 2^n$, let $x_{n,i}$ denote the characteristic function of the interval $[(i-1)2^{-n}, i2^{-n})$. By 2.1, there exists a constant $D = D(K,M,s) < \infty$ so that

$$D^{-1}\Big\|\sum_{i=1}^{2^n} a_i Tx_{n,i}\Big\|_Y \leq \max\Big\{\Big(\sum_\pi \Big\|\max_{1\leq i\leq 2^n} |a_{\pi(i)}Tx_{n,i}|\Big\|_Y^s/2^n!\Big)^{1/s},\ \Big(\sum_{i=1}^{2^n}|a_i|^2/2^n\Big)^{1/2}\Big\}$$

$$\leq D\Big\|\sum_{i=1}^{2^n} a_i Tx_{n,i}\Big\|_Y \quad ,$$

for every sequence of scalars $\{a_i\}_{i=1}^{2^n}$. This means that, for every simple function of the form $\varphi = \sum_{i=1}^{2^n} a_i x_{n,i}$, we get that

$$(+) \qquad K^{-1}D^{-1}\|\varphi\|_X \leq \max\Big\{\Big(\sum_\pi \Big\|\max_{1\leq i\leq 2^n} |a_i Tx_{n,\pi(i)}|\Big\|_Y^s/2^n!\Big)^{1/s},\ \|\varphi\|_2\Big\} \leq DK\|\varphi\|_X$$

and, in particular, that $\|\varphi\|_X \geq K^{-1}D^{-1}\|\varphi\|_2$. Since the simple functions over the dyadic intervals are dense in L_s and hence also in X, by 2.7 and the fact that L_∞ is dense in X, it follows that

$$\|f\|_X \geq K^{-1}D^{-1}\|f\|_2 \quad ,$$

for every $f \in X$.

We distinguigh now between two opposite cases. In the first one we assume that

$$\|x_{n,1}\|_X \leq 2\ DKM\ 2^{-n/2} = 2\ DKM\|x_{n,1}\|_2 \quad ,$$

for every integer n. Let C denote the type 2 constant of X. Then, for $\varphi = \sum_{i=1}^{2^n} a_i x_{n,i}$ as above, we have that

$$\|\varphi\|_X \leq C(\sum_{I=1}^{2^n} \|a_i x_{n,i}\|_X^2)^{1/2} \leq 2\ CDKM\ (\sum_{i=1}^{2^n} \|a_i x_{n,i}\|_2^2)^{1/2} = 2CDKM\|\varphi\|_2 .$$

Consequently, $\|f\|_X \leq 2\ CDKM\ \|f\|_2$ for every $f \in L_2(0,1)$ and, thus, X coincides with $L_2(0,1)$, up to an equivalent norm.

We pass now the opposite case; i.e., when there exist an integer m for which

$$\|x_{m,1}\|_X > 2\ DKM\ \|x_{m,1}\|_2 .$$

The presence of the factor 2 DKM suffices to ensure that when we estimate the norm of $x_{m,1}$ by using the formula (+) the maximum in the inner expression is necessarily attained in the first term. Furthermore, this fact remains true even when we represent $x_{m,1}$ as $\sum_{i=1}^{2^{n-m}} x_{n,i}$ for some $n > m$; i.e.,

$$K^{-1}D^{-1}\|x_{m,1}\|_X \leq (\sum_\pi \| \max_{1\leq i\leq 2^{n-m}} |T\ x_{n,\pi(i)}|\|^s/2^n!)^{1/s} \leq DK\ \|x_{m,1}\|_X .$$

Fix $n > m$, by 2.5, the expression

$$|||a||| = (\sum_\pi \| \max_{1\leq i\leq 2^{n-m}} |a_i Tx_{n,\pi(i)}|\|^s/2^n!)^{1/s}$$

defines an r-convex symmetric structure on $R^{2^{n-m}}$ with r-convexity constant $\leq M$. Hence, by applying 2.7 to the corresponding r.i. structure on [0,1], we get that $|||a||| \geq M^{-1}\|\{1,\ldots,1\}\|(\sum_{i=1}^{2^{n-m}} |a_i|^r/2^{n-m})^{1/r}$. It follows that, for a simple function of the form $\psi = \sum_{i=1}^{2^{n-m}} b_i x_{n,i}$ (i.e., supported entirely by the interval $[0,2^{m-n}]$), we obtain

$$\Big(\sum_\pi \Big\| \max_{1\le i\le 2^{n-m}} |b_i T x_{n,\pi(i)}| \Big\|^s / 2^n! \Big)^{1/s} \ge$$

$$\ge M^{-1} \Big(\sum_\pi \Big\| \max_{1\le i\le 2^{n-m}} |T\, x_{n,\pi(i)}| \Big\|^s / 2^n! \Big)^{1/s} \cdot \Big(\sum_{i=1}^{2^{n-m}} |b_i|^r / 2^{n-m}\Big)^{1/r} \ge$$

$$\ge M^{-1} K^{-1} D^{-1} \|x_{m,1}\|_X \cdot \Big(\sum_{i=1}^{2^{n-m}} |b_i|^2 / 2^{n-m}\Big)^{1/2} \ge$$

$$\ge M^{-1} K^{-1} D^{-1} \cdot 2DKM2^{-m/2} \cdot \Big(\sum_{i=1}^{2^{n-m}} |b_i|^2 / 2^n\Big)^{1/2} \cdot 2\|\psi\|_2 \,.$$

This inequality shows that, for any simple function $\psi \in X$ which is supported by the interval $[0,2^{m-n}]$, the maximum in the inner expression of (+) is always attained in the first term. Since, by 2.5, this expression defines an r-convex structure, it follows that the norm in X is r-convex when restricted to functions supported on $[0,2^{-m}]$. This, of course, implies that X itself is r-convex.

□

3. INVARIANTS OF SOME FINITE DIMENSIONAL BANACH SPACES WITH AN UNCONDITIONAL OR SYMMETRIC BASIS

We start this section by discussing some invariants for a finite dimensional space with an unconditional basis. The main result in this direction is the following:

Theorem 3.1: Let X *be a finite-dimensional Banach space with a normalized K-unconditional basis* $\{x_i\}_{i=1}^n$ *and let* $\{x_i^*\}_{i=1}^n$ *be the sequence of biorthogonal functionsls associated to* $\{x_i\}_{i=1}^n$. *Assume that, for some constant* $M < \infty$ *and for every choice of scalars* $\{a_i\}_{i=1}^n$, *we have that*

$$\left\|\sum_{i=1}^n a_i x_i\right\| \le M\left(\sum_{i=1}^n |a_i|^2\right)^{1/2} .$$

Then,

$$\text{(i)} \qquad \left\|\sum_{i=1}^n x_i\right\|/KM\sqrt{2} \le \gamma_\infty(X) \le d(X,\ell_\infty^n) \le K^2\left\|\sum_{i=1}^n x_i\right\|$$

$$\text{(ii)} \qquad \left\|\sum_{i=1}^n x_i^*\right\|/K \le \pi_1(X) \le KM\sqrt{2}\left\|\sum_{i=1}^n x_i^*\right\| ,$$

where $\gamma_\infty(X)$ *is the projection constant of* X *and* $\pi_1(X)$ *is the* 1*-absolutely summing norm of the identity in* X *(for definitions, see the Introduction).*

Part (i) is due to C. Schütt [75] who proved it by quite a different approach.

The proof of 3.1 requires the use of *diagonal operators;* i.e., operators between two spaces with a Schauder basis whose matrix representation (with respect to these bases) is diagonal. When the bases are fixed, each such diagonal operator Δ is given by a sequence of scalars $\{\delta_i\}_{i=1}^n$. We present first some preliminary lemmas.

Lemma 3.2. Let X *be a finite dimensional Banach space with a Schauder basis* $\{x_i\}_{i=1}^n$ *and biorthogonal functionals* $\{x_i^*\}_{i=1}^n$. *Let* $\{\delta_i\}_{i=1}^n$ *be an arbitrary sequence of scalars and let* Δ *be the diagonal operator from* X *into* ℓ_2^n *determined by* $\{\delta_i\}_{i=1}^n$. *Then,*

$$\pi_1(\Delta) \le \sqrt{2} \int_0^1 \|\sum_{i=1}^n r_i(u)\, \delta_i x_i^*\|_{X^*} du \quad ,$$

where $\pi_1(\Delta)$ *denotes the* 1*-absolutely summing norm of* .

Proof: Let $y_j = \sum_{i=1}^n c_{j,i} x_i$; $j = 1,2,\dots,k$ be arbitrary vectors in X and let $\{e_i\}_{i=1}^n$ denote the unit vector basis of ℓ_2^n. Then, by Khintchine's inequality, we get that

$$\sum_{j=1}^k \|\Delta y_j\|_2 = \sum_{j=1}^k \|\sum_{i=1}^n c_{j,i}\delta_i e_i\|_2 = \sum_{j=1}^k (\sum_{i=1}^n |c_{j,i}\delta_i|^2)^{1/2} \le$$

$$\le A_1^{-1} \sum_{j=1}^k \int_0^1 |\sum_{i=1}^n r_i(u) c_{j,i}\delta_i| du = A_1^{-1} \int_0^1 \sum_{j=1}^k |(\sum_{i=1}^n r_i(u)\delta_i x_i^*)(y_j)| du \le$$

$$\le A_1^{-1} \int_0^1 \|\sum_{i=1}^n r_i(u)\delta_i x_i^*\|_{X^*} du. \ \sup\{\sum_{j=1}^k |y^*(y_j)| ; \|y^*\|_{X^*} \le 1\}$$

and this completes the proof since $A_1^{-1} = \sqrt{2}$.

□

Lemma 3.3. Let X *be a finite dimensional Banach space with a K-unconditional basis* $\{x_i\}_{i=1}^n$ *and biorthogonal functionals* $\{x_i^*\}_{i=1}^n$. *Let* T *be an operator from* X *into some other Banach space* Y *and put* $\delta_i = \|Tx_i\|_Y$; $i = 1,2,\dots,n$. *Then,*

$$\pi_1(T) \ge \|\sum_{i=1}^n \delta_i x_i^*\|_{X^*}/K \ .$$

Proof: Let $\{a_i\}_{i=1}^n$ be an arbitrary sequence of scalars. Then, by the definition of the 1-absolutely summing norm $\pi_1(T)$ of T, it follows that

$$\left|\sum_{i=1}^n a_i\delta_i\right| \le \sum_{i=1}^n \|T(a_ix_i)\|_Y \le \pi_1(T) \sup_{\epsilon_i=\pm 1} \left\|\sum_{i=1}^n \epsilon_i a_i x_i\right\|_X \le$$

$$\le K\pi_1(T) \left\|\sum_{i=1}^n a_i x_i\right\|_X .$$

Hence, since $\{a_i\}_{i=1}^n$ is arbitrarily chosen, we get that

$$\pi_1(T) \ge \sup_{\{a_i\}_{i=1}^n} \left\{\left|\sum_{i=1}^n a_i\delta_i\right| / \left\|\sum_{i=1}^n a_ix_i\right\|_X\right\}/K = \left\|\sum_{i=1}^n \delta_i x_i^*\right\|_{X^*}/K . \qquad \square$$

We are prepared now to present *the proof of 3.1.* We start with Part (i) and observe first that the inequalities

$$\gamma_\infty(X) \le d(X,\ell_\infty^n) \le K^2 \left\|\sum_{i=1}^n x_i\right\|$$

are obvious (use the formal identity mapping between $\{x_i\}_{i=1}^n$ and the unit vector basis of ℓ_∞^n). Let now Δ be the diagonal operator from X into itself which is determined by the sequence of scalars $\{\delta_i\}_{i=1}^n$. By using the fact that the formal identity mapping from ℓ_2^n into X has norm $\le M$ and by applying 3.2 for Δ considered as an operator from X into ℓ_2^n, we get by the ideal property of absolutely summing operators that

$$\pi_1(\Delta) = \pi_1(\Delta\colon X \to X) \le MK\sqrt{2} \left\|\sum_{i=1}^n \delta_i x_i^*\right\|_{X^*}$$

which, by the fact that $\left|\sum_{i=1}^n \delta_i\right| = |\text{trace } \Delta| \le \pi_1(\Delta)\gamma_\infty(X)$, (cf. [66]) implies that

$$\gamma_\infty(X) \ge \sup_{\{\delta_i\}_{i=1}^n} \left\{\left|\sum_{i=1}^n \delta_i\right| / \left\|\sum_{i=1}^n \delta_i x_i^*\right\|\right\}/KM\sqrt{2} = \left\|\sum_{i=1}^n x_i\right\|/KM\sqrt{2} .$$

This completes the proof of (i). The proof of Part (ii) is now very simple. By the argument presented in the proof of Part (i) we deduce that

$$\pi_1(X) \leq MK\sqrt{2}\ \|\sum_{i=1}^{n} x_i^*\|_{X^*}$$

and, by 3.3, it follows that $\pi_1(X) \geq \|\sum_{i=1}^{n} x_i^*\|_{X^*}/K$.

□

In general, the fact that a finite dimensional Banach space X has a normalized unconditional basis $\{x_i\}_{i=1}^n$ which is dominated by the unit vector basis of ℓ_2^n does not imply that every other unconditional basis of X has necessarily the same property. Therefore, in order to be able to compare two unconditional bases of X by using 3.1, we need some additional assumptions.

Corollary 3.4. If $\{x_i\}_{i=1}^n$ *and* $\{y_i\}_{i=1}^n$ *are two normalized K-unconditional bases of a finite dimensional space* X *and* $\{x_i^*\}_{i=1}^n$, $\{y_i^*\}_{i=1}^n$ *denote the respective sequences of biorthogonal functionals then there exists a constant* $0 < C < \infty$ *, depending only on* K *and the cotype 2-constant of* X *, so that*

$$C^{-1}\|\sum_{i=1}^{n} y_i\| \leq \|\sum_{i=1}^{n} x_i\| \leq C\ \|\sum_{i=1}^{n} y_i\|$$

and

$$C^{-1}\|\sum_{i=1}^{n} y_i^*\| \leq \|\sum_{i=1}^{n} x_i^*\| \leq C\ \|\sum_{i=1}^{n} y_i^*\| \ .$$

Proof: If X has cotype 2 constant equal to M then, for every choice of scalars $\{a_i\}_{i=1}^n$, we have that

$$\|\sum_{i=1}^{n} a_i x_i\| \geq (\int_0^1 \|\sum a_i r_i(u) x_i\|^2 du)^{1/2}/K \geq (\sum_{i=1}^{n} |a_i|^2)^{1/2}/KM \ .$$

Hence, by duality, it follows that

$$\|\sum_{i=1}^{n} b_i x_i^*\| \leq KM(\sum_{i=1}^{n} |b_i|^2)^{1/2} ,$$

for every choice of scalars $\{b_i\}_{i=1}^n$. Since a similar inequality is valid for $\{y_i^*\}_{i=1}^n$ we can apply 3.1 for both bases $\{x_i^*\}_{i=1}^n$ and $\{y_i^*\}_{i=1}^n$, and get the desired result.

□

The proof of 3.1 and, subsequently, also that of 3.4, make an essential use of the assumption that the unconditional basis $\{x_i\}_{i=1}^n$ of the space X is dominated by the unit vector basis of ℓ_2^n. However, we do not know if this requirement is really needed. Without these conditions we can prove the following result.

Proposition 3.5. Let $\{x_i\}_{i=1}^n$ *and* $\{y_i\}_{i=1}^n$ *be two K-unconditional bases of a finite dimensional Banach space* X *and let* $\{x_i^*\}_{i=1}^n$, $\{y_i^*\}_{i=1}^n$ *be the respective sequences of biorthogonal functionals. If the matrix* $(a_{i,j})_{i,j=1}^n$ *defined by* $x_i = \sum_{j=1}^n a_{i,j} y_j$; $i = 1,2,\dots,n$ *is orthogonal then*

$$(K^2\sqrt{2})^{-1}\|\sum_{i=1}^n y_i\| \le \|\sum_{i=1}^n x_i\| \le K^2\sqrt{2}\,\|\sum_{i=1}^n y_i\|$$

and

$$(K^2\sqrt{2})^{-1}\|\sum_{i=1}^n y_i^*\| \le \|\sum_{i=1}^n x_i^*\| \le K^2\sqrt{2}\,\|\sum_{i=1}^n y_i^*\| .$$

Proof: The first inequality follows from the second one by duality. In order to prove the second inequality, let I_1 and I_2 be the formal identity mappings from $\{x_i\}_{i=1}^n$, respectively $\{y_i\}_{i=1}^n$, to the unit vector basis $\{e_i\}_{i=1}^n$ of ℓ_2^n. For any $x = \sum_{i=1}^n b_i x_i = \sum_{j=1}^n (\sum_{i=1}^n a_{i,j} b_i) y_j \in X$ we get that

$$\|I_2 x\|_2 = (\sum_{j=1}^n |\sum_{i=1}^n a_{i,j} b_i|^2)^{1/2} = (\sum_{i=1}^n |b_i|^2)^{1/2} = \|I_1 x\|_2$$

since $(a_{i,j})_{j=1}^n$ is an orthogonal matrix. It follows that $\pi_1(I_1) = \pi_1(I_2)$

and the proof of the second inequality can be immediately completed by using 3.2 and 3.3.

□

Proposition 3.5 shows that, in order to construct a finite dimensional Banach space with two normalized unconditional bases $\{x_i\}_{i=1}^n$ and $\{y_i\}_{i=1}^n$ so that $\|\sum_{i=1}^n x_i\|$ differs considerably from $\|\sum_{i=1}^n y_i\|$, we cannot have the bases $\{x_i\}_{i=1}^n$ and $\{y_i\}_{i=1}^n$ related by an orthogonal matrix.

We pass now to some questions involving symmetric bases of finite length. Let X be a finite dimensional Banach space with a normalized K-unconditional basis $\{x_i\}_{i=1}^n$ which satisfies

$$\|\sum_{i=1}^n a_i x_i\| \leq M(\sum_{i=1}^n |a_i|^2)^{1/2} ,$$

for some constant $M < \infty$ and for every choice of scalars $\{a_i\}_{i=1}^n$. Let $\{x_i^*\}_{i=1}^n$ be the sequence of biorthogonal functionals associated to $\{x_i\}_{i=1}^n$ and put $\lambda(n) = \|\sum_{i=1}^n x_i\|$, respectively $\mu(n) = \|\sum_{i=1}^n x_i^*\|$.

Suppose now that X has an additional normalized basis $\{z_i\}_{i=1}^n$ whose symmetry constant will be denoted by $K(\{z_i\}_{i=1}^n)$. This basis need not be dominated by the unit vector basis of ℓ_2^n. Garling and Gordon [30] have shown that in a Banach space X with a 1-symmetric basis $\gamma_\infty(X)\pi_1(X) = n$. It follows easily from their result that in the present case we have

$$n \leq \gamma_\infty(X)\ \pi_1(X) \leq nK(\{z_i\}_i^n)^2 .$$

On the other hand, we know from 3.1 that $\gamma_\infty(X) \geq \lambda(n)/KM\sqrt{2}$ and $\pi_1(X) \geq$ $\geq \mu(n)/K$. Consequently, we get that $K(\{z_i\}_{i=1}^n)^2 \geq \lambda(n)\mu(n)/nK^2M\sqrt{2}$. The discussion above can be summarized as follows.

Proposition 3.6. Let X *be a Banach space with a normalized K-unconditional basis* $\{x_i\}_{i=1}^n$ *and biorthogonal functionals* $\{x_i^*\}_{i=1}^n$ *such that, for some constant* $M < \infty$ *and for every sequence of scalars* $\{a_i\}_{i=1}^n$, *we have that*

$$\|\sum_{i=1}^{n} a_i x_i\| \le M(\sum_{i=1}^{n} |a_i|^2)^{1/2} .$$

Then there exists a constant $c > 0 (c = 2^{-1/4}K^{-1}M^{-1/2})$ *so that every normalized basis* $\{z_i\}_{i=1}^n$ *of* X *has symmetry constant*

$$K(\{z_i\}_{i=1}^n) \ge c \|\sum_{i=1}^{n} x_i\| \cdot \|\sum_{i=1}^{n} x_i^*\|/n .$$

In other words, 3.6 shows that the ratio $\|\sum_{i=1}^{n} x_i\| \cdot \|\sum_{i=1}^{n} x_i^*\|/n$ for *some* unconditional basis of X measures the degree of symmetry of *all* bases of X .

The result 3.1 implies that, in a Banach space X of cotype 2 with a normalized unconditional basis $\{x_i\}_{i=1}^n$ and biorthogonal functionals $\{x_i^*\}_{i=1}^n$, the norms $\|\sum_{i=1}^{n} x_i\|$ and $\|\sum_{i=1}^{n} x_i^*\|$ can be related to some invariants of the space X^* , namely $\pi_1(X^*)$ and, respectively, $\gamma_\infty(X^*)$ or $d(X^*, \ell_\infty^n)$. In the case when $\{x_i\}_{i=1}^n$ is even a symmetric basis of X, the expression $\|\sum_{i=1}^{n} x_i\|$ can be also connected to $d(X, \ell_2^n)$. For reasons which will become clear in the sequel we prefer to state and prove this result within the context of 2-concave symmetric bases rather than in that of spaces of cotype 2.

Proposition 3.7. Let X *be a Banach space with a* K *-symmetric normalized basis* $\{x_i\}_{i=1}^n$. *If the basis* $\{x_i\}_{i=1}^n$ *is 2-concave with 2-concavity constant* M *then*

$$\|\sum_{I=1}^{n} x_i\|/K\sqrt{n} \le d(X, \ell_2^n) \le KM^2 \|\sum_{i=1}^{n} x_i\|/\sqrt{n} .$$

Proof: Let $z = \sum_{i=1}^{n} a_i x_i$ be an arbitrary vector in X and let $\{\sigma_j\}_{j=1}^n$ be the family of all distinct cyclic permutations of the integers $\{1,2,\ldots,n\}$. Using the notion of absolute value of vectors in X as defined in the Introduction (i.e., $|z| = \sum_{i=1}^{n} |a_i| x_i \in X$) and the 2-concavity of the basis $\{x_i\}_{i=1}^n$, we get that

$$\left(\sum_{i=1}^{n} |a_i|^2\right)^{1/2}/M \le \|z\| \le K\left(\sum_{j=1}^{n} \|\sum_{i=1}^{n} a_{\sigma_j(i)}x_i\|^2\right)^{1/2}/\sqrt{n} \le$$

$$\le KM\|\sum_{i=1}^{n}\left(\sum_{j=1}^{n} |a_{\sigma_j(i)}|^2\right)^{1/2} x_i\|/\sqrt{n} = KM\left(\sum_{i=1}^{n} |a_i|^2\right)^{1/2}\|\sum_{i=1}^{n} x_i\|/\sqrt{n} .$$

Hence, by considering the formal identity map between the basis $\{x_i\}_{i=1}^n$ of X and the unit vector basis of ℓ_2^n, it follows that

$$d(X,\ell_2^n) \le KM^2 \|\sum_{i=1}^{n} x_i\|/\sqrt{n} .$$

In order to prove the left-hand side inequality, we first define a new norm $|||\cdot|||$ on X by putting

$$|||z||| = \left(\sum_{\epsilon_i=\pm 1}\sum_{\pi}\|\sum_{i=1}^{n}\epsilon_i a_{\pi(i)}x_i\|^2\right)^{1/2}/(2^n\cdot n!)^{1/2};\ z = \sum_{i=1}^{n} a_i x_i \in X$$

where $\sum_{\pi}$ means the sum over all possible distinct permutations π of the integers $\{1,2,\ldots,n\}$. It is easily checked that, for any $z \in X$,

$$K^{-1}\|z\| \le |||z||| \le K\|z\|$$

and that $d((X,|||\cdot|||),\ell_2^n) \le d(X,\ell_2^n)$, the latter denoting the distance when X is endowed with the original norm $\|\cdot\|$. Since $d(\ell_2^n,\ell_\infty^n) = \sqrt{n}$ and $\{x_i^*\}_{i=1}^n$, the sequence of biorthogonal functionals associated to $\{x_i\}_{i=1}^n$, is 1-unconditional with respect to the new norm $|||\cdot|||$, we have that

$$\sqrt{n} \le d((X^*,|||\cdot|||),\ell_2^n)\cdot d((X^*,|||\cdot|||),\ell_\infty^n) \le d(X,\ell_2^n)\cdot|||\sum_{i=1}^{n} x_i^*||| .$$

But, with respect to the new norm $|||\cdot|||$, $\{x_i\}_{i=1}^n$ is 1-symmetric and, thus,

$$|||\sum_{i=1}^{n} x_i^*||| = n/|||\sum_{i=1}^{n} x_i||| \le Kn/\|\sum_{i=1}^{n} x_i\| .$$

Consequently, $d(X,\ell_2^n) \ge \|\sum_{i=1}^{n} x_i\|/K\sqrt{n}$.

□

Notice that 3.7 generalizes 1.3(i) since, for $1 \le p \le 2$, any monotone unconditional basic sequence in $L_p(0,1)$ has 2-concavity constant at most A_p^{-1}.

After proving that for any 2-concave symmetric basis $\{x_i\}_{i=1}^n$, the expression $\|\sum_{i=1}^n x_i\|$ is, up to a constant, an invariant of the underlying space X it is natural to ask if a space X as above can really have two non-equivalent symmetric bases in the sense of the definition of uniqueness of symmetric bases given in the Introduction.

We shall prove that, under some restrictive conditions, namely bounded q-concavity constant for some $1 < q < 2$, the space X has, up to equivalence, a unique symmetric basis.

Theorem 3.8. Fix $1<q<2$ *and* $M< \infty$, *and let* $C_{q,M}$ *be the family of all finite dimensional Banach spaces having a* 1*-symmetric normalized basis with* q*-concavity constant* $\le M$. *Then each member of* $C_{q,M}$ *has, up to equivalence, a unique symmetric basis.*

The proof of 3.8 requires some preliminary lemmas in which we use frequently the well known notion of p-stable random variables. For the benefit of readers who are not familiar with this notion we mention here a few facts; additional details can be found; e.g., in [23]. Let (Ω,Σ,P) be a probability space and let $0 < p < 2$. A P-measurable function g is called a p-stable random variable if $\int_\Omega e^{itg(\omega)}dP(\omega) = e^{-c|t|^p}$, for some constant $c > 0$ and for every $-\infty < t < \infty$. Any p-stable random variable g belongs to $L_q(\Omega,\Sigma,P)$, whenever $0 < q < p$, but $(\int_\Omega |g(\omega)|^p dP(\omega))^{1/p} = \infty$ for $p < 2$. By choosing properly the above constant c we can ensure that, for some q as above, $\|g\|_q = 1$. The property of p-stable random variables which is mostly used in the context of Banach space theory is the following: every sequence $\{g_i\}_{i=1}^n$ of n independent identically distributed p-stable random variables with $\|g_1\|_q = 1$ for some $1 \leq q < p$ is 1-equivalent to the unit vector basis of ℓ_p^n when considered in $L_q(\Omega,\Sigma,P)$.

Lemma 3.9. Let X *be a finite dimensional Banach space having a* 1*-unconditional basis* $\{x_i\}_{i=1}^n$ *which, for some* $1 < q < 2$ *and* $M < \infty$, *is* q*-concave with* q*-concavity constant* $\leq M$. *Let* $\{g_i\}_{i=1}^n$ *be a sequence of random variables over some probability space* (Ω,Σ,P) *such that for* $1 \leq i \leq n$ $\|g_i\|_q = (\int_\Omega |g_i(\omega)|^q dP(\omega))^{1/q} = 1$. *Then, for every choice of scalars* $\{a_i\}_{i=1}^n$,

$$\min_{1\leq j\leq n} \|g_j\|_1 \cdot \| \sum_{i=1}^n a_i x_i \|_X \leq \int_\Omega \| \sum_{i=1}^n a_i g_i(\omega) x_i \|_X dP(\omega) \leq M \| \sum_{i=1}^n a_i x_i \|_X .$$

Proof: The left-hand side inequality is an immediate consequence of the convexity of the norm since

$$\int_\Omega \| \sum_{i=1}^n a_i g_i(\omega) x_i \|_X dP(\omega) \geq \| \int_\Omega \sum_{i=1}^n |a_i g_i(\omega)| x_i \, dP(\omega) \|_X \geq$$

$$\min_{1\leq j\leq n} \|g_j\|_1 \, \| \sum_{i=1}^n |a_i| x_i \|_X = \min_{1\leq j\leq n} \|g_j\|_1 \cdot \| \sum_{i=1}^n a_i x_i \|_X .$$

On the other hand, by the q-concavity of the basis, we get that

$$\int_\Omega \| \sum_{i=1}^n a_i g_i(\omega) x_i \|_X dP(\omega) \leq (\int_\Omega \| \sum_{i=1}^n a_i g_i(\omega) x_i \|_X^q dP(\omega))^{1/q} \leq$$

$$\leq M \, \| (\int_\Omega | \sum_{i=1}^n a_i g_i(\omega) x_i |^q dP(\omega))^{1/q} \|_X =$$

$$= M \, \| (\int_\Omega \sum_{i=1}^n |a_i g_i(\omega)|^q x_i \, dP(\omega))^{1/q} \|_X = M \, \| \sum_{i=1}^n a_i x_i \|_X .$$

□

The next lemma will be stated in a more general setting in order to facilitate a further use if needed.

Lemma 3.10. Let X *be a finite dimensional Banach space with a* 1*-unconditional basis* $\{x_i\}_{i=1}^n$ *and let* Y *be a Banach lattice which, for some* $1 < q < \infty$ *and some* $M < \infty$, *is* q*-concave with* q*-concavity constant* $\leq M$

Let T *be an isomorphism from* X *into* Y *and, for some* $1 < p < 2$, *let* $\{g_i\}_{i=1}^n$ *be a sequence of independent and identically distributed p-stable variables over a probability space* (Ω,Σ,P) . *Then, for every sequence* $\{b_i\}_{i=1}^n$ *of scalars for which*

$$\int_\Omega \| \sum_{i=1}^n b_i g_i(\omega) x_i \|_X dP(\omega) \leq \| \sum_{i=1}^n b_i x_i \|_X \ ,$$

we have that

$$D^{-1} \| \sum_{i=1}^n b_i x_i \|_X \leq \| \max_{1 \leq i \leq n} |b_i T x_i| \|_Y \leq \| (\sum_{i=1}^n |b_i T x_i|^p)^{1/p} \|_Y \leq$$

$$\leq \|T\| \cdot \|g_1\|_1^{-1} \| \sum_{i=1}^n b_i x_i \|_X \ ,$$

where D *is a constant* $\leq (MB_q \|T^{-1}\|)^{2/(2-p)} (\|T\| \cdot \|g_1\|_1^{-1})^{p/(2-p)}$.

Proof: Since $\{g_i/\|g_1\|_1\}_{i=1}^n$ is, in $L_1(\Omega,\Sigma,P)$, a normalized basis 1-equivalent to the unit vector basis of ℓ_p^n , we get in the lattice Y that

$$(\sum_{i=1}^n |b_i T x_i|^p)^{1/p} = \|g_1\|_1^{-1} \int_\Omega | \sum_{i=1}^n b_i g_i(\omega) T x_i | \, dP(\omega) \ ,$$

where $|y|$ denotes the absolute value of an element $y \in Y$. Hence, by our hypothesis on the sequence $\{b_i\}_{i=1}^n$,

$$\| (\sum_{i=1}^n |b_i T x_i|^p)^{1/p} \|_Y \leq \|g_1\|_1^{-1} \int_\Omega \| \sum_{i=1}^n b_i g_i(\omega) T x_i \|_Y dP(\omega) \leq$$

$$\leq \|T\| \cdot \|g_1\|_1^{-1} \int_\Omega \| \sum_{i=1}^n b_i g_i(\omega) x_i \|_X dP(\omega) \leq \|T\| \cdot \|g_1\|_1^{-1} \| \sum_{i=1}^n b_i x_i \|_X \ .$$

This proves the right-hand side inequality of the statement. Since the inner one is evident it remains to prove only the left-hand side inequality. By using 2.2 with $z_i = b_i Tx_i$; $1 \le i \le n$ and $r = 2$, we get that

$$\|(\sum_{i=1}^n |b_i Tx_i|^2)^{1/2}\|_Y \le \|(\sum_{i=1}^n |b_i Tx_i|^p)^{1/p}\|_Y^{p/2} \cdot \|\max_{1\le i\le n} |b_i Tx_i|\|_Y^{1-p/2} \le$$

$$\le (\|T\|\cdot\|g_1\|_1^{-1})^{p/2} \|\sum_{i=1}^n b_i x_i\|_X^{p/2} \|\max_{1\le i\le n} |b_i Tx_i|\|_Y^{1-p/2} .$$

On the other hand, by the q-concavity of Y and Khintchine's inequality in the lattice Y, it follows that

$$\|\sum_{i=1}^n b_i x_i\|_X = \int_0^1 \|\sum_{i=1}^n b_i r_i(u) x_i\|_X du \le \|T^{-1}\| \int_0^1 \|\sum_{i=1}^n b_i r_i(u) Tx_i\|_Y du \le$$

$$\le \|T^{-1}\| (\int_0^1 \|\sum_{i=1}^n b_i r_i(u) Tx_i\|_Y^q du)^{1/q} \le M\|T^{-1}\| \|(\int_0^1 |\sum_{i=1}^n b_i r_i(u) Tx_i|^q du)^{1/q}\|_Y \le$$

$$\le B_q M \|T^{-1}\| \|(\sum_{i=1}^n |b_i Tx_i|^2)^{1/2}\|_Y .$$

Finally, by combining these inequalities, we get that

$$\|\sum_{i=1}^n b_i x_i\|_X \le B_q M \|T^{-1}\| (\|T\|\cdot\|g_1\|_1^{-1})^{p/2} \|\sum_{i=1}^n b_i x_i\|_X^{p/2} \|\max_{1\le i\le n} |b_i Tx_i|\|_Y^{1-p/2} ;$$

i.e.,

$$\|\sum_{i=1}^n b_i x_i\|_X \le (B_q M \|T^{-1}\|)^{2/(2-p)} (\|T\| \|g_1\|_1^{-1})^{p/(2-p)} \|\max_{1\le i\le n} |b_i Tx_i|\|_Y .$$

□

Lemma 3.11. Let X *and* Y *be two finite dimensional Banach spaces having normalized bases* $\{x_i\}_{i=1}^n$ *, respectively* $\{y_i\}_{i=1}^n$ *, such that* $\{x_i\}_{i=1}^n$ *is* 1-*symmetric and has* q-*concavity constant* $\le M$ *, for some constants* q *and* M *, and* $\{y_i\}_{i=1}^n$ *is* 1-*unconditional. Let* $T: X \to Y$ *be a linear operator with* $\|T\| \le 1$ *and assume that*

$$D^{-1}\|\sum_{i=1}^n y_i\|_Y \le \|\sum_{i=1}^n x_i\|_X \le D\|\max_{1\le i\le n} |Tx_i|\|_Y ,$$

for some constant $0 < D < \infty$. *Then there exist a number* $0 < \beta = \beta(q,M,D) \leq 1$ *independent of* X *and* Y , *and* m *distinct indices* $1 \leq j(h) \leq n$; $h = 1,2,\ldots,m$ *so that* $m \geq \beta n$ *and*

$$\|\sum_{h=1}^{m} a_h y_{j(h)}\|_Y \leq 2D^2 \|\sum_{h=1}^{m} a_h x_h\|_X \quad ,$$

for every choice of scalars $\{a_h\}_{h=1}^m$.

The expression $\max_{1\leq i\leq n} |Tx_i|$ is taken in the sense of the lattice generated by the unconditional basis $\{y_i\}_{i=1}^n$; i.e., coordinatewise.

Proof: Let $\{y_i^*\}_{i=1}^n$ denote the sequence of biorthogonal functionals associated to $\{y_i\}_{i=1}^n$. The key point is to find two sequences of distinct indices $\{j(h)\}_{h=1}^m$ and $\{k(h)\}_{h=1}^m$, with m as above, so that $|y_{j(h)}^* Tx_{k(h)}| \geq 1/2D^2$. Once this is done we have for any choice of scalars $\{a_h\}_{h=1}^m$,

$$\|\sum_{h=1}^{m} a_h y_{j(h)}\|_Y \leq 2D^2 \|\sum_{h=1}^{m} a_h |y_{j(h)}^* Tx_{k(h)}| y_{j(h)}\|_Y$$

$$\leq 2D^2 \|\max_{1\leq h\leq m} |T(a_h x_{k(h)})|\|_Y \leq 2D^2 \|\int_0^1 |\sum_{h=1}^{m} r_h(u) T(a_h x_{k(h)})| du\|_Y \leq$$

$$\leq 2D^2 \|T\| \int_0^1 \|\sum_{h=1}^{m} r_h(u) a_h x_{k(h)}\|_X du \leq 2D^2 \|\sum_{h=1}^{m} a_h x_h\|_X \quad .$$

Set $\eta_o = \{j;\ 1 \leq j \leq n \text{ and } y_j^* (\max_{1\leq i\leq n} |Tx_i|) < 1/2D^2\}$ and let $\eta = \{1,2,\ldots,n\} \sim \eta_o$. Then, by our assumption, we get that

$$\|\chi_{\eta_o} \max_{1\leq i\leq n} |Tx_i|\|_Y \leq \|\sum_{i=1}^{n} y_i\|_Y / 2D^2 \leq \|\sum_{i=1}^{n} x_i\|_X / 2D \quad ,$$

which implies

$$\|\chi_{\eta} \max_{1\leq i\leq n} |Tx_i|\,\|_Y \geq \|\sum_{i=1}^{n} x_i\|_X / 2D \quad .$$

We can construct a partition $\{\eta_k\}_{k=1}^n$ of η into mutually disjoint subsets such that

$$\max_{1\le i\le n} |y_j^* Tx_i| = |y_j^* Tx_k| ,$$

for $j \in \eta_k$. Let $\{\eta_{k(h)}\}_{h=1}^m$ be the set of all those η_k's which are non-empty. We have

$$\|\sum_{i=1}^n x_i\|/2D \le \|\sum_{h=1}^m \chi_{\eta_{k(h)}} \max_{1\le i\le n} |Tx_i|\,\|_Y =$$

$$= \|\sum_{h=1}^m \chi_{\eta_{k(h)}} |Tx_{k(h)}|\,\|_Y \le \|\int_0^1 |\sum_{h=1}^m r_h(u)Tx_{k(h)}|du\|_Y \le$$

$$\le \|T\| \int_0^1 \|\sum_{h=1}^m r_h(u)x_{k(h)}\|_X du \le \|\sum_{h=1}^m x_h\|_X .$$

Thus, by using the 1-symmetricity and the q-concavity constant of $\{x_i\}_{i=1}^n$, it follows that there exists a number $0 < \beta = \beta(q,M,D) \le 1$ so that $\beta n \le m$; i.e., at least βn of the sets $\{\eta_k\}_{k=1}^n$ are non-empty. Now, for each $1 \le h \le m$, choose an arbitrary integer $j(h) \in \eta_{k(h)}$. By the definition of the sets $\{\eta_k\}_{k=1}^n$ and η_o, the indices $\{j(h)\}_{h=1}^m$ have the property that $|y_{j(h)}^* Tx_{k(h)}| \ge 1/2D^2$, as desired.

□

We are prepared now to give the *proof of 3.8.* Let X be a finite dimensional Banach space having a 1-symmetric normalized basis $\{x_i\}_{i=1}^n$ which, for some $1 < q < 2$ and $M < \infty$, belongs to the class $\mathcal{C}_{q,M}$. Let Y be a space with a 1-symmetric normalized basis $\{y_i\}_{i=1}^n$ and assume that, for some $K < \infty$, there is an invertible operator $T\colon X \to Y$ so that $K^{-2}\|x\|_X \le \le \|Tx\|_Y \le \|x\|_X$ for all $x \in X$. Take $p = (q+2)/2$ (or, as a matter of fact, any other fixed p between q and 2) and let $\{g_i\}_{i=1}^n$ be a sequence of independent and identically distributed p-stable random variables over some probability space (Ω,Σ,P) so that $\|g_1\|_q = 1$ (this normalization implies that $\|g_1\|_1$ depends only on q). By 3.9, we know that

$$\int_\Omega \|\sum_{i=1}^n g_i(\omega)x_i\|_X dP(\omega) \le M\|\sum_{i=1}^n x_i\|_X .$$

Therefore, by 3.10, there exists a constant $D_1 = D_1(q,M,K) < \infty$ such that

$$\|\sum_{i=1}^n x_i\|_X \le D_1\| \max_{1\le i\le n} |Tx_i| \|_Y .$$

Since $\{x_i\}_{i=1}^n$ is also 2-concave with 2-concavity constant $\le M$, by [21], the basis $\{y_i\}_{i=1}^n$ of Y is also 2-concave with constant $\le MK^2K_G$, where K_G is the universal constant of Grothendieck. In particular, we have

$$\|\sum_{i=1}^n a_ix_i\|_X \ge (\sum_{i=1}^n |a_i|^2)^{1/2}/M \quad \text{and} \quad \|\sum_{i=1}^n a_iy_i\|_Y \ge (\sum_{i=1}^n |a_i|^2)^{1/2}/MK^2K_G ,$$

for every choice of $\{a_i\}_{i=1}^n$. This implies that each of the corresponding sequences of biorthogonal functionals $\{x_i^*\}_{i=1}^n$ and $\{y_i^*\}_{i=1}^n$ is dominated by the unit vector basis of ℓ_2^n. Hence, by 3.1(ii) applied in X^*, respectively Y^*, we get that

$$\|\sum_{i=1}^n x_i\|_X \le \pi_1(X^*) \le M\sqrt{2}\,\|\sum_{i=1}^n x_i\|_X$$

and

$$\|\sum_{i=1}^n y_i\|_Y \le \pi_1(Y^*) \le MK^2K_G\sqrt{2}\,\|\sum_{i=1}^n y_i\|_Y .$$

Consequently, since $K^{-2}\pi_1(Y^*) \le \pi_1(X^*) \le K^2\pi_1(Y^*)$, it follows that there exists a constant $D_2 = D_2(K,M) < \infty$ such that

$$D_2^{-1}\|\sum_{i=1}^n x_i\|_X \le \|\sum_{i=1}^n y_i\|_Y \le D_2\|\sum_{i=1}^n x_i\|_X .$$

Hence, by putting $D = \max(D_1,D_2)$, we are in a position to apply 3.11 and conclude the existence of a number $0 < \beta = \beta(q,M,K) \le 1$, independent of $\{x_i\}_{i=1}^n$ and $\{y_i\}_{i=1}^n$, and of an integer $m \ge \beta n$ such that

$$\|\sum_{h=1}^{m} a_h y_h\|_Y \le 2D^2 \|\sum_{h=1}^{m} a_h x_h\|_X ,$$

for every choice of scalars $\{a_h\}_{h=1}^m$.

By splitting an arbitrary vector into at most β^{-1}-blocks, each of length $\le m$, and by using the previous inequality, we easily get that

$$\|\sum_{i=1}^{n} b_i y_i\|_Y \le 2D^2\beta^{-1} \|\sum_{i=1}^{n} b_i x_i\|_X ,$$

for every sequence of scalars $\{b_i\}_{i=1}^n$.

In order to complete the proof we would like to reverse the roles of X and Y . This is not a-priori possible since the hypotheses on X , respectively Y , are not identical. We however point out that Y is also 2-concave (with 2-concavity constant $\le MK^2K_G$) and, by what we have already proved above, that

$$\int_\Omega \|\sum_{i=1}^{n} g_i(\omega) y_i\|_Y dP(\omega) \le 2D^2\beta^{-1} \int_\Omega \|\sum_{i=1}^{n} g_i(\omega) x_i\|_X dP(\omega) \le$$

$$\le 2D^2\beta^{-1}M\|\sum_{i=1}^{n} x_i\|_X \le 2D^3\beta^{-1}M\|\sum_{i=1}^{n} y_i\|_Y .$$

Therefore, by applying 3.10 and 3.11, with the roles of X and Y being reversed (and with the sequence $\{g_i/2D^3\beta^{-1}M\}_{i=1}^n$ instead of $\{g_i\}_{i=1}^n$), we conclude the existence of a constant $C < \infty$, independent of n , so that

$$\|\sum_{i=1}^{n} b_i x_i\|_X \le C\|\sum_{i=1}^{n} b_i y_i\|_Y ,$$

for every choice of scalars $\{b_i\}_{i=1}^n$.

□

Remarks 1. The proof of 3.8 implies more than was stated: it actually follows that there exist a number $\gamma > 0$, independent of n , and permutations $\{\pi_j\}_{j=1}^4$ of the integers $\{1,2,\dots,n\}$ such that $|y^*_{\pi_1(h)} T x_{\pi_2(h)}| \ge \gamma$ and

$|x^*_{\pi_3}(h) T^{-1} y_{\pi_4}(h)| \geq \gamma$ for $h = 1,2,\ldots,m$ and $m \geq \beta n$ (see also the proof of 3.11). Clearly, this kind of equivalence is false in spaces which are "too close" to ℓ_2^n ; $n = 1,2,\ldots$ (take e.g., the unit vector basis and the Walsh system in ℓ_2^n).

2. The infinite statement corresponding to 3.8. is false: it is possible to find an infinite dimensional Banach space X with two non-equivalent symmetric bases so that both of them are q-concave for some $1 < q < 2$ (cf. [46]).

4 PERFECTLY HOMOGENEOUS BASES OF FINITE LENGTH

We begin this section by presenting a localization of Zippin's [82] characterization of perfectly homogeneous bases i.e. of those normalized bases which are equivalent to each of their normalized block bases. Perfectly homogeneous bases of infinite length are clearly unconditional and even subsymmetric, but in the finite dimensional case the subsymmetry does not seem to suffice. Since the results in the finite case are necessarily of a quantitative nature we introduce first the following definition.

Definition 4.1. Let $\{x_i\}_{i=1}^n$ *be a normalized basis of a Banach space* X. *The perfectly homogeneous constant* $PH(\{x_i\}_{i=1}^n)$ *of* $\{x_i\}_{i=1}^n$ *is the smallest constant* K *such that every sequence of the form* $y_j = \sum_{i\in\sigma_j} \pm x_i / \|\sum_{i\in\sigma_j} x_i\|$; $j = 1,2,\ldots,k$, *where* $\sigma_j \cap \sigma_h = \emptyset$ *whenever* $j \neq h$, *is* K-*equivalent to the basic sequence* $\{x_i\}_{i=1}^k$.

Since with this definition of the perfectly homogeneous constant the symmetry constant $K(\{x_i\}_{i=1}^n)$ of $\{x_i\}_{i=1}^n$ does not exceed $PH(\{x_i\}_{i=1}^n)$, we may as well restrict our attention to 1-symmetric bases.

The main result of this section is the following.

Theorem 4.2. Let $\{x_i\}_{i=1}^n$ *be a* 1-*symmetric normalized basis of a Banach space* X *and set* $K = PH(\{x_i\}_{i=1}^n)$. *Then* $\{x_i\}_{i=1}^n$ *is* $160K^2$ -*equivalent to the unit vector basis of* ℓ_p^n , *where* $p = \log n/\log\|\sum_{i=1}^n x_i\|$ *(with the convention* $1/0 = \infty$) .

In the proof of 4.2, we make use of the following elementary lemmas.

Lemma 4.3. Let g *be a real function on the integers* $\{0,1,2,\ldots,N\}$ *such that* $g(N) = 0$ *and* $|g(j+k) - g(j) - g(k)| \leq C$, *whenever* $j + k \leq N$. *Then*
$$\max_{0\leq j\leq N} |g(j)| \leq 2C .$$

Proof: Let i be the first integer for which $\max_{0\le j\le N} |g(j)| = |g(i)|$. Since $-g$ also satisfies the hypotheses of the lemma we may assume without loss of generality that $g(i) \geq 0$.

We first consider the case when $2i \leq N$. Under this assumption, it follows that

$$-C \leq g(2i) - 2g(i) \leq g(i) - 2g(i) = -g(i)$$

i.e., $g(i) \leq C \leq 2C$, as desired. In the other case; i.e., when $2i > N$ we have $g(i-(N-i)) < g(i)$. Thus, using the fact that $g(i) + g(N-i) = g(i) + g(N-i) - g(N) \leq C$, we get $g(i) \leq C - g(N-i) < C + g(i) - g(i-(N-i)) - g(N-i) \leq 2C$ and this completes the proof.

□

Lemma 4.4. Let $\{x_i\}_{i=1}^n$ *be a* 1*-symmetric normalized basis of a Banach space* X *and let* $\{x_i^*\}_{i=1}^n$ *be the corresponding sequence of biorthogonal functionals. Then*

$$PH(\{x_i^*\}_{i=1}^n) = PH(\{x_i\}_{i=1}^n) .$$

Proof: Suppose that $\{\sigma_j\}_{j=1}^k$ is a sequence of pairwise disjoint subsets of the integers $\{1,2,\dots,n\}$ and put

$$y_j = (\sum_{i\in\sigma_j} x_i)/\|\sum_{i\in\sigma_j} x_i\|;\ y_j^* = (\sum_{i\in\sigma_j} x_i^*)/\|\sum_{i\in\sigma_j} x_i^*\|;\ j = 1,2,\dots,k .$$

By the 1-symmetry of $\{x_i\}_{i=1}^n$ it follows that $y_j^* y_j = \bar{\bar{\sigma}}_j/\|\sum_{i\in\sigma_j} x_i\|\cdot\|\sum_{i\in\sigma_j} x_i^*\| = 1$ for all $1 \leq j \leq k$ and, in addition, the averaging operation $P = \sum_{j=1}^k y_j^* \otimes y_j$ is a norm one projection from X onto $Y = \operatorname{span}\{y_j\}_{j=1}^k$. Thus, Y^* is isometric to $P^*X^* = \operatorname{span}\{y_j^*\}_{j=1}^k$ in the canonical way which means that $\{y_j^*\}_{j=1}^k$ is 1-equivalent to the sequence of biorthogonal functionals associated to $\{y_j\}_{j=1}^k$ in Y^*. Hence, since $\{\sigma_j\}_{j=1}^k$ was arbitrarily chosen, we get that $PH(\{x_i\}_{i=1}^n) = PH(\{x_i^*\}_{i=1}^n)$.

□

Proof of 4.2: Let N be an integer such that $2^N \le n < 2^{N+1}$. We first show that, for all integers j and k with $0 \le j+k \le N$, we have the multiplicative inequality

$$K^{-1}\lambda(2^{j+k}) \le \lambda(2^j)\lambda(2^k) \le K\lambda(2^{j+k}),$$

where $\lambda(m) = \|\sum_{i=1}^m x_i\|$; $m = 1,2,\ldots,n$. In order to prove this inequality it suffices to notice that, by the perfect homogeneity of $\{x_i\}_{i=1}^n$, the normalized block basis $y_h = (\sum_{i=m(h)+1}^{m(h+1)} x_i)/\lambda(2^j)$; $h = 1,2,\ldots,k$, where $m(h) = 2^{(h-1)j}$, is K-equivalent to $\{x_i\}_{i=1}^k$ and then to use this equivalence for the vector $\sum_{h=1}^k y_h = (\sum_{i=1}^{2^{k+j}} x_i)/\lambda(2^j)$.

Instead of working with this inequality it is more convenient for us, as for Zippin in [82], to use a linearized version of it. Let $\{x_i^*\}_{i=1}^n$ be the biorthogonal functionals corresponding to $\{x_i\}_{i=1}^n$ and set

$$\alpha = (\log_2\|\sum_{i=1}^{2^n} x_i\|)/N\ ;\quad \alpha^* = (\log_2\|\sum_{i=1}^{2^N} x_i^*\|)/N\ .$$

Since $\alpha + \alpha^* = 1$ we may assume without loss of generality that $1/2 \le \alpha \le 1$; otherwise, by using 4.4, we work with X^* instead of X. Put $g(j) = \log_2\lambda(2^j) - \alpha j$; $0 \le j \le N$ and observe that $g(N) = \log_2\lambda(2^N) - \alpha N = 0$ and $|g(j+k) - g(j) - g(k)| \le \log_2 K$, whenever $0 \le j+k \le N$. Hence, by 4.3, we conclude that $\max_{0\le j\le N} |g(j)| \le 2\log_2 K$; i.e.,

$$K^{-2}2^{\alpha j} \le \lambda(2^j) \le K^2 2^{\alpha j}\ ;\ 0 \le j \le N\ .$$

By the fact that $\{x_i\}_{i=1}^n$ is a 1-unconditional basis, we further get that

$$(+)\qquad 2^{-1}K^{-2}j^{\alpha} \le \lambda(j) \le 2K^2 j^{\alpha}\ ;\ 1 \le j \le n .$$

The next step in the proof is to show that $\{x_i\}_{i=1}^{N+2}$ is $10K^3$-equivalent to the unit vector basis of ℓ_q^{N+2}, where $q = 1/\alpha$. Suppose that $\{a_j\}_{j=1}^{N+2}$ are non-negative reals such that $\sum_{j=1}^{N+2} a_j = 2^N + N + 2$. In view of this condition it is possible to choose non-negative integers $m_j \le a_j$; $j=1,2,\dots,N+2$ such that $\sum_{j=1}^{N+2} m_j = 2^N$ and $\max_{1\le j\le N+2} (a_j - m_j) < 2$. A simple computation shows that

$$\|\sum_{j=1}^{N+2} m_j^{\alpha} x_j\| \le \|\sum_{j=1}^{N+2} a_j^{\alpha} x_j\| \le \|\sum_{j=1}^{N+2} m_j^{\alpha} x_j\| + 2(N+2) .$$

Let now $\{\sigma_j\}_{j=1}^{N+2}$ be a partition of the integers $\{1,2,\dots,2^N\}$ into pairwise disjoint subsets such that $\bar{\bar{\sigma}}_j = m_j$ for all $1 \le j \le N+2$. Then, by (+) and our hypotheses, it follows that

$$\|\sum_{j=1}^{N+2} m_j^{\alpha} x_j\| \le K\|\sum_{j=1}^{N+2} m_j^{\alpha}(\sum_{i\in\sigma_j} x_i)/\lambda(m_j)\| \le 2K^3\|\sum_{j=1}^{N+2} \sum_{i\in\sigma_j} x_i\| =$$

$$= 2K^3\lambda(2^N) = 2K^3 2^{N\alpha} .$$

Since the opposite inequality can be proved in a similar way we obtain that

$$2^{-1}K^{-3}2^{N\alpha} \le \|\sum_{j=1}^{N+2} m_j^{\alpha} x_j\| \le 2K^3 2^{N\alpha}$$

which, if we return to the scalars $\{a_j\}_{j=1}^{N+2}$, implies that

$$2^{-1}K^{-3}2^{N\alpha} \le \|\sum_{j=1}^{N+2} a_j^{\alpha} x_j\| \le 2K^3 2^{N\alpha} + 2(N+2) .$$

On the other hand, by the assumption that $1/2 \le \alpha \le 1$, we have

$$(\sum_{j=1}^{N+2} a_j)^{1/q} = (2^N + N + 2)^{\alpha} \le 3 \cdot 2^{N\alpha}$$

and $2(N+2) \leq 2(N+2)^{2\alpha} \leq 8(2^N + N + 2)^{\alpha}$. Thus,

$$6^{-1}K^{-3}(\sum_{j=1}^{N+2} a_j)^{1/q} \leq \| \sum_{j=1}^{N+2} a_j^{1/q} x_j \| \leq 10K^3(\sum_{j=1}^{N+2} a_j)^{1/q}$$

and this completes the proof of this step since the above formula is homogeneous and $\{x_i\}_{i=1}^n$ is 1-unconditional.

In the final step of the proof we will show that $\{x_i\}_{i=1}^n$ is $80K^6$-equivalent to the unit vector basis of ℓ_q^n. Suppose that $\| \sum_{i=1}^n b_i x_i \| = 1$ and assume that $b_1 \geq b_2 \geq \dots \geq b_n \geq 0$. Let k be the smallest integer for which $b_{k+1} \leq 2^{-N-2}$. Then $\| \sum_{i=k+1}^n b_i x_i \| \leq n2^{-N-2} \leq 2^{-1}$ which implies that $\| \sum_{i=1}^k b_i x_i \| > 2^{-1}$. For $1 \leq j \leq N+2$ let $\delta_j = \{i;\ 2^{-j} < b_i \leq 2^{-j+1}\}$ and $h_j = \bar{\bar{\delta}}_j$. Since $\| \sum_{j=1}^{N+2} 2^{-j} \sum_{i \in \delta_j} x_i \| \geq 4^{-1}$ we get, by the perfect homogeneity of $\{x_i\}_{i=1}^n$, that

$$\| \sum_{j=1}^{N+2} 2^{-j} \lambda(h_j) x_j \| \geq 4^{-1}K^{-1} .$$

Hence, by the previous step of the proof, we have that

$$(\sum_{j=1}^{N+2} 2^{-jq} \lambda(h_j)^q)^{1/q} \geq 40^{-1}K^{-4}$$

or, by (+), that

$$(\sum_{j=1}^{N+2} 2^{-jq} h_j)^{1/q} \geq 80^{-1}K^{-6} .$$

But, of course,

$$(\sum_{i=1}^{n} b_i^q)^{1/q} \geq (\sum_{j=1}^{N+2} 2^{-jq} h_j)^{1/q}$$

so we have

$$\left(\sum_{i=1}^{n} |b_i|^q\right)^{1/q} \geq 80^{-1}K^{-6} = 80^{-1}K^{-6} \left\| \sum_{i=1}^{n} b_i x_i \right\| ,$$

whenever $\| \sum_{i=1}^{n} b_i x_i \| = 1$ and $b_1 \geq b_1 \geq \ldots \geq b_n \geq 0$. Evidently, this inequality remains valid for all choices of the scalars $\{b_i\}_{i=1}^n$ and, exactly in the same way, we prove the opposite inequality. In conclusion, $\{x_i\}_{i=1}^n$ is $80K^6$-equivalent to the unit vector basis of ℓ_q^n. In order to complete the proof, we have to compare the values of p and q. We recall that $\lambda(2^N) = 2^{N/q}$, $\lambda(n) = n^{1/p}$ and $2^N \leq n < 2^{N+1}$. Thus, by a simple computation, it follows that

$$2^{-1}n^{1/q} \leq n^{1/p} \leq 2n^{1/q} ,$$

and this certainly implies that the unit vector basis of ℓ_q^n is 2-equivalent to that of ℓ_p^n. This evidently completes the proof.

□

Remark: The isometric version of 4.2 is false in the finite dimensional case. Indeed, let $n > 1$ and define a norm on R^n by putting

$$\left\| \sum_{i=1}^{n} a_i e_i \right\| = \max_{\pi} [(|a_{\pi(1)}| + |a_{\pi(2)}|)^2/2 + |a_{\pi(3)}|^2 + \ldots + |a_{\pi(n)}|^2]^{1/2} ,$$

where the maximum is taken over all possible permutations π of the integers $\{1,2,\ldots,n\}$. It is easily checked that $\{e_i\}_{i=1}^n$ is a 1-symmetric basis with $PH(\{e_i\}_{i=1}^n) = 1$. Moreover, for every $x = \sum_{i=1}^{n} a_i e_i \in R^n$, we have that

$$\left(\frac{n-1}{n}\right)^{1/2} \left(\sum_{i=1}^{n} |a_i|^2\right)^{1/2} \leq \|x\| \leq \left(\sum_{i=1}^{n} |a_i|^2\right)^{1/2}$$

but $\|x\| = (\sum_{i=1}^{n} |a_i|^2)^{1/2}$ if and only if $|a_i| = |a_j|$ for some $i \neq j$. This shows that R^n endowed with the above norm $\|\cdot\|$ is not isometric to ℓ_2^n.

We pass now to the characterization of those Banach spaces which have a unique unconditional basis of finite length.

As is well known, in the infinite-dimensional case, Lindenstrauss and Zippin [51] showed that ℓ_1, ℓ_2 and c_0 are the only spaces with a unique normalized unconditional basis. This result does not admit a completely satisfactory local analogue since 1.9 implies via a simple limiting argument that, for any $1 < p < \infty$, there exists a constant K_p so that every normalized basis of ℓ_p^n with unconditional constant $< \gamma_p(\ell_2)/\sqrt{2}$ ($\approx \max\{(\sqrt{p/2}, \sqrt{p/2(p-1)}\})$ is K_p-equivalent to the unit vector basis of ℓ_p^n. However, by a careful analysis and under suitable restrictions, it is possible to prove some satisfactory results of a positive nature.

The first step in the study of spaces with a unique unconditional basis is to reduce the problem to the case of ℓ_p^n-spaces. This is achieved by using 4.2; however, 4.2 can be applied only if first we are able to show that a normalized block basis with constant coefficients of a given unconditional basis $\{x_i\}_{i=1}^n$ extends to an unconditional basis for the whole space. For infinite-dimensional spaces this is essentially a simple consequence of Pelczynski's decomposition method [63] (cf. [49] for a simple proof or [51] for the original proof which actually gives a slightly more general result). In the finite-dimensional case we have to prove first a localization of the decomposition method. Before stating our decomposition result we point out that if a Banach space has (in some reasonable sense) a unique unconditional basis then this basis is necessarily symmetric. Therefore, it suffices to consider only 1-symmetric bases.

Proposition 4.5. There exists an absolute constant $C < \infty$ *such that if* X *is a finite dimensional Banach space with a* 1*-symmetric normalized basis* $\{x_i\}_{i=1}^m$ *and* $y_j = \sum_{i \in \sigma_j} x_i$; $j = 1,2,\ldots,k$, *where* $\{\sigma_j\}_{j=1}^k$ *are pairwise disjoint subsets of the integers* $\{1,2,\ldots,m\}$, *is a block basis with constant coefficients of* $\{x_i\}_{i=1}^m$ *whose closed linear span is denoted by* Y *then*

$$d(X, Y \oplus_\infty \operatorname{span}\{x_i\}_{i=1}^{m-k}) \le C^2 .$$

(Here, the direct sum is taken in the sense of ℓ_∞^2 ; i.e., the maximum of the respective norms.)

Proof: For simplicity, we shall assume that there exists an integer N so that $m = 2^{N+1}-1$ and that all the blocks $\{y_j\}_{j=1}^k$ are supported by the first 2^N vectors of the basis $\{x_i\}_{i=1}^m$. Notice that there is no loss of generality in imposing these conditions since, when we pass from this case to the case of a general m and of an arbitrary block basis $\{y_j\}_{j=1}^k$ with constant coefficients, we might have to increase the value of the constant C by a factor of 4 , at most. In view of the assumption made on m we can rewrite the basis $\{x_i\}_{i=1}^m$ of X as $\{x_i^h ; 0 \le h \le N , 1 \le i \le 2^h\}$. By putting $X_h = [x_i^{N-h}]_{i=1}^{2^{N-h}}$; $0 \le h \le N$ we can represent the space X as the internal direct sum

$$X = X_0 + X_1 + \dots + X_N .$$

By the assumption made on $\{y_j\}_{j=1}^k$ we consider Y as a subspace of X_0 . The idea of the proof is to split each vector y_j in half and consider the span of these halves of vectors as a subspace Y_1 of X_1 . Then we split each basis vector of Y_1 in half and consider the span of the halves of the vectors as a subspace Y_2 of X_2 . We continue in this way in order to build subspaces Y_h of X_h so that Y_h is spanned by blocks with constant coefficients and the natural backward shift from $Y_1 + Y_2 + \dots + Y_N$ into $Y_0 + Y_1 + \dots + Y_{N-1}$ (where $Y_0 = Y$) is a good isomorphism. In this construction we have to be careful since $\dim X_N = 1$ and thus the dimensions of the Y_h's must shrink.

We present now this construction in a precise manner. Put $\sigma_j^0 = \sigma_j$, $y_j^0 = \sum_{i \in \sigma_j^0} x_i^0$ for $1 \le j \le k$ and $Y_0 = [y_j^0]_{j=1}^k$. We may assume that $\bar{\bar{\sigma}}_j^0 \ge \bar{\bar{\sigma}}_{j+1}^0$, and let k_1 be the last index j , for which $\bar{\bar{\sigma}}_j^0 \ge 2$. For

$1 \leq j \leq k_1$, choose pairwise disjoint subsets $\{\sigma_j^1\}_{j=1}^{k_1}$ of the integers $\{1,2,\ldots,2^{N-1}\}$ such that $\bar{\bar{\sigma}}_j^1 = [\bar{\bar{\sigma}}_j^0/2]$. Set $y_j^1 = \sum_{i\in\sigma_j^1} x_i^1$; $1 \leq j \leq k_1$ and let $Y_1 = [y_j^1]_{j=1}^{k_1}$. Obviously, Y_1 is a subset of X_1.

We continue inductively and build vectors $\{y_j^{h+1}\}_{j=1}^{k_{h+1}}$ from $\{y_j^h\}_{h=1}^{k_h}$ in the same way $\{y_j^1\}_{j=1}^{k_1}$ was built from $\{y_j^0\}_{j=1}^{k}$. In other words, if $y_j^h = \sum_{i\in\sigma_j^h} x_i^h$; $1 \leq j \leq k_h$ then, for $1 \leq j \leq k_h$ such that $\bar{\bar{\sigma}}_j^h > 1$ (say, for $1 \leq j \leq k_{h+1}$), we choose mutually disjoint subsets $\{\sigma_j^{h+1}\}_{j=1}^{k_{h+1}}$ of the integers $\{1,2,\ldots,2^{N-h-1}\}$ so that $\bar{\bar{\sigma}}_j^{h+1} = [\bar{\bar{\sigma}}_j^h/2]$, and define $y_j^{h+1} = \sum_{i\in\sigma_j^{h+1}} x_i^{h+1}$.

Observe that $Y_{h+1} = [y_j^{h+1}]_{j=1}^{k_{h+1}}$ is a subspace of X_{h+1}. Once this construction has been completed, we notice that the internal direct sum $\widetilde{Y} = Y_0 + Y_1 + \ldots + Y_N$ is spanned by a block basis with constant coefficients of $\{x_i\}_{i=1}^m$ and, hence, there is a norm one averaging projection P from X onto $\widetilde{Y}$. Now we want to define a "good" isomorphism from X, considered as the internal direct sum $\ker P + Y$, into $X \oplus_\infty Y$ so that the range of T is isometric to $[x_i]_{i=1}^{m-k} \oplus_\infty Y$. The operator T is defined as follows: for $x \in \ker P$ put $Tx = (x,0)$ while in $\widetilde{Y}$ let $Ty_j^0 = (0,y_j)$ if $1 \leq j \leq k$ and $Ty_j^h = (y_j^{h-1},0)$ if $1 \leq j \leq k_h$ and $h = 1,2,\ldots,N$. Elsewhere in X extend T linearly. It is quite obvious that T acts as an isometry on $\ker P$ as well as on Y_0. The main point of the construction above is that T is also a "good" isomorphism on $Z = Y_1 + \ldots + Y_N$. Indeed, since $(\overline{\overline{\text{supp}\, y_j^{h-1}}}) \geq \geq \overline{\overline{\text{supp}}}\, y_j^h)$ we immediately get that $\|Tz\| \geq \|z\|$ for all $z \in Z$. On the other hand,

$$\tfrac{1}{2}(\overline{\overline{\text{supp}}}\; y_j^{h-1}) \geq (\overline{\overline{\text{supp}}}\; y_j^h) > \tfrac{1}{3}(\overline{\overline{\text{supp}}}\; y_j^{h-1})$$

so if $z \in Z$ then Tz can be decomposed as the sum of 3 vectors with mutually disjoint supports, two of which have the same distribution and therefore also the same norm as z and the third of which has norm at most $\|z\|$. Thus,

$$\|Tz\| \leq 3\,\|z\| \quad \text{for all } z \in Z.$$

Hence, T is a good isomorphism on Z, and whence also on $Z + Y_0 = \text{range } P$. Since T maps $\ker P$ into $(\ker P) \oplus \{0\}$ and range P into $(\text{range } P) \oplus Y$, there exists an absolute constant $C < \infty$ such that $\|T\|$, $\|T^{-1}\| \leq C$.

Furthermore, it is easily checked that the range of T is $[x_i]_{i\in\sigma} \oplus Y$, where $i \notin \sigma$ iff x_i is equal to y_j^h for some j and h. Of course, $\bar{\bar{\sigma}} = m-k$, so the proof is complete.

□

We return now to the study of spaces with a unique unconditional basis. We first remark that by combining 4.2 and 4.5, we get immediately the following result.

Proposition 4.6. For every K *there exists a number* $M(K) < \infty$ *such that if a finite dimensional Banach space* X *has a normalized* 1*-symmetric basis* $\{x_i\}_{i=1}^m$ *and every other normalized basis of* X, *whose unconditional constant is* $\leq C$ *(with* $C > 1$, *as given in the statement of 4.5), is* K*-equivalent to* $\{x_i\}_{i=1}^m$, *then the basis* $\{x_i\}_{i=1}^m$ *is* $M(K)$*-equivalent to the unit vector basis of* ℓ_p^m, *where* $p = \log m / \log \|\sum_{i=1}^m x_i\|$.

In view of 4.6, we have to investigate now unconditional bases in the spaces ℓ_p^n; $1 \leq p \leq \infty$; $n = 1,2,\ldots$. By duality, it is clearly enough to consider only the case $p > 2$. A way of producing in ℓ_p^n unconditional bases which are distant from the usual unit vector basis is contained in the following lemma proved in [6] (by means similar to those used in the proof of 4.5).

Lemma 4.7. Let $k = k(p,n)$ *denote the maximal dimension of a subspace of* ℓ_p^n; $p > 2$ *which is 2-isomorphic to a Hilbert space. There exists a constant* $D < \infty$ *such that, for every integer* n *and every* $p > 2$, *the space* ℓ_p^{nk+k}, *where* $k = k(p,n)$, *has a normalized basis* $\{x_i\}_{i=1}^{nk+k}$, *with unconditoional constant* $\leq pD$, *such that* $\{x_i\}_{i=1}^k$ *is 2-equivalent to the unit vector basis of* ℓ_2^k.

Proof: Fix $p > 2$ and an integer n. Let H_k ; $k = k(p,n)$ be a subspace of ℓ_p^n which is 2-isomorphic to ℓ_2^k. In [6] it was proved that there exist a constant $D_p < \infty$, independent of n, and an isomorphism T between ℓ_p^{nk+k} and $\ell_2^k \oplus_p \ell_p^{nk}$ such that $\|T\|$, $\|T^{-1}\| \leq D_p$ and T maps H_k, considered as a subspace of ℓ_p^{nk+k}, onto ℓ_2^k. Moreover, the constant D_p behaves like the square of the type 2-constant of the space ℓ_p ; i.e., it is of order of magnitude p. More precisely, there is a constant D, independent of p, so that $D_p \leq pD$. This obviously completes the proof. □

By estimating the value of $k(p,n)$ for a given n we are able to apply 4.7 in order to prove that Hilbert spaces are the only smooth spaces having a unique normalized unconditional basis. More exactly, we have the following result.

Theorem 4.8: Let $1 < r < s < \infty$ *and let*

$$\mathfrak{F}_{r,s} = \{\ell_p^n;\ r \leq p \leq s,\ n = 1,2,\dots\}$$

There exists a constant $C = C(r,s)$ *which satisfies the following: Given any* $K < \infty$ *there exists a number* $f(K) = f(K,r,s) < \infty$ *so that if* $X \in \mathfrak{F}_{r,s}$ ($\dim X = n$, say) *and every normalized basis of* X, *whose unconditional constant is at most* C, *is* K*-equivalent to the unit vector basis of* X *then*

$$d(X,\ell_2^n) \leq f(K) .$$

Proof: As we have already said, the main point in this proof is to estimate the value of $k(p,n)$ from below. This can be done by using e.g. the arguments in [27], which yield that, there is a constant $c > 0$ such that

$$k(p,n) \geq cn^{2/p} ,$$

for every n and every $p > 2$.

Now, set $C = C(r,s) = D \max\{s, r/(r-1)\}$, where D is the absolute constant appearing in the statement of 4.7. This choice of the constant C implies that each of the spaces ℓ_p^{nk+k} , $r \le p \le s$, $n = 1,2,\ldots,k = k(p,n)$ has, by 4.7, a normalized basis $\{x_i\}_{i=1}^{nk+k}$ whose unconditional constant is $\le C$ and such that $\{x_i\}_{i=1}^{k}$ is 2-equivalent to the unit vector basis of ℓ_2^k . If, for some $K < \infty$, this basis is K-equivalent to the unit vector basis of ℓ_p^{nk+k} then, by using only the first k vectors, we get that $k^{1/2-1/p} \le 2K$. On the other hand, it dollows from the estimate given above for $k = k(p,n)$ that

$$d(\ell_p^{nk+k},\ell_2^{nk+k}) = (nk+k)^{1/2-1/p} \le 2\, c^{(2-p)/4} k^{(1+p/2)(1/2-1/p)} \le$$
$$\le 2\, c^{(2-p)/4} (2K)^{1+p/2} .$$

This completes the proof in view of the bounds imposed on p . □

The conjunction of 4.6 and 4.8 justifies our assertion that Euclidean spaces are the only smooth finite dimensional spaces which have a unique unconditional basis.

Corollary 4.9. Let an integer N *and a function* $g:[1,\infty] \to [1,\infty)$ *be given. Let* F *be the family of all finite dimensional spaces* X *with monotonely unconditional basis which satisfy the smoothness condition*

(N) $d(E,\ell_1^N) \ge 2$ *for each subspace* E *of* X

and the uniqueness condition

{g} For every $1 \le K < \infty$ *, every* K*-unconditional normalized basis for* X *is* $g(K)$ *equivalent to the given monotonely unconditional basis for* X .

Then there is a constant $M = M(N,g) < \infty$ *such that for all* X *in* F ($\dim X = n$, say), $d(X,\ell_2^n) \le M$.

Sketch of proof: In view of 4.8, we only need to show that there are $1 < r < s < \infty$ and a constant $D = D(N,g)$ so that each space X in F of sufficiently large dimension ($\dim X = m$, say) satisfies $d(X,\ell_p^m) \le D$ for some $r \le p \le s$. Given any X in F, the specified monotonely unconditional basis for X is $g(1)$-

symmetric, hence X is $g(1)$ isomorphic to a space X_1 satisfies the smoothness condition (N_1), where $N_1 = N^k$ and k is the smallest positive integer for which $[2g(1)^{1/k}] < 2$. Also, X_1 satisfies the uniqueness condition $\{g_1\}$, where $g_1(K) = g(1)g(g(1)K)$.

Set $F_1 = \{X_1 : X \in F\}$, and let X_1 be a generic member of F_1 (with $\dim X = m$, say). Notice that the 1-symmetric normalized basis $\{x_i\}_{i=1}^m$ for X_1 is, by 4.6, $M = M(g_1(C))$ equivalent to the unit vector basis for ℓ_p^m , where C , M , and $p = \log m/\log\|\sum_{i=1}^m x_i\|$ are as in 4.6. Since X_1 satisfies (N_1) , it is intuitively obvious and easily computable that there are $1 < r < s < \infty$ depending only on N_1 and g_1 such that p must be in (r,s) when m is sufficiently large. This completes the sketch of the proof. □

It is known [49] that a K-unconditional normalized basis for a Euclidean space is K-equivalent to an orthonormal basis. It is also known [44] that a normalized K-unconditional basis for ℓ_1^n or ℓ_∞^n is $K_G K^2$ equivalent to the unit vector basis for the space. We ask whether this common property of ℓ_1^n , ℓ_∞^n , and ℓ_2^n gives a joint characterization of these spaces. More generally, given a family F of finite dimensional Banach spaces each of which has a monotonely unconditional basis and given a function $g: [1,\infty) \to [1,\infty)$, we ask:

Problem 4.10. Does there exist a constant $M = M(g) < \infty$ *such that if* X *is in* F ($\dim X = n$ *, say) and, for each* $1 \le K \le \infty$ *, every* K*-unconditional normalized basis for* X *is* $g(K)$ *-equivalent to the given monotonely unconditional normalized basis for* X *then* $d(X, \ell_p^n) \le M$ *for some* $p \in \{1,2,\infty\}$?

Of course, we are mainly interested in the case where F is the family of all finite dimensional spaces which have a monotonely unconditional basis or, which is essentially the same by 4.6 and 4.8, in the case where F is $\{\ell_p^n : 1 \le p \le \infty,\ n = 1,2,\dots\}$. For this family F it would be interesting to have a positive answer for all functions g of the form $g(t) = Bt^2$, but there may be a much stronger positive result. One can use the construction in the proof of 4.7 to show, for example, that if $F = \{\ell_p^n;\ n = 1,2,\ldots,2,$ $2 \le p \le (\log n)^{\alpha}\}$, and $0 < \alpha < 1$, then 4.10 has an affirmative answer for functions g which satisfy $g(t) \le B \exp \frac{1}{2} t^{1/\alpha - 1}$. The real problem occurs when $p(n) \to \infty$ fairly fast, but more slowly than $\log n$ (since ℓ_p^n is uniformly isomorphic to ℓ_∞^n when $p \ge \log n$) . Our calculation suggests that $p(n) = [\log(\log n)]^{-1} \log n$ may be the "worst" case.

5. UNIQUENESS OF THE REARRANGEMENT STRUCTURE

The main result of this section is a generalization to the "continuous" case of r.i. function spaces on $[0,1]$ of Theorem 3.8 on the uniqueness of symmetric bases of finite length. We start by studying isomorphic embeddings of a q-concave $(1 \leq q < 2)$ r.i. space of functions into another r.i. function space (in the next section, we shall present a more general result of this sort). This theorem, besides being one of the main tools in the proof of the uniqueness of the r.i. structure, has itself some interesting consequences which will be presented in the sequel.

Theorem 5.1: Let X *be a r.i. function space on* $[0,1]$ *so that* $\|f\|_X \leq C\|f\|_q$ *, for some* $1 \leq q < 2$ *, some constant* $C < \infty$ *and for every* $f \in L_q(0,1)$. *Let* Y *be a r.i. function space on* $[0,1]$ *or on* $[0,\infty)$ *which does not contain uniformly isomorphic copies of* ℓ_∞^n *for all* n *. If* X *is isomorphic to a subspace of* Y *then one of the following conditions holds:*

(i) There exists a constant $V < \infty$ *such that*

$$\|f\|_Y \leq V\|f\|_X$$

for every $f \in X$ *, or,*

(ii) Y *contains a sequence of normalized functions with mutually disjoint supports which, in the norm of* Y *, is equivalent to the usual Haar system in* X .

We begin with the following lemma.

Lemma 5.2: Let Y *be a r.i. space of functions on a finite interval* $[0,s]$ *with* $s \geq 1$ *which is not, even up to an equivalent norm, equal to* $L_\infty(0,s)$. *Suppose that there exist a constant* $R < \infty$ *and a bounded operator* T *from a r.i. function space* X *on* $[0,1]$ *into* Y *such that if, for* $n = 0,1,2,\ldots$ *and* $1 \leq i \leq 2^n$ *, we put*

$$x_{n,i} = \chi_{[(i-1)2^{-n},i2^{-n})}, \quad y_n = \max_{1\le i\le 2^n} |Tx_{n,i}|$$

and $\delta_n = \{t \in [0,s];\ y_n(t) \le R\}$, $\qquad \inf_n \|y_n\chi_{\delta_n}\|_Y > 0$.

Then, there exists a constant $V<\infty$ *so that*

$$\|f\|_Y \le V\|f\|_X \quad ,$$

for every $f \in X$.

Proof. The main point is to find positive numbers α, β and γ having the property that, for each n there exist disjoint subsets $\{\eta_j\}_{j=1}^{2^n}$ of $[0,s]$ such that $|(Tx_{n,j})(t)| \ge \alpha/(2\|\chi_{[0,s]}\|_Y)$ for $t \in \eta_j$ and $\mu(\eta_j) \ge \gamma 2^{-n}$ for at least $\beta 2^n$ values of j .

To this end, set

$$\alpha = \inf_n \|y_n\ \chi_{\delta_n}\|_Y \quad ,$$

fix n , and put

$$\eta = \{t \in [0,s];\ \alpha/(2\|\chi_{[0,s]}\|_Y) \le y_n(t) \le R\}$$

so that

$$\|y_n\ \chi_\eta\|_Y \ge \alpha/2 \quad .$$

Let $\{\eta_j\}_{j=1}^{2^n}$ be a partition of η into mutually disjoint measurable subsets such that, for $t \in \eta_j$, we have

$$|(T\ x_{n,j})(t)| = y_n(t) \ .$$

We also assume, as we certainly may, that

$$\mu(\eta_1) \ge \mu(\eta_2) \ge \dots \ge \mu(\eta_{2^n}) \ge 0,$$

Then, for each integer $1 \le m \le 2^n$, we get that

$$\|y_n \sum_{j=1}^{m} \chi_{\eta_j}\|_Y \leq \|\int_0^1 |\sum_{j=1}^{m} r_j(u)\ Tx_{n,j}|du\|_Y \leq$$

$$\leq \|T\| \cdot \int_0^1 \|\sum_{j=1}^{m} r_j(u) x_{n,j}\|_X\ du = \|T\| \cdot \|\sum_{j=1}^{m} x_{n,j}\|_X = \|T\| \cdot \|\chi_{[0,m2^{-n}]}\|_X$$

which implies that

$$\alpha/2 \leq \|y_n \chi_n\|_Y \leq \|T\| \cdot \|\chi_{[0,m2^{-n}]}\|_X + \|y_n \sum_{j=m+1}^{2^n} \chi_{\eta_j}\|_Y \leq$$

$$\leq \|T\| \cdot \|\chi_{[0,m2^{-n})}\|_X + R\|\chi_{[0,2^n\mu(\eta_{m+1}))}\|_Y .$$

If X is equal to $L_\infty(0,1)$, up to an equivalent norm, then the assertion of the Lemma is trivially satisfied. Otherwise, it is easily checked that $\lim_{t \to \infty} \|\chi_{[0,t)}\|_X = 0$ which, in particular, implies the existence of a number $\beta > 0$ so that $\|\chi_{[0,\beta)}\|_X \leq \alpha/4\|T\|$.

Let m be the largest integer for which $m2^{-n} \leq \beta$. Then $\alpha/4 \leq R\|\chi_{[0,2^n\mu(\eta_{m+1}))}\|_Y$. Hence, by the fact that Y is not equal to $L_\infty(0,s)$, even up to an equivalent norm, it follows that there exists a number $\gamma > 0$ such that

$$2^n\mu(\eta_{m+1}) \geq \gamma ;$$

i.e., that $\mu(\eta_1) \geq \mu(\eta_2) \geq \dots \geq \mu(\eta_{m+1}) \geq \gamma 2^{-n}$.

In other words, we have just shown that a percentage $(m+1)2^{-n} > \beta$ of the 2^n sets $\{\eta_j\}_{j=1}^{2^n}$ have measure $\geq \gamma 2^{-n}$, where both β and γ are independent of n .

Let N be a fixed integer such that $N\gamma \geq 1$. Then, for every step function of the form

$$g = \sum_{j=1}^{m+1} a_j\ \chi_{[(j-1)N^{-1}2^{-n},\,j\,N^{-1}2^{-n})} ,$$

where m has been fixed above (of course, depending on n) and $\{a_j\}_{j=1}^{m+1}$ are arbitrary scalars, we have

$$\|g\|_Y \leq \|\sum_{j=1}^{m+1} a_j \chi_{\eta_j}\|_Y \leq 2\alpha^{-1}\|\chi_{[0,s]}\|_Y \cdot \| \max_{1\leq j\leq m+1} |a_j \, Tx_{n,j}|\|_Y \leq$$

$$\leq 2\alpha^{-1}\|\chi_{[0,s]}\|_Y \|\int_0^1 |\sum_{j=1}^{m+1} r_j(u) a_j Tx_{n,j}| du\|_Y \leq 2\|T\|\alpha^{-1}\|\chi_{[0,s]}\|_Y \|\sum_{j=1}^{m+1} a_j x_{n,j}\|_X \leq$$

$$\leq 2\|T\| \, N \, \alpha^{-1}\|\chi_{[0,s]}\|_Y \|g\|_X \ .$$

For a general step function of the form

$$f = \sum_{i=1}^{N2^n} a_i \chi_{[(i-1)N^{-1}2^{-n}, iN^{-1}2^{-n})} \ ,$$

we split the interval $[0,1]$ into (at most $N2^n(m+1)^{-1} + 1 < N\beta^{-1} + 1$) intervals each having measure $\leq (m+1)N^{-1}2^{-n}$ and then apply the previous inequality in each of these intervals. It follows that, for every dyadic step function f ,

$$\|f\|_Y \leq V\|f\|_X \ ,$$

where $V = (2\|T\|N\alpha^{-1}\|\chi_{[0,s]}\|_Y) \ (N\beta^{-1} + 1)$ is independent of n . This completes the proof in view of the discussion on r.i. spaces in the introduction.

□

Remark: Note that the same proof remains valid if (a) we replace the $x_{n,i}$'s in the statement of Lemma 5.2 by functions $u_{n,i}$ such that $\mu(\text{supp } u_{n,i}) = 2^{-n}$ and $|\sum_{i=1}^{2^n} u_{n,i}| = 1$ and (b) if Y is a r.i. space except that $L_\infty(0,s)$ is not assumed to be dense in Y .

Proof of 5.1: For each $n = 0,1,2,\ldots$ and $1 \leq i \leq 2^n$, denote by $x_{n,i}$ the characteristic function of the interval $[(i-1)2^{-n}, i2^{-n})$. Let T be an isomorphism from X into Y .

We distinguish now between several cases.

Case I: There exist a constant $R < \infty$, and a number $1 \le s < \infty$ such that if for $n = 0,1,2,\ldots$, we put

$$z_n = \Big(\sum_{i=1}^{2^n} |Tx_{n,i}|^2\Big)^{1/2} \quad \text{and} \quad \sigma_n = \{t \in [0,s];\ z_n(t) \le R\} ,$$

then $\inf_n \|\chi_{\sigma_n} z_n\|_Y > 0$.

Fix $q < p < 2$ and consider a sequence $\{g_i\}_{i=1}^{\infty}$ of independent and identically distributed p-stable random variables over some probability space (Ω,Σ,P) so that $\|g_1\|_q = 1$. Put

$$w_n = \Big(\sum_{i=1}^{2^n} |Tx_{n,i}|^p\Big)^{1/p} ,\ y_n = \max_{1\le i\le 2^n} |Tx_{n,i}| \text{ and } \delta_n = \{t \in [0,s];\ y_n(t) \le R\},$$

and observe that

$$\|w_n\|_Y = \|g_1\|_1^{-1} \Big\| \int_\Omega \Big| \sum_{i=1}^{2^n} g_i(w)\, Tx_{n,i}\Big| dP(w)\Big\|_Y \le$$

$$\le \|g_1\|_1^{-1}\cdot\|T\| \int_\Omega \Big\| \sum_{i=1}^{2^n} g_i(w) x_{n,i}\Big\|_X dP(w) \le \|g_1\|_1^{-1}\|T\| C \Big(\int_\Omega \Big\| \sum_{i=1}^{2^n} g_i(w) x_{n,i}\Big\|_q^q dP(w)\Big)^{1/q} =$$

$$= \|g_1\|_1^{-1}\, \|T\| C \Big(\int_0^1 \Big(\sum_{i=1}^{2^n} |x_{n,i}(t)|^p\Big)^{q/p} dt\Big)^{1/q} = \|g_1\|_1^{-1}\, \|T\| C ,$$

for all n. Thus, by 2.2(ii), we get that

$$\|\chi_{\sigma_n} z_n\|_Y \le \|\chi_{\sigma_n} w_n\|_Y^{p/2} \|\chi_{\sigma_n} y_n\|_Y^{1-p/2} \le \big(\|g_1\|_1^{-1}\, \|T\| C\big)^{p/2} \cdot \|\chi_{\sigma_n} y_n\|_Y^{1-p/2}$$

and this clearly implies that

$$0 < \inf_n \|\chi_{\sigma_n} y_n\|_Y \le \inf_n \|\chi_{\delta_n} y_n\|_Y .$$

Hence, by 5.2, there exists a constant $V < \infty$ so that $\|f\|_Y \le V\|f\|_X$, for all $f \in X$, and this completes the proof in Case I.

In order to study the next case we first point out that for every measurable subset E of $[0,1]$ with non-zero Lebesgue measure, the subspace X_E of X, consisting of those functions in X which are supported by E, is linearly and lattice isomorphic to X. Indeed, let τ be a one-to-one measurable transformation from $[0,1]$ onto E such that $\mu(\tau^{-1}F) = \mu(F)/\mu(E)$, whenever F is a measurable subset of E. Then, the mapping $f \to f(\tau)$ defines an isomorphism S_E, from X_E onto X, which preserves the lattice operations.

Now, define functions $x^E_{n,i}$ and Z^E_n relative to E by putting

$$x^E_{n,i} = S_E^{-1} x_{n,i} \quad \text{and} \quad z^E_n = \Big(\sum_{i=1}^{2^n} |Tx^E_{n,i}|^2\Big)^{1/2} .$$

Case II: There exist a measurable subset E of $[0,1]$ with non-zero Lebesgue measure, a constant $R < \infty$ and a number $1 \leq s < \infty$ such that if we set

$$\sigma^E_n = \{t \in [0,s];\ z^E_n(t) \leq R\}$$

then $\inf_n \|\chi_{\sigma_n} z^E_n\|_Y > 0$.

This case reduces immediately to the preceding one by using, instead of T, the isomorphism $T_E = TS_E^{-1}$ from X into Y. It follows again that there exists a constant $V_E < \infty$ so that $\|f\|_Y \leq V_E\|f\|_X$, for all $f \in X$, and this completes the proof for Case II.

In the remaining case, namely Case III, we suppose that, for every measurable subset E of $[0,1]$ with $\mu(E) > 0$, every constant $R < \infty$, and every $1 \leq s < \infty$,

$$\liminf_{n \to \infty} \|\chi_{\sigma^E_n} z^E_n\|_Y = 0 ,$$

where z^E_n and σ^E_n have the meanings assigned to them in Case II.

We want to show in this case that the usual Haar basis of X is equivalent to a sequence of pairwise disjoint elements of Y. Since X is r.i. this is a simple consequence of the following lemma.

Lemma 5.3. *Suppose that the assumption of Case III above is satisfied and let* $\{f_j\}_{j=1}^m$ *be a finite sequence of pairwise disjoint elements of* Y. *Then for every measurable subset* E *of* $[0,1]$ *with* $\mu(E)>0$ *and for every* $\epsilon>0$, *there exist a sequence* $\{g_j\}_{j=0}^m$ *of mutually disjoint functions in* Y *and a partition of* E *into two disjoint measurable subsets with* $\mu(E_1) = \mu(E_2)$ *so that* $\|f_j-g_j\| \leq \epsilon$ *for* $1\leq j\leq m$ *and* $\|T(\chi_{E_1}-\chi_{E_2}) - g_0\|_Y \leq \epsilon$.

Proof: Since Y does not contain isomorphic copies of ℓ_∞^n for all n it follows from [54] that Y is a r-concave with r-concavity constant $\leq M$, for some $r < \infty$ and some $M < \infty$. In particular, Y is a complete and order continuous lattice and thus, for every $\epsilon > 0$, there exists $1 \leq s < \infty$ and a number $\rho > 0$ so that

$$\|\chi_{[s,\infty)}f_j\|_Y < \epsilon/2 \, , \quad \text{for} \quad 1 \leq j \leq m \, ,$$

and for every measurable subset F of $[0,s]$ with $\mu(F) < \rho$, we have

$$\|\chi_F \, f_j\|_Y \leq \epsilon/2 \quad \text{for} \quad 1 \leq j \leq m \, .$$

On the other hand, by using the assumption of Case III for a given set $E \subset [0,1]$ with $\mu(E) > 0$, we can find a subset G of $[0,s]$ with $\mu(G) > s-\rho$ and an integer n such that $\|\chi_G \, z_n^E\|_Y \leq \epsilon/4MB_r$.

Now, for $u \in [0,1]$, put

$$h_1(u) = \sum_{i=1}^{2^{n-1}} r_i(u)x_{n,i}^E, \quad h_2(u) = \sum_{i=1}^{2^{n-1}} r_i(u)x_{n,2^{n-1}+i}^E$$

and observe that, by the r-concavity of Y, we have for $j = 1,2$

$$(\int_0^1 \|\chi_G T\, h_j(u)\|_Y^r du)^{1/r} \le M\|(\int_0^1 |\chi_G T\, h_j(u)|^r du)^{1/r}\|_Y \le$$

$$\le MB_r \|\chi_G(\sum_{i=1}^{2^n} |Tx_{n,i}^E|^2)^{1/2}\|_Y = MB_r\|\chi_G\, z_n^E\|_Y < \epsilon/4 \ .$$

It follows that, for u taking values in more than half of the interval $[0,1]$ (or, equivalently, for more than half of all possible choices of signs), we get $\|\chi_G\, Th_j(u)\|_Y \le \epsilon/2$ for both $j = 1,2$. It is therefore possible to choose $u_0 \in [0,1]$ so that for $j = 1,2$ $\|\chi_G T\, h_j(u_0)\|_Y \le \epsilon/2$. Put $h = h_1(u_0) - h_2(u_0)$, $g_0 = Th - \chi_G\, Th$ and $g_j = \chi_G f_j$ for $1 \le j \le m$. Then $\{g_j\}_{j=1}^m$ forms a sequence of disjoint elements of Y such that

$$\|f_j - g_j\|_Y = \|f_j - \chi_G f_j\|_Y \le \epsilon \ ; \ 1 \le j \le m$$

(since $\mu([0,s] - G) < \rho$) and

$$\|g_0 - Th\|_Y = \|\chi_G Th\|_Y \le \epsilon \ .$$

The proof can now be completed by setting

$$E_1 = \{t \in E;\ h(t) = 1\} \quad\text{and}\quad E_2 = \{t \in E;\ h(t) = -1\} \ .$$

□

Remarks: 1. Case (ii) of 5.1 implies that

(iii) Y *contains a sequence of normalized functions with mutually disjoint supports which, in the norm of* Y , *is equivalent to the unit vector basis of* ℓ_2 .

Indeed, since the norm in X is between that of $L_1(0,1)$ and that of $L_q(0,1)$ it follows from Khintchine's inequality that

$$A_1(\sum_{i=1}^{\infty} |a_i|^2)^{1/2} \le \|\sum_{i=1}^{\infty} a_i r_i\|_X \le C\, B_q(\sum_{i=1}^{\infty} |a_i|^2)^{1/2} \quad ,$$

for every choice of scalars $\{a_i\}_{i=1}^{\infty}$; i.e., that the Rademacher functions in X , which obviously form a block basis of the Haar basis of X , are equivalent to the unit vector basis of ℓ_2 .

2. Notice that 5.1 and Remark 1 imply that if X and Y are two r.i. spaces on [0,1] which are q-concave for some $q < \infty$ and each embeds isomorphically into the other, then they are identical up to an equivalent norm.

3. A weaker version of 5.1 was proved in [47] for Orlicz function spaces.

Corollary 5.4: A r.i. function space Y *on* [0,1] *which does not contain uniformly isomorphic copies of* ℓ_∞^n *for all* n *and which contains an isomorphic copy of* $L_1(0,1)$ *coincides, up to an equivalent norm, with the space* $L_1(0,1)$.

Proof: Observe first that Case (ii) of 5.1 is excluded in our situation since it would imply that the Haar basis of $L_1(0,1)$ is unconditional and this is false. On the other hand, Case (i) of 5.1 implies the existence of a constant $C < \infty$ such that $\|f\|_Y \le C\|f\|_1$, for every $f \in L_1(0,1)$. This completes the proof since we also have $\|f\|_1 \le \|f\|_Y$ for every $f \in Y$. □

Remark: N. J. Kalton [83] proved recently the same result under the weaker assumption that Y does not contain an isomorphic copy of c_0 .

We are ready to state our main uniqueness result.

Theorem 5.5: Let X *be a r.i. function space on* [0,1] *which is q-concave for some* $1 \le q < 2$. *Then* X *has unique r.i. structure on* [0,1] *; i.e. every r.i. function space* Y *on* [0,1] *which is isomorphic to* X *is already equal to* X , *up to an equivalent norm.*

Proof of 5.5 under the additional assumption that X *is* B-convex; i.e., that it does not contain uniformly isomorphic copies of ℓ_1^n for all n .

Suppose that a r.i. function space Y on [0,1] is linearly isomorphic to X . If Y were q-concave for the same value of q then, by Remark 2 following 5.1 we would conclude the existence of a constant $V < \infty$ so that

$$V^{-1}\|f\|_X \le \|f\|_Y \le V\|f\|_X .$$

In order to prove that Y is q-concave, we use the fact that X is assumed to be B-convex. Since B-convexity is clearly an isomorphic invariant it follows that Y is B-convex too and that, by [54], there exists an $r > 1$ so that both X and Y are r-convex. By duality, we get that their duals X^* and Y^* are r'-concave with $r' = r/(r-1)$ which implies that they are proper r.i. function spaces in the sense of the definition given in the Introduction; i.e., that $L_\infty(0,1)$ is dense in both X^* and Y^*. We also have that X^* is q'-convex with $q' = q/(q-1)$ and thus, by 2.6, Y^* is either q'-convex too or Y^* is equal to $L_2(0,1)$, up to an equivalent norm. Since the latter possibility clearly contradicts our q-concavity assumption on X it follows that Y^* is q'-convex; i.e., Y is q-concave. □

The *proof of 5.5 in the general case* is more complicated and is given as a particular case of the following result.

Theorem 5.6: Let X *be a r.i. function space on* $[0,1]$ *or on* $[0,\infty)$ *which is q-concave for some* $1 \le q < 2$. *Assume that the restriction* $X_{|[0,1]} = \{f \in X;\ f\chi_{[0,1)} = 0\}$ *of* X *to the interval* $[0,1]$ *is not equal to* $L_1(0,1)$ *up to an equivalent norm. If a r.i. function space* Y *on* $[0,1]$ *is isomorphic to a complemented subspace of* X *then, up to an equivalent norm,* Y *is equal either to* $L_2(0,1)$ *or to* $X_{|[0,1]}$.

When X is r.i. on $[0,1]$, the condition $X \neq L_1[0,1]$ is not really a restriction, since a result of Lindenstrauss and Pelczynski [44] (based on Grothendieck's work [33]) yields that the only r.i. space $[0,1]$ which embeds into L_1 as a complemented subspace is $L_1[0,1]$ itself. When X is r.i. on $[0,\infty)$, we did not try to check whether the condition imposed on X is necessary.

The proof of 5.6 is rather involved and, therefore, we shall try first to explain its main ideas. The hypothesis of 5.6 implies the existence of two bounded linear operators $T: Y \to X$ and $S: X \to Y$ such that $ST =$ identity on Y. We want to pass to duals and use the classification theorem 2.1 in order to study the map $S^*: Y^* \to X^*$ but this is not a-priori possible since, in general, X is not B-convex and, therefore, X^* is not of type 2 as required by 2.1. In order to replace X by a B-convex space we make use of interpolation. Before interpolating we prove that there exist two probability measures λ and ν on $[0,1]$ or $[0,\infty)$, both continuous with respect to Lebesgue measure μ, such that T is continuous when considered as an operator from $L_2(\lambda)$ to $L_2(\nu)$ and S is continuous as an operator from $L_2(\nu)$ to $L_2(\lambda)$. Now, the interpolation spaces X and Y defined by $\tilde{X} = [X,L_2(\nu)]_{1/2,2}$, $\tilde{Y} = [Y,L_2(\lambda)]_{1/2,2}$ are 2-concave and B-convex and T is continuous as an operator from $\tilde{Y}$ into $\tilde{X}$ and S is continuous as an operator from $\tilde{X}$ into $\tilde{Y}$. We also get that S^* is an isomorphism from $\tilde{Y}^*$ into the lattice $\tilde{X}^*$, which is of type 2. The space $\tilde{Y}^*$ is not necessarily a r.i. function space. However, if we restrict our attention to subsets E of $(0,1]$ on which the Radon-Nikodym derivative $d\lambda/d\mu$ is essentially constant, we get that $\tilde{Y}^*_E$ (= the set of all $f \in \tilde{Y}^*$ which are supported on E) is, up to a constant, a r.i. function space. The final step is to show, by using 2.1, that if Y is not equal to $L_2(0,1)$, up to an equivalent norm, then the norm in Y^* dominates that in X^*; i.e., the norm in X dominates the norm in Y. In particular, the norm in Y is dominated by that in $L_q(0,1)$ and the proof is then completed by applying 5.1 with the roles of X and Y being reversed.

The actual proof of 5.6 requires some preliminary results.

Lemma 5.7: Let X *be a Banach lattice of measurable functions on* $[0,1]$ *which has 2-concavity constant one.*

1. *Assume that* $L_2(0,1) \subseteq X$ *with the inclusion being contractive and having dense range. Let* T *be a bounded linear operator from a Hilbert space* H *into* X. *Then, for every* $\epsilon > 0$ *there exists a strictly positive (Lebesgue) integrable function* ρ *on* $[0,1]$ *such that*

(i) $\left(\int_0^1 |Tz|^2 \rho d\mu\right)^{1/2} \le (1+\varepsilon)\sqrt{2}\,\|T\|\,\|z\|_H$; $z \in H$

(ii) $\|x\|_X \le \left(\int_0^1 |x|^2 \rho d\mu\right)^{1/2}$; $x \in L_2(\rho d\mu)$.

Note that 5.7 (and 5.8 below) can, in view of the representation theorem for p-concave Banach lattices mentioned in the Introduction, be applied to any separable purely non-atomic 2-concave Banach lattice.

The result 5.7 is actually a factorization theorem: it asserts that T factors through $L_2(\rho d\mu)$ in the sense that $T = i\,T_0$, where T_0 is, in fact, equal to T considered as an operator from H into $L_2(\rho d\mu)$ (with $\|T_0\| \le (1+\varepsilon)\sqrt{2}\,\|T\|$, by 5.7(i)) and i is the formal identity map from $L_2(\rho d\mu)$ into X (with $\|i\| \le 1$ by 5.7(ii)). This result is a special case of Krivine [41] Cor. 3 and its proof follows from a separation argument.

Proof of 5.7: Fix $\varepsilon > 0$ and suppose that

$$\|T\| = (1-\varepsilon)/\sqrt{2} .$$

We consider now the following convex sets.

$$C_1 = \operatorname{conv}\{f^2;\ f \in C(0,1),\ |f| \le |Tz| + h \ \text{ for some } \ z \in H \ \text{ and } \ h \in L_2(0,1)$$
$$\|z\|_H < 1 \ \text{ and } \ \|h\|_2 \le \varepsilon\}$$

$$C_2 = \operatorname{conv}\{g^2;\ g \in C(0,1)\ ,\ \|g\|_X \ge 1\} .$$

Assume that the convex set $C_0 = C_1 - C_2$ contains a non-negative function k . Then

$$k = \sum_{i=1}^m \alpha_i f_i^2 - \sum_{i=1}^m \beta_i g_i^2 ,$$

where $\alpha_i, \beta_i \ge 0$, $\sum_{i=1}^m \alpha_i = \sum_{i=1}^m \beta_i = 1$, $|f_i| \le |Tz_i| + h_i$ for some $z_i \in H$ and $h_i \in L_2(0,1)$ with $\|z_i\|_H < 1$ and $\|h_i\|_2 \le \varepsilon$ for $1 \le i \le m$. Since $k \ge 0$ we have

$$1 \le \left(\sum_{i=1}^{m} \beta_i \|g_i\|_X^2\right)^{1/2} \le \left\|\left(\sum_{i=1}^{m} \beta_i g_i^2\right)^{1/2}\right\|_X \le \left\|\left(\sum_{i=1}^{m} \alpha_i f_i^2\right)^{1/2}\right\|_X$$

$$\le \left\|\left(\sum_{i=1}^{m} \alpha_i |Tz_i|^2\right)^{1/2} + \left(\sum_{i=1}^{m} \alpha_i h_i^2\right)^{1/2}\right\|_X \le A_1^{-1} \int_0^1 \left\|\sum_{i=1}^{m} r_i(u)\sqrt{\alpha_i}\, Tz_i\right\|_X du +$$

$$+ \left\|\left(\sum_{i=1}^{m} \alpha_i h_i^2\right)^{1/2}\right\|_2$$

$$\le \sqrt{2}\, \|T\| \int_0^1 \left\|\sum_{i=1}^{m} r_i(u)\sqrt{\alpha_i}\, z_i\right\|_H du + \varepsilon < \sqrt{2}\, \|T\| + \varepsilon = 1$$

and this is impossible. This proves that the convex set C_0 is disjoint from the open convex set $C_+(0,1)$ of all $f \in C(0,1)$ which are positive on $[0,1]$. Hence, by the Hahn-Banach theorem, there is a regular probability measure ν on $[0,1]$ such that $\int_0^1 k\, d\nu \le 0$ for every $k \in C_0$. Put

$$\gamma^2 = \inf\{\int_0^1 g^2\, d\nu\ ,\ g \in C(0,1)\ ,\ \|g\|_X \ge 1\}\ .$$

First note that $\gamma \ge \varepsilon$, since the constant function ε^2 belongs to C_1. Now, by the definition of γ and the property of ν, we get that

$$\int_0^1 f^2 d\nu \le \gamma^2 \quad \text{for every} \quad f^2 \in C_1\ .$$

In particular, $\int_0^1 h^2\, d\nu \le \gamma^2$ for every $h \in C(0,1)$ such that $\|h\|_2 \le \varepsilon$. This immediately yields that the measure ν is absolutely continuous with respect to the Lebesgue measure μ and also that the Radon-Nikodym derivative $\psi = d\nu/d\mu$ is bounded by γ/ε.

We want to show that $\varphi = \gamma^{-2}\psi$ satisfies the statement in lemma 5.7. First notice that the inequality

$$\left(\int_0^1 g^2 \psi\, d\mu\right)^{1/2} \ge \gamma\, \|g\|_X\ ,$$

which holds for all $g \in C(0,1)$, can be clearly extended to hold for every $g \in L_2(\psi d\mu)$. This proves the inequality (ii) in 5.7 and also shows that φ is strictly positive on $[0,1]$.

Now, let $z \in H$ be such that $\|z\|_H < 1$. Let n be an arbitrary integer and let η be a positive real. We may clearly find $f \in C(0,1)$ so that

$$\|f - |Tz| \wedge n\|_2 \le \epsilon \quad \text{and} \quad \|f - |Tz| \wedge n\|_{L_2(\psi d\mu)} \le \eta .$$

Since $|f| \le |Tz| + (f - |Tz| \wedge n)$, $f^2 \in C_1$ and $\int_0^1 f^2 \psi d\mu \le \gamma^2$, it follows that

$$\int_0^1 |Tz \wedge n|^2 \psi d\mu \le (\gamma+\eta)^2 .$$

Since n and η are arbitrary, we finally get

$$\int_0^1 |Tz|^2 \psi d\mu \le \gamma^2 ,$$

which completes the proof of 5.7.

□

The next proposition is a generalization of a result from [36].

Proposition 5.8: Let X *and* Y *be two Banach lattices of meaurable functions on* $[0,1]$ *with 2-concavity constant one. Assume that* $L_2(0,1) \subseteq X$, $L_2(0,1) \subseteq Y$ *with inclusions being contractive and having dense range. Let* $S: X \to Y$ *and* $T: Y \to X$ *be bounded linear operators so that* $ST =$ *identity on* Y. *Then there exist two positive integrable functions* φ *and* ψ *such that*

(i) The natural injections i_X, *of* $L_2(\varphi d\mu)$ *into* X, *and* i_Y *of* $L_2(\psi d\mu)$ *into* Y, *are continuous (even contractive).*

(ii) S *and* T *are bounded as operators from* $L_2(\varphi d\mu)$ *into* $L_2(\psi d\mu)$, *respectively from* $L_2(\psi d\mu)$ *into* $L_2(\varphi d\mu)$ *(where the bounds are given by, respectively,* $2\|S: X\to Y\|$ *and* $4\|T: Y\to X\|$ *).*

The result 5.8 can be better understood by using the following diagram

$$\begin{array}{ccccc} Y & \longrightarrow & X & \longrightarrow & Y \\ {\scriptstyle i_Y}\uparrow & & {\scriptstyle i_X}\uparrow & & {\scriptstyle i_Y}\uparrow \\ L_2(\psi d\mu) & \xrightarrow{T} & L_2(\rho d\mu) & \xrightarrow{S} & L_2(\psi d\mu) \end{array}$$

Proof: Put $\rho_0 \equiv 1$ and notice that, by 2.7, the injection i_0 from $L_2(\rho_0 d\mu) = L_2(0,1)$ into X has norm 1 and, therefore, $S_0 = Si_0$ is a bounded linear operator from $L_2(\rho_0 d\mu)$ into Y. By 5.7, there exists a positive integrable function ψ_0 such that

(a) $(\int_0^1 |Sf|^2 \psi_0 d\mu)^{1/2} \leq 2\|S\|(\int_0^1 |f|^2 \rho_0 d\mu)^{1/2}$,

(b) $\|f\|_Y \leq (\int_0^1 |f|^2 \psi_0 d\mu)^{1/2}$,

at least for every $f \in L_\infty(0,1)$. Condition (b) shows that the injection j_0 from $L_2(\psi_0 d\mu)$ into Y has also norm 1 and, thus, $T_0 = Tj_0$ is a bounded linear operator from $L_2(\psi_0 d\mu)$ into X. By applying again 5.7, we can find a positive integrable function ρ_1 so that

(c) $(\int_0^1 |Tf|^2 \rho_1 d\mu)^{1/2}) \leq 2\|T\|(\int_0^1 |f|^2 \psi_0 d\mu)^{1/2}$,

(d) $\|f\|_X \leq (\int_0^1 |f|^2 \rho_1 d\mu)^{1/2}$,

for every $f \in L_\infty(0,1)$. In this manner, we define inductively two sequences $\{\rho_n\}_{n=0}^\infty$ and $\{\psi_n\}_{n=0}^\infty$ of strictly positive functions such that, for every $n \geq 0$ and $f \in L_\infty(0,1)$,

$$(\int_0^1 |Sf|^2 \psi_n \, d\mu)^{1/2} \leq 2\|S\|(\int_0^1 |f|^2 \rho_n \, d\mu)^{1/2} \quad ,$$

$$\|f\|_Y \leq (\int_0^1 |f|^2 \psi_n d\mu)^{1/2}$$

and

$$\left(\int_0^1 |Tf|^2 \rho_{n+1} d\mu\right)^{1/2} \le 2\|T\| \left(\int_0^1 |f|^2 \psi_n d\mu\right)^{1/2} ,$$

$$\|f\|_X \le \left(\int_0^1 |f|^2 \rho_{n+1} d\mu\right)^{1/2} .$$

A simple duality argument in $L_2(0,1)$ shows that the above inequalities are equivalent to having, for every $n \ge 0$ and every $g \in L_\infty(0,1)$, that

$$\left(\int_0^1 |S^* g|^2 \rho_n^{-1} d\mu\right)^{1/2} \le 2\|S\| \left(\int_0^1 |g|^2 \psi_n^{-1} d\mu\right)^{1/2} ,$$

$$\|g\|_{Y*} \ge \left(\int_0^1 |g|^2 \psi_n^{-1} d\mu\right)^{1/2}$$

and

$$\left(\int_0^1 |T^* g|^2 \psi_n^{-1} d\mu\right)^{1/2} \le 2\|T\| \left(\int_0^1 |g|^2 \rho_{n+1}^{-1} d\mu\right)^{1/2} ,$$

$$\|g\|_{X*} \ge \left(\int_0^1 |g|^2 \rho_n^{-1} d\mu\right)^{1/2} .$$

If we set

$$\varphi^{-1} = \sum_{n=0}^{\infty} 2^{-n-1} \varphi_n^{-1} \quad \text{and} \quad \psi^{-1} = \sum_{n=0}^{\infty} 2^{-n-1} \psi_n^{-1}$$

then we obtain

$$\left(\int_0^1 |S^* g|^2 \varphi^{-1} d\mu\right)^{1/2} \le 2\|S\| \left(\int_0^1 |g|^2 \psi^{-1} d\mu\right)^{1/2} ,$$

$$\|g\|_{Y*} \ge \left(\int_0^1 |g|^2 \psi^{-1} d\mu\right)^{1/2}$$

and

$$\left(\int_0^1 |T^* g|\psi^{-1} d\mu\right)^{1/2} \le 4\|T\| \left(\int_0^1 |g|^2 \varphi^{-1} d\mu\right)^{1/2},$$

$$\|g\|_{X*} \ge \left(\int_0^1 |g|^2 \varphi^{-1} d\mu\right)^{1/2} .$$

The proof can now be completed by another duality argument and by the fact that $L_\infty(0,1)$ is dense in $X, Y, L_2(\varphi d\mu)$ and $L_2(\psi d\mu)$. □

Proof of 5.6: We treat the case where X is a r.i. space on $[0,1]$ and indicate later the modifications necessary for the $[0,\infty)$ case. So let X and Y be two r.i. function spaces on $[0,1]$ as in the statement of 5.6 and let $S: X \to Y$ and $T: Y \to X$ be bounded linear operators so that ST is the identity on Y. Since X is q-concave for $1 \leq q < 2$ it follows from [21] that Y is 2-concave. By passing to equivalent r.i. norms on X and Y, we may assume that the 2-concavity constants of X and Y are both one. Thus by the comments in the Introduction, we are in a position to apply 5.8 and to conclude that there exist two positive Lebesgue integrable functions φ and ψ on $[0,1]$ such that

(a) S and T are continuous as operators from $L_2(\varphi d\mu)$ into $L_2(\psi d\mu)$, respectively, from $L_2(\psi d\mu)$ into $L_2(\varphi d\mu)$, and

(b) $\|f\|_X \leq (\int_0^1 |f|^2 \varphi d\mu)^{1/2}$, $\|f\|_Y \leq (\int_0^1 |f|^2 \psi d\mu)^{1/2}$, for every $f \in L_\infty(0,1)$.

Consider now the Lions-Peetre interpolation spaces [52]

$$\tilde{X} = [X, L_2(\varphi d\mu)]_{1/2,2} \quad , \quad \tilde{Y} = [Y, L_2(\psi d\mu)]_{1/2,2}$$

whose precise definition is given in the Introduction. Since the Hilbert space is B-convex it follows from [3] that both $\tilde{X}$ and $\tilde{Y}$ are B-convex. Similarly, since both X and Y are 2-concave we get that $\tilde{X}$ and $\tilde{Y}$ are 2-concave Banach lattices, (cf. e.g., the proof of 6.6 below). Moreover, the fact that $\tilde{X}$ and $\tilde{Y}$ are interpolation spaces implies that S and T are continuous as operators from $\tilde{X}$ into $\tilde{Y}$, respectively, from $\tilde{Y}$ into $\tilde{X}$.

The lattice $\tilde{Y}$ need not be a r.i. function space on $[0,1]$, but if we restrict our attention to a measurable subset E of $[0,1]$ with $\mu(E) > 0$ such that

$$M^{-1} \leq \psi(u) \leq M ,$$

for all $u \in E$ and for some constant $M < \infty$, then the space $\widetilde{Y}_E$, of those functions in $\widetilde{Y}$ which vanish outside E, is linearly and lattice isomorphic to a r.i. function space on E and, therefore, also to a r.i. space of functions on $[0,1]$.

Observe now that, since ST is the identity on Y, S^* is an isomorphism from Y^* into X^* and, in particular, from Y^*_E into X^*. The spaces X^*, Y^* and Y^*_E need not be r.i. function spaces in the sense of our definition given in the Introduction since $L_\infty(0,1)$, respectively, $L_\infty(E)$ is not necessarily norm dense in them. However, the axioms (a), (b) and (d) of the definition of r.i. function spaces are satisfied. In order to avoid this problem we shall work with the closures Y^*_0 and $Y^*_{E,0}$ of $L_\infty(0,1)$, respectively, $L_\infty(E)$ in Y^*, respectively, Y^*_E.

Let $x^E_{n,i}$; $i = 1,2,\ldots,2^n$; $n = 1,2,\ldots$ have the same meaning as in the proof of 5.1 and put

$$y^E_n = \max_{1 \le i \le 2^n} |S^* x^E_{n,i}| \ ; \ n = 1,2,\ldots$$

(recall that $x^E_{n,i}$ is the characteristic function of the image, under a one to one map from $[0,1]$ onto E, of the interval $[(i-1)2^{-n}, i2^{-n}))$.

We distinguish between two cases.

Case I: There exists a constant $R < \infty$ such that if we set $\delta^E_n = \{t \in [0,1] \ ; \ y^E_n(t) \le R\}$ then

$$\inf_n \|\chi_{\delta^E_n} y^E_n\|_{X^*} > 0 \quad .$$

Since we plan to use 5.2 for the isomorphism $S^*: Y^*_{E,0} \to X^*$, it is important for us that X^* is not equal to $L_\infty(0,1)$, even up to an equivalence renorming. Another observation is that $Y^*_{E,0}$ is lattice isomorphic and linearly isometric to Y^*_0 which is a proper r.i. function space on $[0,1]$ and, under this isomorphism, $x^E_{n,i}$ stands in correspondence with $x_{n,i}$. This means that we may indeed use 5.2 (see Remark (b) following the proof of 5.2) and conclude that there exists a constant $V_1 < \infty$ so that $\|g\|_{X^*} \le V_1\|g\|_{Y_0^*}$, for every $g \in Y_0^*$. Since Y_0^* is norming over Y we further get that

$$\|f\|_Y \le V_1\|f\|_X \quad ,$$

for every $f \in X$. Hence, by the q-concavity of X and 2.7, it follows that there exists a constant $V_0 < \infty$ so that $\|f\|_Y \le V_0\|f\|_q$, $f \in L_q(0,1)$. This puts us in a position to use 5.1 for the isomorphism T of Y into X. Since Case II of 5.1 would imply (see Remark 1 following the proof of 5.1) that X contains a sequence of disjointly supported functions which is equivalent to the unit vector basis of ℓ_2 and this would contradict the q-concavity of X, we conclude that there is a constant $V_2 < \infty$ so that

$$\|f\|_X \le V_2\|f\|_Y \ ,$$

for every $f \in Y$. This, of course, shows that, in Case I, X and Y are equal, up to an equivalent norm.

In Case II, i.e., when there exists no R as required by Case I, the sequence $\{y_n^E\}_{n=1}^\infty$ clearly contains a subsequence $\{y_{n'}^E\}_{n'=1}^\infty$ which converges to zero a.e. in $[0,1]$ (with respect to the Lebesgue measure μ). Suppose that

$$\limsup_{n'\to\infty} \|y_{n'}^E\|_{L_2(\varphi^{-1}d\mu)} = \eta > 0 \ .$$

Then there exists a further subsequence $\{y_{n''}^E\}_{n''=1}^\infty$ and a sequence $\{G_{n''}\}_{n''=1}^\infty$ of pairwise disjoint subsets of $[0,1]$ so that

$$\|\chi_{G_{n''}} y_{n''}^E\|_{L_2(\varphi^{-1}d\mu)} \ge \eta/2$$

for all n''. Hence, it follows from condition (b) above, by duality, that

$$K\|\sum_{n''=1}^\infty a_{n''}\chi_{G_{n''}} y_{n''}^E\|_{X^*} \ge \|\sum_{n''=1}^\infty a_{n''}\chi_{G_{n''}} y_{n''}^E\|_{L_2(\varphi^{-1}d\mu)} \ge \eta(\sum_{n''=1}^\infty |a_{n''}|^2)^{1/2}/2 \ ,$$

for every choice of scalars $\{a_{n''}\}_{n''=1}^{\infty}$. This, however, contradicts the q-concavity of X which is equivalent to the fact that X* is q'-convex for $q' = q/(q-1) > 2$. This means that we have just shown that

$$\lim_{n'\to\infty} \|y_{n'}^E\|_{L_2(\varphi^{-1}d\mu)} = 0 .$$

Thus, by the dual interpolation inequality (cf. [52])

$$\|y_{n'}^E\|_{\tilde{X}^*} \leq \|y_{n'}^E\|_{X^*}^{1/2} \cdot \|y_{n'}^E\|_{L_2(\varphi^{-1}d\mu)}^{1/2} \quad ; \; n' = 1,2,\ldots$$

We get that also

$$(+) \qquad \lim_{n'\to\infty} \|y_{n'}^E\|_{\tilde{X}^*} = 0 .$$

In order to complete the proof we use the classification theorem 2.1 for the map $S^*: \tilde{Y}_E^* \to \tilde{X}^*$. We have already seen that, by the choice of E , $\tilde{Y}_E^*$ is linearly and lattice isomorphic to a r.i. function space on [0,1] ($L_\infty(E)$ is norm dense in $\tilde{Y}_E^*$ since $\tilde{Y}_E^*$ is B-convex, but we don't use this in the sequel.) Moreover, S* is an isomorphism from $\tilde{Y}^*$ onto $\tilde{X}^*$ (since T*S* = identity on $\tilde{Y}^*$) and $\tilde{X}^*$ is 2-convex and s-concave for some even integer s (since $\tilde{X}$ is B-convex and 2-concave). Thus, all the requirements of 2.1 are satisfied by S* and, consequently, there exists a constant $D < \infty$, independent of n , so that, whenever $f = \sum_{i=1}^{2^n} b_i x_{n,i}^E$ for some n and some coefficients $\{b_i\}_{i=1}^{2^n}$, we have

$$\|f\|_{\tilde{Y}_E^*} \leq D \max\{(\sum_\pi \| \max_{1\leq i\leq 2^n} |b_i S^* x_{n,\pi(i)}^E| \|_{\tilde{X}^*}^s /2^n!)^{1/s}, \|f\|_2\}$$

from which it follows that

$$\|f\|_{\tilde{Y}_E^*} \leq D \max\{\|f\|_\infty \|y_n^E\|_{\tilde{X}^*} , \|f\|_2\} .$$

But any fixed f as above can be also expressed as $f = \sum_{i=1}^{2^{n'}} c_i x_{n',i}^E$ with $n' \geq n$ and suitable $\{c_i\}_{i=1}^{2^{n'}}$. Hence, by (+), we get that

$$\|f\|_{\tilde{Y}_E^*} \leq D\|f\|_2 .$$

This implies that, for every function $g \in L_\infty(E)$, we have

$$\|g\|_2 \leq D\|g\|_{\tilde{Y}_E} = D\|g\|_{\tilde{Y}} \leq D\|g\|_Y^{1/2} \cdot \|g\|_{L_2(\psi\, d\mu)}^{1/2} \leq DM^{1/4}\|g\|_Y^{1/2} \cdot \|g\|_2^{1/2} ;$$

that is,

$$\|g\|_2 \leq D^2 M^{1/2} \|g\|_Y .$$

This inequality remains true, with a larger constant, for every $g \in L_\infty(0,1)$, and, thus, also for every $g \in Y$. On the other hand, since Y is 2-concave it follows from 2.7 that

$$\|g\|_Y \leq D_1\|g\|_2 ,$$

for some $D_1 < \infty$ and for every $g \in L_2(0,1)$. This shows that in Case II, Y is equal to $L_2(0,1)$, up to an equivalent norm.

Now we discuss the case when X is a r.i. space on $[0,\infty)$. Again we can assume that the 2-concavity constants of X and Y are both one.

The first case can be written as

Case I': There exist $E \subseteq [0,1]$ with $\mu(E) > 0$, a constant $R < \infty$, and a number $1 \leq s < \infty$ such that if we set $\delta_n^E = \{t \in [0,s] ; y_n^E(t) \leq R\}$ then

$$\inf_n \|\chi_{\delta_n^E}\, y_n^E\|_{X^*} > 0 .$$

In this case the argument for Case I shows that Y is $\{x \in X;\ x\chi_{[1,\infty)} = 0\}$ up to an equivalent norm.

Now assume that Case I' fails. Let f be a bounded, decreasing, strictly positive, norm one function in X, so that, in particular, f is a weak order unit in X. By the remarks in the Introduction, there is an isometry U from X onto a Banach lattice X_1 of measurable functions on $[0,1]$ such that $Uf = \chi_{[0,1]}$ and $L_2(0,1) \subseteq X_1$ with the injection contractive and having dense range. Let $S_1: X_1 \to Y$ and $T_1: Y \to X_1$ be defined by $S_1 = SU^{-1}$ and $T_1 = US$, so that S_1T_1 is the identity on Y. We apply 5.8 to find positive Lebesgue measurable functions φ and ψ on $[0,1]$ so that (a') and (b') hold, where (a') and (b') are (a) and (b), respectively, on page 119, except that X,S,T are replaced by X_1,S_1,T_1. We then build the interpolation spaces $\tilde{X}_1 = [X_1, L_2(\varphi d\mu)]_{1/2,2}$, and $\tilde{Y}$ as in the case when X is r.i. on $[0,1]$. Of course, X_1 is not r.i., but this was not used in the argument we presented. It is important that Y is r.i. on $[0,1]$, since, as before, we need to restrict attention to the functions in $\tilde{Y}_E$ where E is a subset of $[0,1]$ of positive measure so that

$$M^{-1} \le \psi(u) \le M \quad \text{for} \quad u \in E ,$$

for some positive constant M. The earlier proof of Case II shows that $Y = L_2[0,1]$, once we verify that the remaining case can be expressed as

Case II': Given any $R < \infty$, if

$$\delta_{1n}^E = \{t \in [0,1]: y_{1n}^E = \max_{1 \le i \le 2^n} |S_1^* x_{n,i}^E| \le R\} ,$$

then
$$\lim_{n \to \infty} \|\chi_{\delta_{1n}^E} y_{1n}^E\|_{X_1^*} = 0 .$$

Notice that $\chi_{\delta_{1n}^E} y_{1n}^E$ is just the infimum in the lattice X_1^* of Y_{1n}^E and $\chi_{\delta_{1n}^E}$. Since U is a lattice isomorphism, $\chi_{\delta_{1n}^E} = U(\chi_{\Delta_n^E} f)$, where

$$\Delta_n^E = \{t \in [0,\infty): y_n^E(t) \le Rf(t)\} ;$$

and we can rewrite Case II' as

Case II": For any $R < \infty$,

$$\|y_n^E \chi_{\Delta E_n} f\|_{X^*} \to 0 \quad \text{as} \quad n \to \infty .$$

Since for any R,

$$\lim_{s \to \infty} \|Rf \chi_{[s,\infty)}\|_X = 0 ,$$

it is clear that Case II" is the complementary case to Case I'.

□

Remarks 1. An examination of the proof of 5.6 shows that the only way in which we used the q-concavity of X was to ensure that X^* contains no copy of ℓ_2 spanned by disjoint functions. Therefore, Theorems 5.6 and 5.5 remain valid when X is a 2-concave r.i. function space such that no sequence of disjoint functions in X is equivalent to the unit vector basis of ℓ_2. (Recall that, by 2.4, X is q-concave for some $q < 2$ if there is some n so that X contains no 2-isomorph of ℓ_2^n spanned by disjoint elements.)

2. Notice that the Hilbert spaces $L_2(\varphi d\mu)$ and $L_2(\psi d\mu)$ and the interpolation spaces $\tilde{X}$ and $\tilde{Y}$ are not used in the analysis of Case I. This machinery is introduced only to show that if 5.2 does not apply to the embedding $S^*: Y_E^* \to X^*$ for some set E, then Y is $L_2(\mu)$ with an equivalent norm. In Case II, the interpolation scheme plays a double role:

1) it allows 2.1 to be applied even when X^* is not r-concave for some r;

2) it facillitates the computation $\|g\|_2 \le$ constant $\|g\|_Y$. Of course, 1) is not important when X is B-convex, but even in this case the computation $\|g\|_2 \le$ constant $\|g\|_Y$ does not appear to be easy. A different, but even more complicated, method for making this calculation is given in the next section.

6. A CLASSIFICATION THEOREM FOR EMBEDDINGS OF REARRANGEMENT INVARIANT FUNCTION SPACES

We now present an attempt to classify all the possible isomorphic embeddings of a r.i. function space X on $[0,1]$ with an unconditional basis into a r.i. function space Y on $[0,1]$ or on $[0,\infty)$ which does not contain uniformly isomorphic copies of ℓ_∞^n for all n. In the previous section, namely Theorem 5.1, we treated this question under the additional assumption that X is q-concave for some $q < 2$ but without requiring that X have an unconditional basis.

Let us discuss first some special cases of isomorphic embeddings of X into Y. One such case occurs when there exists a constant $C < \infty$ so that $\|f\|_Y \le C\|f\|_X$, for every $f \in X$. This can occur, for example, if we embed the space $L_p(0,1)$ into $L_r(0,1)$ for $1 \le r < p \le 2$ by using a suitable family of independent p-stable random variables (cf. [11], [71]). A second case happens when the Haar system in X is mapped to a sequence of disjoint functions in Y (recall that, by [62], X has an unconditional basis if and only if the usual Haar basis forms an unconditional basis in X). This kind of embedding will be used in Section 10 below to construct Example 10.1. A third case occurs when $L_2(0,1)$ is embedded isomorphically in Y (e.g. take $Y = L_p(0,1)$, with $1 \le p < \infty$).

The main result of this section asserts that, essentially speaking, the cases described above are the only possible ones.

Theorem 6.1: Let X *be a r.i. function space on* $[0,1]$ *such that the Haar system* $\{h_n\}_{n=1}^\infty$ *is an unconditional basis for* X. *Let* Y *be a r.i. function space on* $[0,1]$ *or on* $[0,\infty)$ *which does not contain uniformly isomorphic copies of* ℓ_∞^n *for all* n. *If* X *embeds isomorphically into* Y *then one of the following three (non-exclusive) possibilities holds:*

(i) *There exists a constant* $C < \infty$ *such that*

$$\|f\|_Y \le C\|f\|_X$$

for every $f \in X$.

(ii) *The Haar system in* X *is equivalent to a sequence of disjoint functions in* Y .

(iii) X *is isomorphic to* $L_2(0,1)$.

One application of 6.1 is 6.12, which shows that if X is a super-reflexive r.i. function space on $[0,1]$ for which the Haar system is not equivalent to a sequence of disjoint functions in X , then X has a unique representation as a r.i. function space on $[0,1]$.

The proof of 6.1 is quite long and involved but we believe that some of the arguments used in it could be of intrinsic value. Before starting the actual proof of 6.1 we fix our notations and make some definitions.

As in Section 5, for $i = 1,2,\dots,2^n$ and $n = 0,1,\dots,$ $x_{n,i}$ denotes the characteristic function of the interval $E_{n,i} = [(i-1)2^{-n}, i2^{-n})$. The Haar functions $h_{n,i}$; $i = 1,2,\dots,2^n$, $n = 0,1,\dots,$ with the $L_\infty(0,1)$-normalization, are as usual defined by

$$h_0 = x_{0,1} \quad \text{and} \quad h_{n,i} = x_{n+1,2i-1} - x_{n+1,2i} \ .$$

For fixed n , the collection of all the dyadic intervals $[(i-1)2^{-n}, i2^{-n})$ $i = 1,2,\dots,2^n$ is denoted by π_n while $\mathcal{S}_n$ stands for the finite algebra of sets generated by π_n . We also set $\pi = \bigcup_{n=0}^{\infty} \mathcal{S}_n$. The (incorrect) notation $h_{n,i} \subset E$, when $E \in \mathcal{S}$, simply means that the support of $h_{n,i}$ is contained in the set E . The supports of the Haar functions; i.e., the sets in $\mathcal{S}$, form what will be called the *fundamental tree* $\{E_{n,i}\}_{i=1,n=0}^{2^n\ \infty}$. In general, a *tree* $\{F_{n,i}\}_{i=1,n=0}^{2^n\ \infty}$ of sets is a family of measurable subsets of $[0,1]$ (with respect to the Lebesgue measure μ) such that, for every $n \geq 0$, we have

$$F_{n,i} = F_{n+1,2i-1} \cup F_{n+1,2i} \quad \text{and} \quad \mu(F_{n,i}) = 2^{-n} \quad \text{for all } i = 1,2,\ldots,2^n.$$

It is quite obvious that to every tree $\{F_{n,i}\}_{i=1,n=0}^{2^n\ \infty}$ one can associate a Haar system $\{h'_{n,i}\}_{i=1,n=0}^{2^n\ \infty}$ so that the support of $h'_{n,i}$ is exactly $F_{n,i}$.

There is a certain class of disjoint blocks of the Haar system which will be of particular interest in the sequel. These block bases, which will be called *gaussian Haar systems*, are defined as follows:

Let $\{G_{k,j}\}_{j=1,k=0}^{2^k\ \infty}$ be a tree such that $G_{k,j} \in \mathcal{E}$ for $k = 0,1,\ldots$; $j = 1,\ldots,2^k$ (e.g., $G_{k,j}$ may be $E_{k,j}$, but we will also need other trees) and let $\{N_{k,j}\}_{j=1,k=0}^{2^k\ \infty}$ be a family of subsets of the integers such that

(a) $\inf N_{k,j} > \inf\{\ell;\ G_{k,j} \in \mathcal{E}_\ell\}$ and

(b) $N_{k,j} \cap N_{m,i} = \emptyset$ whenever $G_{k,j} \subset G_{m,i}$.

Let α_n be reals so that $\sum_{n \in N_{k,j}} \alpha_n^2 = 1$, for every $j = 1,2,\ldots,2^k$ and $k = 0,1,\ldots$, and, finally, let $\epsilon_{n,i}$ be equal to either $+1$ or -1 for $i = 1,2,\ldots,2^n$ and $n = 0,1,\ldots$. A gaussian Haar system $\{\tilde{h}_{k,j}\}_{j=1,k=0}^{2^k\ \infty}$ is then defined by putting

$$\tilde{h}_{k,j} = \sum_{n \in N_{k,j}} \alpha_n \sum_{h_{n,i} \subset G_{k,j}} \epsilon_{n,i} h_{n,i} .$$

Observe that condition (b) above implies that the vectors $\tilde{h}_{k,j}$; $j = 1,2,\ldots,2^k$, $k = 0,1,\ldots$ are disjoint blocks of the usual Haar basis such that $\operatorname{supp} \tilde{h}_{k,j} = G_{k,j}$. (The reason we use the adjective "gaussian" is that $\tilde{h}_{k,j}$ has, by the central limit theorem, approximately gaussian distribution relative to the set $G_{k,j}$ if the cardinality $\overline{\overline{N}}_{k,j}$ of $N_{k,j}$ is large enough and $\alpha_n = (\overline{\overline{N}}_{k,j})^{-1/2}$ for $n \in N_{k,j}$).

Lemma 6.2: Let X *be a r.i. space of functions on* [0,1] *which is s-concave for some* $s < \infty$. *If the Haar system* $\{h_{n,i}\}_{i=1,n=0}^{2^n\ \ \infty}$ *is unconditional in* X *then any gaussian Haar system* $\{\tilde{h}_{k,j}\}_{j=1,k=0}^{2^k\ \ \infty}$ *is also unconditional and, moreover,* $\{\tilde{h}_{k,j}\}_{j=1,k=0}^{2^k\ \ \infty}$ *is equivalent to* $\{h_{n,i}\}_{i=1,n=0}^{2^n\ \ \infty}$.

Proof: The unconditionality of the gaussian Haar systems is evident since each such system consists of disjoint blocks of $\{h_{n,i}\}_{i=1,n=0}^{2^n\ \ \infty}$. Now let M be the s-concavity constant of X and let K be the unconditionality constant of the Haar system in X . Then, by (*) (of the Introduction), we have, for every choice of reals $\{a_{k,j}\}_{j=1,k=0}^{2^k\ \ \infty}$, that

$$\|\sum_{k=0}^{\infty}\sum_{j=1}^{2^k} a_{k,j}\tilde{h}_{k,j}\|_X = \|\sum_{k=0}^{\infty}\sum_{j=1}^{2^k} a_{k,j}\sum_{n\in N_{k,j}}\alpha_n\sum_{h_{n,i}\subset G_{k,j}}\epsilon_{n,i}h_{n,i}\|_X \le$$

$$\le KMB_s\|(\sum_{k=0}^{\infty}\sum_{j=0}^{2^k} a_{k,j}^2\sum_{n\in N_{k,j}}\alpha_n^2\sum_{h_{n,i}\subset G_{k,j}}h_{n,i}^2)^{1/2}\|_X =$$

$$= KMB_s\|(\sum_{k=0}^{\infty}\sum_{j=1}^{2^k} a_{k,j}^2\chi_{G_{k,j}})^{1/2}\|_X .$$

On the other hand, we get in a similar way that

$$\|\sum_{k=0}^{\infty}\sum_{j=1}^{2^k} a_{k,j}h_{k,j}\|_X \ge K^{-1}2^{-1/2}\|(\sum_{k=0}^{\infty}\sum_{j=1}^{2^k} a_{k,j}^2h_{k,j}^2)^{1/2}\|_X =$$

$$= K^{-1}2^{-1/2}\|(\sum_{k=0}^{\infty}\sum_{j=1}^{2^k} a_{k,j}^2x_{k,j})^{1/2}\|_X =$$

$$= K^{-1}2^{-1/2}\|(\sum_{k=0}^{\infty}\sum_{j=1}^{2^k} a_{k,j}^2\chi_{G_{k,j}})^{1/2}\|_X$$

i.e.

$$\|\sum_{k=0}^{\infty}\sum_{j=1}^{2^k} a_{k,j}\tilde{h}_{k,j}\|_X \le 2^{1/2}K^2MB_s\|\sum_{k=0}^{\infty}\sum_{j=1}^{2^k} a_{k,j}h_{k,j}\|_X .$$

The opposite inequality is proved in exactly the same manner.

□

As in 6.2, the main tool in the proof of 6.1 will be the so called square function which has the advantage of reducing many problems to questions concerning positive functions.

Since some parts of the proof of 6.1 do not use the r.i. properties of Y, we prefer to state the lemmas in a setting of lattices.

By a "good lattice Y of functions on $[0,1]$", we shall mean a σ-complete and σ-order continuous Banach lattice Y of (Lebesgue classes) of measurable functions so that $L_\infty(0,1) \subset Y \subset L_1(0,1)$, with continuous injections.

Lemma 6.3: Let X *be a r.i. function space on* $[0,1]$, Y *a good lattice of functions on* $[0,1]$, *and let* $\{h_{n,i}\}_{i=1,n=0}^{2^n \ \infty}$ *denote the Haar system in* X. *Assume that* X *does not contain uniformly isomorphic copies of* ℓ_∞^n *for all* n. *Let* T *be an operator from* X *into* Y. *For* $n = 0,1,2,\ldots$ *set*

$$v_n = \left(\sum_{i=1}^{2^n} |Th_{n,i}|^2\right)^{1/2} .$$

Then there exists a constant $C_1 < \infty$ *such that, for every choice of reals* $\{a_n\}_{n=0}^\infty$, *we have*

$$\left\|\left(\sum_{n=0}^\infty a_n^2 v_n^2\right)^{1/2}\right\|_Y \leq C_1\left(\sum_{n=0}^\infty a_n^2\right)^{1/2} .$$

In particular, the convex hull $\operatorname{conv}\{v_n^2;\ n \geq 0\}$ *is bounded in probability. More precisely, for every* $\rho > 0$ *and* $\beta_n \geq 0$ *satisfying* $\sum_{n=0}^\infty \beta_n = 1$, *we have*

$$\mu(\{t \in [0,1],\ \sum_{n=0}^\infty \beta_n v_n^2(t) > \rho\}) \leq C_1 \rho^{-1/2} .$$

Proof: The fact that X does not contain uniformly isomorphic copies of ℓ_∞^n for all n implies, by [54], that X is an s-concave Banach lattice for some $s < \infty$. Let M denote the s-concavity constant of X. By Khintchine's inequality, it follows that, for every choice of reals $\{a_n\}_{n=0}^\infty$, we have

$$\|(\sum_{n=0}^{\infty} a_n^2 v_n^2)^{1/2}\|_Y = \|(\sum_{n=0}^{\infty} a_n^2 \sum_{i=1}^{2^n} |Th_{n,i}|^2)^{1/2}\|_Y \le$$

$$\le A_1^{-1} \| \int_0^1 \int_0^1 | \sum_{n=0}^{\infty} a_n r_n(u) \sum_{i=1}^{2^n} r_{n,i}(t) Th_{n,i} | du dt \|_Y \le$$

$$\le \|T\| \sqrt{2} \int_0^1 \int_0^1 \| \sum_{n=0}^{\infty} a_n r_n(u) \sum_{i=1}^{2^n} r_{n,i}(t) h_{n,i} \|_X du dt ,$$

where $\{r_n\}_{n=0}^{\infty}$ and $\{r_{n,i}\}_{i=1,n=0}^{2^n \ \ \infty}$ are two sequences of independent Rademacher functions. Hence, by the s-concavity of X we get that

$$\|(\sum_{n=0}^{\infty} a_n^2 v_n^2)^{1/2}\|_Y \le \|T\|\sqrt{2} \int_0^1 (\int_0^1 \| \sum_{n=0}^{\infty} a_n r_n(u) \sum_{i=1}^{2^n} r_{n,i}(t) h_{n,i} \|_X^s du)^{1/s} dt \le$$

$$\le M \|T\| \sqrt{2} \int_0^1 \| (\int_0^1 | \sum_{n=0}^{\infty} a_n r_n(u) \sum_{i=1}^{2^n} r_{n,i}(t) h_{n,i} |^s du)^{1/s} \|_X dt \le$$

$$\le MB_s \|T\| \sqrt{2} \int_0^1 \| (\sum_{n=0}^{\infty} a_n^2 | \sum_{i=1}^{2^n} r_{n,i}(t) h_{n,i} |^2)^{1/2} \|_X dt =$$

$$= MB_s \|T\| \sqrt{2} \, (\sum_{n=0}^{\infty} a_n^2)^{1/2} .$$

In order to prove the second assertion, observe that

$$\mu(\{t \in [0,1], \sum_{n=0}^{\infty} \beta_n v_n^2(t) > \rho\}) \le \rho^{-1/2} \int_0^1 (\sum_{n=0}^{\infty} \beta_n v_n^2(t))^{1/2} d\mu \le$$

$$\le \rho^{-1/2} \|(\sum_{n=0}^{\infty} \beta_n v_n^2)^{1/2}\|_Y \le MB_s \|T\| \sqrt{2} \, \rho^{-1/2} (\sum_{n=0}^{\infty} \beta_n)^{1/2} ,$$

for every ρ and $\beta_n \ge 0$ for all n.

□

Lemma 6.4: Suppose that for each n , λ_n *is an additive measure on* $\mathcal{S}_n$ *taking values in the set* L_0^+ *of non-negative meaurable real functions on* $[0,1]$ *such that* $\text{conv}\{\lambda_n(0,1)\}_{n=1}^{\infty}$ *is bounded in probability. Then there is an additive measure* Λ *on* $\mathcal{S}$ *taking values in* L_0^+ *and a sequence* $\{\sigma_m\}_{m=1}^{\infty}$ *of disjoint finite subsets of the integers such that the measures*

$$\nu_m = (\sum_{n\in\sigma_m} \lambda_n)/\bar{\bar{\sigma}}_m$$

defined on $\mathcal{S}_k$ $k = \min \sigma_m$ *have the property that for every* $E \in \mathcal{S}$, $\{\nu_m(E)\}_{m=1}^{\infty}$ *converges* μ-a.e. *to* $\Lambda(E)$.

Proof: Set $v_n = \lambda_n(0,1)$. By a result of E.M. Nikishin [61] (see also [53]), it follows that, for every $\epsilon > 0$, there exists a measurable subset D_ϵ of $[0,1]$ so that $\mu(D_\epsilon) > 1-\epsilon$ and

$$\sup_n \int_{D_\epsilon} v_n d\mu < \infty \ .$$

This fact shows that it suffices to prove the lemma under the assumption that $\{v_n\}_{n=1}^{\infty}$ is a bounded sequence in $L_1(0,1)$ (notice that if, for each $\epsilon > 0$, one can find measures $\{\nu_{m,\epsilon}(\cdot)\}_{m=1}^{\infty}$ so that $\{\nu_{m,\epsilon}(F)\}_{m=1}^{\infty}$ converges μ-a.e. in D_ϵ then, by a diagonal argument one finds measures $\{\nu_m\}_{m=1}^{\infty}$ which converge μ-a.e. in all of $[0,1]$). Therefore, there is no loss of generality in assuming that

$$\sup_n \int_0^1 v_n d\mu < \infty \ .$$

By a known argument (see e.g. [16]), the sequence $\{v_n\}_{n=1}^{\infty}$ can be decomposed as a sum of a sequence of "peaks" and a sequence of equi-integrable functions. More precisely, there exist a subsequence of $\{v_n\}_{n=1}^{\infty}$, which for simplicity of notation will still be denoted by $\{v_n\}_{n=1}^{\infty}$, and subsets $\{G_n\}_{n=1}^{\infty}$ and $\{H_n\}_{n=1}^{\infty}$ of $[0,1]$ so that

$$v_n = \chi_{G_n} v_n + \chi_{H_n} v_n \ ,$$

for all n , the sets $\{G_n\}_{n=1}^{\infty}$ are pairwise disjoint and the functions $\{\chi_{H_n} v_n\}_{n=1}^{\infty}$ are equi-integrable. The last fact implies that, for every fixed set $E \in \pi$, the sequence $\{\chi_{H_n} \lambda_n(E)\}_{n=1}^{\infty}$ is also equi-integrable; i.e., weakly compact in $L_1(0,1)$. Therefore, this sequence contains a subsequence which is weakly convergent in $L_1(0,1)$. Since π is countable it is clearly possible to construct by a simple diagonal argument, an additive function Λ on $\mathcal{S}$, whose values are μ-measurable functions on $[0,1]$, and a subsequence $\{\chi_{H_{n_i}}\lambda_{n_i}(\cdot)\}_{i=1}^{\infty}$ of $\{\chi_{H_n}\lambda_n(\cdot)\}_{n=1}^{\infty}$ such that, for every $F \in \mathcal{S}$, the sequence $\{\chi_{H_{n_i}}\lambda_{n_i}(F)\}_{i=1}^{\infty}$ converges weakly in $L_1(0,1)$ to $\Lambda(F)$. It follows from the Banach-Saks property of $L_1(0,1)$ (cf. [78]) that there exists a sequence $\{\sigma_m\}_{m=1}^{\infty}$ of disjoint finite subsets of the sequence $\{n_i\}_{i=1}^{\infty}$ with $\min \sigma_{m+1} > \max \sigma_m$ so that

$$\|\bar{\bar{\sigma}}_m^{-1} \sum_{n\in\sigma_m} \chi_{H_n}\lambda_n(F) - \Lambda(F)\|_1 \leq 1/2^m$$

for every m and every $F \in \mathcal{S}_m$. Put $\nu_m(\cdot) = (\sum_{n\in\sigma_m} \lambda_n(\cdot))/\bar{\bar{\sigma}}_m$ and $\Lambda_m(\cdot) = \sum_{n\in\sigma_m} \chi_{H_n}\lambda_n(\cdot)/\bar{\bar{\sigma}}_m$ and observe that, for $F \in \mathcal{S}$, $\|\Lambda_m(F) - \Lambda(F)\|_1 \to 0$ as $m \to \infty$. In particular, we get that, for every $F \in \mathcal{S}$, the sequence $\{\Lambda_m(F)\}_{m=1}^{\infty}$ converges μ-a.e. to $\Lambda(F)$. On the other hand, since all the functions $\nu_m(F) - \Lambda_m(F)$; $F \in \mathcal{S}_m$ are supported by the set $\bigcup_{n\in\sigma_m} G_n$ and $\mu(\bigcup_{n\in\sigma_m} G_n) \to 0$ as $m \to \infty$, it follows that the sequence $\{\nu_m(F) - \Lambda_m(F)\}_{m=1}^{\infty}$ tends μ-a.e. to 0 for every $F \in \mathcal{S}$. Consequently, for any $F \in \mathcal{S}$, $\{\nu_m(F)\}_{m=1}^{\infty}$ converges μ-a.e. in $[0,1]$ to $\Lambda(F)$ and this completes the proof.

□

The proof of 6.4 puts in evidence the role played by sequences of functions in $L_1(0,1)$ which are equi-integrable. Actually, the proof of 6.1 requires the use of equi-integrability in general r.i. function spaces on $[0,1]$ or even in Banach lattices. The meaning of this notion is made precise in the following definition.

Definition 6.5: Let Y *be a* σ*-complete and* σ*-order continuous Banach lattice. If* y *is a positive element in* Y *let* P_y *denote the band projection generated by* y . *A subset* M *of* Y *is said to be equi-integrable in* Y *if it is bounded and if, for every sequence* $\{y_n\}_{n=0}^{\infty}$ *of disjoint elements in* Y , *we have*

$$\lim_{n\to\infty} \sup_{x\in M} \|P_{y_n}(x)\|_Y = 0 .$$

In the case where Y is a good lattice of functions on [0,1] , the band projections are of the form $P_x = \chi_A \cdot x$, where A is a measurable subset of [0,1] . Therefore, in this case, the equi-integrability of a subset M is expressed by the condition:

(+) For every sequence $\{A_n\}_{n=0}^{\infty}$ of disjoint measurable subsets of [0,1] , we have

$$\lim_{n\to\infty} \sup_{x\in M} \|\chi_{A_n} \cdot x\|_Y = 0 .$$

(Notice that (+) already yields the boundedness of M .)

The usual Kadec-Pelczynski [38] disjointification procedure in σ-complete and σ-order continuous lattices (see, for example, [37]) shows that (+) is equivalent to

($\ddagger$) For every sequence $\{A_n\}_{n=0}^{\infty}$ of measurable subsets of [0,1] such that $\lim_{n\to\infty} \mu(A_n) = 0$, we have

$$\lim_{n\to\infty} \sup_{x\in M} \|\chi_{A_n} \cdot x\|_Y = 0 .$$

Observe that the unit ball $B_{L_\infty(0,1)}$ of $L_\infty[0,1]$ is equi-integrable in every good lattice Y of functions on [0,1] . Clearly, this is equivalent to saying that $\lim_n \|\chi_{A_n}\|_Y = 0$ for every sequence $\{A_n\}_{n=0}^{\infty}$ of measurable sets such that $\lim_n \mu(A_n) = 0$. If this were false, we could choose a sequence

$\{A_n\}_{n=0}^\infty$ with $\sum_{n=0}^\infty \mu(A_n) < \infty$ and $\inf_n \|\chi_{A_n}\| > 0$. But now, if we set $B_k = \bigcup_{n=k}^\infty A_n$, the sequence $\{\chi_{B_k}\}_{k=0}^\infty$ is decreasing a.e. to zero, and thus $\lim_k \|\chi_{B_k}\|_Y = 0$, by the σ-completeness and σ-order continuity of Y, thus contradicting $\inf_n \|\chi_{A_n}\|_Y = 0$.

The condition ($\ddagger$) is also equivalent to the following one, again in the case where Y is a good lattice of functions on $[0,1)$.

($\ddagger$) For every $\epsilon > 0$, there exists a constant R_ϵ so that if f is an arbitrary function in M and

$$E(R_\epsilon, f) = \{t \in (0,1);\ |f(t)| \geq R_\epsilon\}$$

then

$$\|\chi_{E(R_\epsilon, f)} f\|_Y \leq \epsilon .$$

(If ($\ddagger$) is false for some $\epsilon > 0$, we can choose a sequence $\{f_n\}_{n=0}^\infty$ in M such that $\|\chi_{E(2^n, f_n)} f_n\| > \epsilon$. But, since M is bounded in Y, $\mu(E(2^n, f_n)) \leq 2^{-n} \int_0^1 f_n \, d\mu \leq C \cdot 2^{-n}$, and thus ($\ddagger$) is not satisfied. The other direction is also easy, and follows from the fact that $\lim_{n \to \infty} \mu(A_n) = 0$ implies $\lim_n \|\chi_{A_n}\|_Y = 0$, as we have already seen.)

If $\{f_n\}_{n=1}^\infty$ is a bounded sequence in a good lattice Y, then there is a subsequence $\{n_k\}_{k=1}^\infty$ and a sequence $\{A_k\}_{k=1}^\infty$ of disjoint measurable subsets of $[0,1]$ such that the sequence $\{\chi_{\sim A_k} \cdot f_{n_k}\}_{k=1}^\infty$ is equi-integrable. This observation, due to Dacunha-Castelle [16] in the context of Orlicz spaces, generalizes easily, with essentially the same proof:

Let δ be the supremum of all ρ's such that there exists a sequence $\{A_k\}_{k=0}^\infty$ satisfying $\lim_{k \to \infty} \mu(A_k) = 0$ and a subsequence $\{n_k\}_{k=1}^\infty$ for which

$\|\chi_{A_k} f_{n_k}\|_Y \geq \rho$ for every k. With this definition, it is clear that we can find $\{A_k\}_{k=1}^\infty$ satisfying $\lim_{k\to\infty} \mu(A_k) = 0$ and a subsequence $\{n_k\}_{k=1}^\infty$ such that

$$\lim_{k\to\infty} \|\chi_{A_k} f_{n_k}\|_Y = \delta .$$

By using again the disjointification procedure and a further subsequence $\{n_k'\}_{k=1}^\infty$ we can find a sequence $\{B_k\}_{k=1}^\infty$ of disjoint measurable sets such that

$$\lim_{k\to\infty} \|\chi_{B_k} f_{n_k'}\|_Y = \delta .$$

We leave to the reader the verification that $\{\chi_{\sim B_k} f_{n_k'}\}_{k=1}^\infty$ is equi-integrable in Y.

Another observation to be used in the sequel is the following: if Y is a good lattice of functions on $[0,1]$, $\{y_n\}_{n=1}^\infty$ an equi-integrable sequence in Y and if $\{y_n\}_{n=1}^\infty$ converges μ-a.e. to a limit y, then $y \in Y$ and $\lim_{n\to\infty} \|y - y_n\|_Y = 0$. Again, this simple fact is left to the reader.

It is known [57] that if M is an equi-integrable set in $L_1(0,1)$ then there exists an Orlicz function space $L_M(0,1)$ which is different from $L_1(0,1)$ (i.e., $M(t)/t \to \infty$ as $t \to \infty$) such that M is a bounded subset of $L_M(0,1)$. This fact is, in a certain sense, true for any good lattice on $[0,1]$.

Proposition 6.6: Let Y *be a good lattice of functions on* $[0,1]$. *Then for every equi-integrable set* M *in* Y, *there exists a lattice of functions* Z *on* $[0,1]$ *so that*

(i) The set M *is equi-integrable in* Z, *and in particular* $\sup\{\|f\|_Z ; f \in M\} < \infty$.

(ii) The unit ball B_Z *of* Z *is an equi-integrable set in* Y *(and therefore is bounded in* Y*).*

Furthermore, if Y *is 2-convex or it does not contain uniformly isomorphic copies of* ℓ_∞^n *for all* n *or is a r.i. space then* Z *has the same property.*

Proof: The space Z will be constructed by the factorization method described in [18] (or, equivalently, the Lions-Peetre interpolation approach [52]).

Let M be an equi-integrable set in Y ; then, by the definition 6.5, we can find an increasing sequence $\{R_n\}_{n=0}^{\infty}$ so that, for every n and each $f \in M$ we have $\|\chi_{E(R_n,f)}f\|_Y \le 4^{-n}$. This means that f can be decomposed as the sum of two disjoint functions $f = g_n + h_n$ with $\|g_n\|_\infty \le R_n$ and $\|h_n\|_Y \le 4^{-n}$. Fix $n \le 0$ and consider the convex set

$$C_n = 2^{-n}B_Y + 2^n R_n B_{L_\infty(0,1)} \quad ,$$

where B_Y and $B_{L_\infty(0,1)}$ denote the closed unit balls of Y , respectively, $L_\infty(0,1)$. Notice that

$$2^{-n}B_Y \subset C_n \subset (2^{-n} + 2^n R_n)B_Y$$

i.e. the gauge function $|||\cdot|||_n$ of C_n is a norm on Y equivalent to $\|\cdot\|_Y$ and invariant under all automorphisms of the measure [0,1] when Y is r.i. Z is defined to be the "diagonal" of the direct sum $(\sum_{n=1}^{\infty} \oplus(Y, |||\cdot|||_n))_2$; i.e., the set of all $f \in Y$ for which

$$\|f\|_Z = (\sum_{n=0}^{\infty} |||f|||_n^2)^{1/2}/K < \infty ,$$

where $K = (\sum_{n=0}^{\infty} |||1|||_n^2)^{1/2}$. Notice that, since $\|1\|_Z = 1$, Z is a r.i. space of functions on [0,1] in the case when Y is r.i.

If $f \in M$ we have $|||f|||_n \le 2^{-n}$ for every n (since $f = g_n + h_n$ with $\|g_n\|_\infty \le R_n$ and $\|h_n\|_Y \le 4^{-n}$) . For $n \le k$ we can write

$$|||\chi_{E(R_n,f)}f|||_n \le 2^n\|\chi_{E(R_k,f)}f\|_Y \le 2^n 4^{-k} \quad .$$

Therefore,

$$\|\chi_{E(R_k,f)}f\|_Z \le K^{-1}\left(\sum_{n=0}^{k-1} (2^n 4^{-k})^2 + \sum_{n=k}^{\infty} 4^{-n}\right)^{1/2} \le$$

$$\le K^{-1}(5/3 \cdot 4^{-k})^{1/2} = K_1 \cdot 2^{-k} \quad ,$$

which shows that M is equi-integrable in Z.

Let now f be an arbitrary function in the unit ball of Z and let n be an integer. Set

$$F_n = \{t \in [0,1], \ |f(t)| \ge 2^{n+1}R_n K\}$$

and observe that $|||\chi_{F_n} f|||_n \le K$. It follows that

$$\chi_{F_n} f = 2^{-n}Kg + 2^n R_n Kh \quad ,$$

where $g \in B_Y$ and $h \in B_{L_\infty(0,1)}$. Hence, for each $t \in F_n$, we obtain

$$2^{-n}K|g(t)| \ge |f(t)| - 2^n R_n K|h(t)| \ge$$

$$\ge |f(t)| - 2^n R_n K \ge |f(t)|/2 \quad .$$

This shows that

$$\|\chi_{F_n} f\|_Y \le 2^{-n+1}K\|g\|_Y \le 2^{-n+1}K \quad ,$$

for all n or, in other words, the unit ball of Z is equi-integrable in Y.

If Y is 2-convex, then for each $n \ge 0$, Y endowed with the norm $|||\cdot|||_n$ is also 2-convex with 2-convexity constant $\le \sqrt{2}$-times the 2-convexity constant of Y (use the fact that the dual of $(Y, |||\cdot|||_n)$ is isometric to a sublattice of $(Y^* \oplus L^*_\infty(0,1))_\infty$.)

Suppose now that Y does not contain uniformly isomorphic copies of ℓ_∞^n for all n. This implies, by [35], that Y has a lower q-estimate for some $q < \infty$. Of course, the same is true for Y endowed with the norm $|||\cdot|||_0$ since this norm is equivalent to $\|\cdot\|_Y$. Therefore, there exists an integer k so that, for every sequence $\{y_i\}_{i=1}^k$ of disjoint elements in Y, we have

$$\sum_{i=1}^{k} \|y_i\|_Y^2 \le k\|\sum_{i=1}^{k} y_i\|_Y^2/16 \quad \text{and} \quad \sum_{i=1}^{k} |||y_i|||_0^2 \le k|||\sum_{i=1}^{k} y_i|||_0^2/16 .$$

Let $\{f_i\}_{i=1}^k$ be a sequence of mutually disjoint functions in Z such that $f = \sum_{i=1}^{k} f_i$ has norm $\|f\|_Z \le 1$. By the definition of the norms $|||\cdot|||_n$, we may write for each $n \ge 0$

$$f = u_n + v_n ,$$

where $\|u_n\|_Y \le 2^{-n}|||f|||_n$ and $\|v_n\|_\infty \le 2^n R_n |||f|||_n$. Let E_i denote the support of f_i and put

$$u_{n,i} = u_n \chi_{E_i} \; ; \; v_{n,i} = v_n \chi_{E_i} .$$

Then, for each $1 \le i \le k$ and every $n \ge 1$, we have

$$f_i = u_{n-1,i} + v_{n-1,i}$$

and $|||f_i|||_n \le \max(2^n\|u_{n-1,i}\|_Y , 2^{-n}R_n^{-1}\|v_{n-1,i}\|_\infty)$. Using all these facts, we get that

$$\sum_{i=1}^{k} \|f_i\|_Z^2 = \sum_{i=1}^{k}\sum_{n=0}^{\infty} |||f_i|||_n^2/K^2 = \sum_{i=1}^{k} |||f_i|||_0^2/K^2 + \sum_{i=1}^{k}\sum_{n=1}^{\infty} |||f_i|||_n^2/K^2 \le$$

$$\le k|||\sum_{i=1}^{k} f_i|||_0^2/16K^2 + \sum_{n=1}^{\infty}\sum_{i=1}^{k} \{\max(K^{-1}2^n\|u_{n-1,i}\|_Y, K^{-1}2^{-n}R_n^{-1}\|v_{n-1,i}\|_\infty)\}^2 \le$$

$$\le k/16 + \sum_{n=1}^{\infty}\sum_{i=1}^{k} \max(K^{-2}4^n\|u_{n-1,i}\|_Y^2 , K^{-2}2^{-2}|||f|||_{n-1}^2) \le$$

$$\le k/16 + \sum_{n=1}^{\infty} (K^{-2}4^n k 16^{-1}\|u_{n-1}\|_Y^2 + K^{-2}2^{-2}k|||f|||_{n-1}^2) \le$$

$$\le k/16 + \sum_{n=1}^{\infty} (K^{-2}4^n k 16^{-1}4^{-n+1}|||f|||_{n-1}^2 + K^{-2}2^{-2}k|||f|||_{n-1}^2) =$$

$$= k/16 + k\|f\|_Z^2/2 \le 9k/16 .$$

This means that we have just proved that no sequence $\{z_i\}_{i=1}^k$ of mutually disjoint elements in Z is $1+\epsilon$-equivalent to the unit vector basis of ℓ_∞^k for $\epsilon < 7/9$. Consequently, by [35], Z does not contain uniformly isomorphic copies of ℓ_∞^n for all n.

□

Remarks: 1. If Y is p-convex then, by a suitable modification of the construction in 6.6, Z can be built to be also p-convex.

2. Proposition 6.6 has two consequences:

a) If M is equi-integrable in Y so is conv M.

b) Assume that Y does not contain uniformly isomorphic copies of ℓ_∞^n, for all n. Let $\{y_{k,i}\}_{i=1,k=0}^{i_k\ \ \infty}$ be a family of elements in Y such that $V_k = (\sum_{i=1}^{i_k} y_{k,i}^2)^{1/2}$ is an equi-integrable sequence in Y. It is then possible to select signs $\{\epsilon_{k,i}\}_{i=1,k=0}^{i_k\ \ \infty}$ in such a way that the sequence $s_k = \sum_{i=1}^{i_k} \epsilon_{k,i} y_{k,i}$ will be equi-integrable in Y.

The last step in the preparation needed to prove 6.1 involves a result which is most convenient to state and to prove by using the 2-concavification $Y_{1/2}$ of the lattice of functions Y. The space $Y_{1/2}$ is the set of all measurable functions f on $[0,1]$ such that $|f|^{1/2} \in Y$. The expression

$$\|f\|_{Y_{1/2}} = \| \, |f|^{1/2} \|_Y^2 ,$$

which is defined for every $f \in Y_{1/2}$, is a homogeneous function on $Y_{1/2}$, having the property that

$$\|f + g\|_{Y_{1/2}}^{1/2} \leq \|f\|_{Y_{1/2}}^{1/2} + \|f\|_{Y_{1/2}}^{1/2} \ ; \ f,g \in Y_{1/2} .$$

In other words, $Y_{1/2}$ is 1/2-normed lattice and it is easily seen that $Y_{1/2}$ is a Banach lattice if and only if the 2-convexity constant of Y is equal to 1.

We prove now the following lemma.

Lemma 6.7: Let Y *be a good lattice of functions on* [0,1] *which is q-concave with q-concavity constant* $\leq M$ *for some* $2 \leq q < \infty$ *and some* $M < \infty$. *Then, for every sequence* $\{y_i\}_{i=1}^{2N}$ *of positive elements in* Y *with* $\sum_{i=1}^{2N} y_i \in Y_{1/2}$, *there exists a subset* $\sigma \subset \{1,2,\dots,2N\}$ *such that* σ *contains exactly* N *integers and*

$$\|\sum_{i\in\sigma} y_i - \sum_{i\notin\sigma} y_i\|_{Y_{1/2}} \leq M^2 B^2_{q/2} \|\sum_{i=1}^{2N} y_i\|^{1/2}_{Y_{1/2}} \|\max_{1\leq i\leq 2N} y_i\|^{1/2}_{Y_{1/2}} .$$

Proof: Suppose that $\|\sum_{i=1}^{2N} y_i\|_{Y_{1/2}} = 1$ and put

$$\theta = \|\max_{1\leq i\leq 2N} y_i\|_{Y_{1/2}} .$$

Then by the q-concavity of Y and Khintchine's inequality it follows that

$$\left(\int_0^1 \|\sum_{i=1}^{n} (y_{2i}-y_{2i-1}) r_i(u)\|^{q/2}_{Y_{1/2}} du\right)^{2/q} = \left(\int_0^1 \|\,|\sum_{i=1}^{N} (y_{2i}-y_{2i-1}) r_i(u)|^{1/2}\|^{q}_{Y} du\right)^{2/q}$$

$$\leq M^2 \|\left(\int_0^1 |\sum_{i=1}^{N} (y_{2i}-y_{2i-1}) r_i(u)|^{q/2} du\right)^{1/q}\|^2_Y \leq M^2 B^2_{q/2} \|\left(\sum_{i=1}^{N} |y_{2i}-y_{2i-1}|^2\right)^{1/4}\|^2_Y .$$

Hence by 2.2 (ii) we obtain

$$\left(\int_0^1 \|\sum_{i=1}^{N} (y_{2i}-y_{2i-1}) r_i(u)\|^{q/2}_{Y_{1/2}} du\right)^{2/q} \leq M^2 B^2_{q/2} \|\left(\sum_{i=1}^{2N} y_i^2\right)^{1/4}\|^2_Y$$

$$\leq M^2 B^2_{q/2} \|\left(\sum_{i=1}^{2N} y_i\right)^{1/2}\|_Y \cdot \|\left(\max_{1\leq i\leq 2N} y_i\right)^{1/2}\|_Y = M^2 B^2_{q/2} \|\sum_{i=1}^{2N} y_i\|^{1/2}_{Y_{1/2}} \|\max_{1\leq i\leq 2N} y_i\|^{1/2}_{Y_{1/2}}$$

$$= M^2 B^2_{q/2}\, \theta^{1/2} .$$

It is therefore possible to find signs $\{\epsilon_i\}_{i=1}^{n}$ so that

$$\|\sum_{i=1}^{N} \epsilon_i (y_{2i}-y_{2i-1})\|_{Y_{1/2}} \leq M^2 B^2_{q/2}\, \theta^{1/2} .$$

Set $\sigma = \{2i;\ 1 \leq i \leq N\ ,\ \epsilon_i = +1\} \cup \{2i-1;\ 1 \leq i \leq N\ ,\ \epsilon_i = -1\}$. Then, clearly σ consists of precisely N integers and

$$\|\sum_{i\in\sigma} y_i - \sum_{i\notin\sigma} y_i\|_{Y_{1/2}} \leq M^2 B^2_{q/2}\ \theta^{1/2} .$$

□

The next proposition really contains theorem 6.1; it is formulated in the more general case where Y is a lattice of functions on $[0,1]$.

Proposition 6.8: Let X *be a r.i. function space on* $[0,1]$ *and* Y *be a good lattice of functions on* $[0,1]$. *Assume that the Haar system* $\{h_{n,i}\}_{i=1,n=0}^{2^n\ \ \infty}$ *is an unconditional basis for* X *and that* Y *does not contain uniformly isomorphic copies of* ℓ_∞^n *for all* n . *If* X *embeds into* Y , *one (or both) of the following holds:*

(i) There is another embedding $\tilde{T}$ *from* X *into* Y , *and there are constants* $\delta > 0$, R *such that*

$$\|\chi_{A_n} \max_{1\leq i\leq 2^n} |\tilde{T}h_{n,i}|\ \|_Y \geq \delta ,$$

$$A_n = \{t \in [0,1];\ \max_{1\leq i\leq 2^n} |\tilde{T}h_{n,i}(t)| \leq R\} ,$$

(ii) There exist constants D *and* w , $0\leq w\leq 1$, *and a sequence* $\{g_{n,i}\}_{n=0,i=1}^{\infty\ \ 2^n}$ *of disjoint functions in* Y *so that for every* $f = \sum_{n,i} a_{n,i}h_{n,i} \in X$, *we have*

$$D^{-1}\|f\|_X \leq \max\{\|\sum_{n,i} a_{n,i}g_{n,i}\|_Y\ ,\ w\|f\|_2\} \leq D\|f\|_X .$$

Before proving 6.8, let us deduce 6.1 from 6.8. Suppose that T is an isomorphism from X into Y , where X and Y satisfy the conditions of 6.1. We want to construct a good lattice of functions Y' on $[0,1]$, which isomorphic to Y , and an embedding T' from X into Y' . If Y is a r.i. function space on $[0,1]$, we simply put $Y' = Y$ and $T' = T$.

If Y is a r.i. function space on $[0,\infty)$ we construct a lattice Y' of functions on $[0,1]$ in the following way: let f_o be a norm one decreasing function in Y such that $1 \geq f_o > 0$ everywhere and φ be a homeomorphism from $[0,\infty)$ onto $[0,1)$. The space Y' will be the space of functions g on $[0,1]$ such that $f_o \cdot g(\varphi)$ belongs to Y, endowed with the norm

$$\|g\|_{Y'} = \|f_o \cdot g(\varphi)\|_Y .$$

Clearly, the mapping V defined by $Vg = f_o \cdot g(\varphi)$ is an isometry and a lattice isomorphism from Y' onto Y. We now define the isomorphism T' of X into Y' by $T' = V^{-1}T$.

If Y is a r.i. function space on $[0,1]$ and if the first case in 6.8 is satisfied for $T' = T$ it follows from 5.2 and the remark thereafter that there is a constant C such that

$$\|f\|_Y \leq C\|f\|_X ,$$

for every $f \in X$.

In the case where Y is a r.i. function space on $[0,\infty)$ and the first case in 6.8 occurs for the embedding $T' = V^{-1}T$ from X into Y' we get that there are $\delta > 0$, R and an isomorphic embedding $\tilde{T}$ from X into Y' such that

$$\|\chi_{A_n} \max_{1\leq i\leq 2^n} |\tilde{T}h_{n,i}| \,\|_{Y'} \geq \delta ;$$

where $A_n = \{t \in [0,1];\ \max_{1\leq i\leq 2^n} |\tilde{T}h_{n,i}(t)| \leq R\}$, for every n. Now, since Y, and hence also Y', do not contain uniformly isomorphic copies of ℓ_∞^n, it follows that $\lim_{u\to 1} \|\chi_{[u,1]}\|_{Y'} = 0$. If we choose $u < 1$ such that $\|\chi_{[u,1]}\|_{Y'} \leq \delta/2R$, we get

$$\|\chi_{A_n}\chi_{[0,u]} \max_{1\leq i\leq 2^n} |\tilde{T}h_{n,i}| \,\|_{Y'} \geq \delta/2 \quad ,$$

for every n. It is easily seen that, for every $0 < u < 1$, the space Y'_u, of functions in Y' supported on $[0,u]$ is, up to an equivalent renorming, the r.i. function space $Y(0,u)$ of functions in Y supported on $[0,u]$. We may thus apply 5.2 to the mapping $\chi_{[0,u]}\tilde{T}$ from X into Y'_u, and this yields (i) of 6.1.

We shall show now that if T' satisfies the second case of 6.8 then we get the case (ii) or (iii) in 6.1. Since Y' is isometric and lattice isomorphic to Y it is clear that we can find disjoint functions $\{g_{n,i}\}_{n=0,i=1}^{\infty \quad 2^n}$ in Y such that, for every $f = \sum_{n,i} a_{n,i} h_{n,i} \in X$, we have

$$(\%) \quad D^{-1}\|f\|_X \le \max\{\|\sum_{n,i} a_{n,i} g_{n,i}\|_Y, w\|f\|_2\} \le D\|f\|_X .$$

It is obvious that $w = 0$ implies the case (ii) in 6.1. By our assumption on Y, both X and Y are s-concave for some $s < \infty$. If $w > 0$ and if X is not isomorphic to $L_2(0,1)$ then there exists a function $f \in X$ such that $1 = \|f\|_X \ge 2MB_s A_1^{-1} Dw\|f\|_2$, where M is the s-concavity constant of X. We may assume without loss of generality that

$$f = \sum_{i=1}^{2^n} a_i x_{n,i}$$

for a suitable n. Put

$$f_m = f \cdot r_{m+n} = f \cdot (\sum_{i=1}^{2^{m+n-1}} h_{m+n-1,i}) \; ; \; m = 1,2,\ldots$$

Observe that $\{f_m\}_{m=1}^{\infty}$ is a block basis of the Haar system in X which is 1-unconditional and K-equivalent to the unit vector basis of ℓ_2, with $K \le MB_s A_1^{-1}$. It follows that, for every $g = \sum_{m=1}^{\infty} b_m f_m \in [f_m]_{m=1}^{\infty}$, we have

$$\|g\|_X \ge (MB_s A_1^{-1})^{-1} \cdot (\sum_{m=1}^{\infty} b_m^2)^{1/2} = (MB_s A_1^{-1}\|f\|_2)^{-1}\|g\|_2 \ge 2Dw\|g\|_2 .$$

Therefore, for every $g \in [f_m]_{m=1}^{\infty}$, the maximum in the inner term of (%) is not given by $w\|g\|_2$. This means that the sequence $\{f_m\}_{m=1}^{\infty}$ is D equivalent to a sequence of disjoint blocks of the vectors $\{g_{n,i}\}_{n=0,i=1}^{\infty\ \ 2^n}$, or in other words, ℓ_2 embeds on disjoint vectors in Y. Hence, we can find disjoint vectors $\{k_{n,i}\}_{n=0,i=1}^{\infty\ \ 2^n}$ in Y, all disjoint from the $g_{n,i}$'s, and a constant D_1 such that if we put $g'_{n,i} = g_{n,i} + k_{n,i}$, then for every $f = \sum_{n,i} a_{n,i}h_{n,i} \in X$, we have

$$D_1^{-1}\|\sum_{n,i} a_{n,i}g'_{n,i}\|_Y \le \max\{\|\sum_{n,i} a_{n,i}g_{n,i}\|_Y, w\|f\|_2\} \le D_1\|\sum_{n,i} a_{n,i}g'_{n,i}\|_Y .$$

This completes the deduction of 6.1 from 6.8.

We begin now the actual *proof of* 6.8. Let X be a r.i. function space on [0,1] and Y be a good lattice of functions on [0,1] such that the Haar system $\{h_{n,i}\}_{n=0,i=1}^{\infty\ \ 2^n}$ (normalized in $L_\infty(0,1)$) is an unconditional basis of X and Y does not contain uniformly isomorphic copies of ℓ_∞^n for all n. As we already pointed out above, this last assumption imposed on Y and the fact that X is assumed to be isomorphic to a subspace of Y imply, by [54], that both X and Y are s-concave for some $s < \infty$. We shall suppose that the s-concavity constants of X and Y are $\le$ a constant $M < \infty$.

Let T be an isomorphism from X into Y and, for $E \in \pi_k$ and $n \ge k \ge 0$, put

$$v_n(E) = \Big(\sum_{h_{n,i} \subset E} |Th_{n,i}|^2\Big)^{1/2} .$$

The function $v_n^2(\cdot)$ can be, of course, extended to an additive function on $\mathcal{B}_k$; for simplicity, we shall denote $v_n([0,1])$ by v_n (recall that π_k denotes the collection of all the intervals $E_{k,i} = [(i-1)2^{-k}, i2^{-k})$; $i = 1,2,\dots,2^k$, $\mathcal{B}_k$ stands for the finite algebra generated by π_k and the notation $\sum_{h_{n,i} \subset E}$ means the sum over all indices $1 \le i \le 2^n$ for which the support of the Haar function $h_{n,i}$ is contained in E). By 6.4, there exist an additive measure Λ on $\mathcal{S} = \bigcup_{k=0}^{\infty} \mathcal{S}_k$ which takes values in the linear space of all measurable functions on $[0,1]$, a sequence $\{\sigma_m\}_{m=1}^{\infty}$ of mutually disjoint finite subsets of the integers and positive reals $\{\alpha_n\}_{n=1}^{\infty}$ with $\sum_{n\in\sigma_m} \alpha_n^2 = 1$ for all m such that the functions

$$w_m(E) = \Big(\sum_{n\in\sigma_m} \alpha_n^2 v_n^2(E)\Big)^{1/2} \; ; \; E \in \mathcal{S}_k \, , \; k \le \min \sigma_n \, , \; m = 1,2,\ldots$$

have the property that, for each $F \in \mathcal{S}$, the sequence $\{w_m^2(F)\}_{m=1}^{\infty}$ (which possibly is not defined for a finite number of indices) converges μ-a.e. to $\Lambda(F)$. Observe also that $w_m(E) \le w_m(F)$ (as functions in Y) whenever both expressions are defined and $E \subseteq F$.

Put $w_m = w_m([0,1])$; $m \ge 1$. It was observed after 6.5 that there exist a subsequence of $\{w_m\}_{m=1}^{\infty}$, which for sake of simplicity is still denoted in the same manner, and subsets $\{C_m\}_{m=1}^{\infty}$ and $\{D_m\}_{m=1}^{\infty}$ of $[0,1]$ so that

$$w_m = \chi_{C_m} w_m + \chi_{D_m} w_m$$

for all m , the sets $\{C_m\}_{m=1}^{\infty}$ are pairwise disjoint and the functions $\{\chi_{D_m} w_m\}_{m=1}^{\infty}$ are equi-integrable in Y . It follows that, for each $F \in \mathcal{S}$, the sequence $\{\chi_{C_m} w_m(F)\}_{m=1}^{\infty}$ converges μ-a.e. to 0 , and, thus, the sequence $\{\chi_{D_m} w_m(F)\}_{m=1}^{\infty}$ converges μ-a.e. to $\sqrt{\Lambda(F)}$. Moreover, the sequence $\{\chi_{D_m} w_m(F)\}_{m=1}^{\infty}$ is equi-integrable in Y for every $F \in \mathcal{S}$, by the monotonicity of $w_m(\cdot)$. It follows that the sequence $\{\chi_{D_m} w_m(F)\}_{m=1}^{\infty}$ converges to $\sqrt{\Lambda(F)}$ in the norm of Y . Also, the sequence $\{|\chi_{D_m} w_m^2(f) - \Lambda(F)|^{1/2}\}_{m=1}^{\infty}$ is equi-integrable in Y and converges μ-a.e. to 0 , and hence, by using the definition of the quasi-norm in the q-concavification $Y_{1/2}$ of Y (introduced prior to this proof) we conclude that, for every $F \in \mathcal{S}$,

$$(*) \qquad \lim_{m\to\infty} \|\chi_{D_m} w_m^2(F) - \Lambda(F)\|_{Y_{1/2}} = 0 \, .$$

Set

$$\Lambda_n = \max\{\Lambda(F);\ F \in \pi_n\} \ ;\ n = 0,1,2,\ldots$$

and notice that $\{\Lambda_n\}_{n=0}^{\infty}$ is a non-increasing sequence of measurable functions (since $\Lambda(\cdot)$ is additive and its values are non-negative functions). Let $\tilde{\Lambda}$ denote the pointwise limit of $\{\Lambda_n\}_{n=0}^{\infty}$. We shall distinguish now between two cases according to the behavior of the function $\tilde{\Lambda}$.

Case I: $\|\tilde{\Lambda}\|_{Y_{1/2}} \neq 0$. We first apply 6.6 for the subset

$$M = \{\chi_{D_m} w_m \ ;\ m = 1,2,\ldots\}$$

of Y and conclude the existence of a lattice of functions Z on $[0,1]$ which satisfies the condition (ii) of 6.6, does not contain uniformly isomorphic copies of ℓ_∞^n for all n and for which

$$Q = \sup_n \|\chi_{D_m} w_m\|_Z < \infty \ .$$

Furthermore, since the exact value of s for which Y is supposed to be s-concave is not really important in this proof we may assume without loss of generality that Z is also s-concave with s-concavity constant $\leq M$. Since the injection of Z into $L_1(0,1)$ is continuous, we can normalize so that it has norm one.

The assumption imposed in Case I implies that there exists a $\delta > 0$ so that

$$\|\max\{\Lambda(F);\ F \in \pi_n\}\|_{Y_{1/2}} > \delta \quad ,$$

for all n. Fix n and observe that, by (*) and the continuity of the maximum of a finite number of elements in $Y_{1/2}$, it follows that there is an integer $m = m(n)$ for which

$$\|\max\{\chi_{D_m} w_m^2(F) \ ;\ F \in \pi_n\}\|_{Y_{1/2}} > \delta$$

or equivalently,

$$\|\max\{\chi_{D_m} w_m(F)\ ;\ F \in \pi_n\}\|_Y > \sqrt{\delta}\ .$$

Let $\{G_i\}_{i=1}^{2^n}$ be a family of pairwise disjoint measurable subsets of the interval $[0,1]$ so that $\max\{\chi_{D_m} w_m(F);\ F \in \pi_n\} = \sum_{i=1}^{2^n} \chi_{D_m \cap G_i} w_m(E_{n,i})$.

Let σ_m (with $m = m(n)$) be the subset of integers and $\{\alpha_k\}_{k\in\sigma_m}$ the positive reals used in the definition of $w_m(\cdot)$. Set

$$y(u) = \sum_{i=1}^{2^n} \chi_{D_m \cap G_i} \sum_{k\in\sigma_m} \alpha_k \sum_{h_{k,j}\subset E_{n,i}} r_{k,j}(u) Th_{k,j};\ u \in [0,1]$$

and notice that, by Khintchine's inequality, we have

$$\int_0^1 \|y(u)\|_Y dy \geq \left\|\int_0^1 |y(u)|\, dy\right\|_Y \geq$$

$$\geq A_1 \Big\|\Big(\sum_{i=1}^{2^n} \chi_{D_m\cap G_i} \sum_{k\in\sigma_m} \alpha_k^2 \sum_{h_{k,j}\subset E_{n,i}} |Th_{k,j}|^2\Big)^{1/2}\Big\|_Y =$$

$$= A_1 \Big\|\Big(\sum_{i=1}^{2^n} \chi_{D_m\cap G_i} \sum_{k\in\sigma_m} \alpha_k^2 v_k^2(E_{n,i})\Big)^{1/2}\Big\|_Y = A_1 \Big\|\Big(\sum_{i=1}^{2^n} \chi_{D_m\cap G_i} w_m^2(E_{n,i})\Big)^{1/2}\Big\|_Y =$$

$$= A_1 \Big\|\sum_{i=1}^{2^n} \chi_{D_m\cap G_i} w_m(E_{n,i})\Big\|_Y = A_1 \|\max\{\chi_{D_m} w_m(F);\ F \in \pi_n\}\|_Y > \sqrt{\delta/2}\ .$$

But, by a result of J.P. Kahane [39], there exists a real η ; $0 < \eta < 1$, such that if $\{x_i\}_{i=1}^{\ell}$ are arbitrary vectors in a general Banach space and $\int_0^1 \|\sum_{i=1}^{\ell} r_i(u) x_i\| du \geq 1$ then

$$\mu(\{u \in [0,1];\ \|\sum_{i=1}^{\ell} r_i(u) x_i\| \geq \eta\}) \geq \eta\ .$$

It therefore follows from the previous computation that

$$\mu(\{u \in [0,1];\ \|y(u)\|_Y \geq \eta\sqrt{\delta/2}\}) \geq \eta\ .$$

On the other hand, if we put

$$y_i(u) = \chi_{D_m} \sum_{k\in\sigma_m} \alpha_k \sum_{h_{k,j}\subset E_{n,i}} r_{k,j}(u)Th_{k,j};\ u \in [0,1],\ i = 1,2,\ldots,2^n,$$

then, by using the s-concavity of the space Z introduced above and Khintchine's inequality in $L_s([0,1] \times [0,1])$, we get that

$$\int_0^1 \| \max_{1\le i\le 2^n} |y_i(u)| \|_Z du \le \int_0^1 \| (\int_0^1 | \sum_{i=1}^{2^n} r_i(t)y_i(u)|^s dt)^{1/s} \|_Z du \le$$

$$\le (\int_0^1 \| (\int_0^1 | \sum_{i=1}^{2^n} r_i(t)y_i(u)|^s dt)^{1/s} \|_Z^s du)^{1/s} \le M \| (\int_0^1\int_0^1 | \sum_{i=1}^{2^n} r_i(t)y_i(u)|^s dt du)^{1/s} \|_Z \le$$

$$\le B_s M \| (\sum_{i=1}^{2^n} \chi_{D_m} \sum_{k\in\sigma_m} \alpha_k^2 \sum_{h_{k,j}\subset E_{n,i}} |Th_{k,j}|^2)^{1/2} \|_Z = B_s M \| (\sum_{i=1}^{2^n} \chi_{D_m} w_m^2(E_{n,i}))^{1/2} \|_Z =$$

$$= B_s M \| \chi_{D_m} w_m \|_Z \le B_s MQ \ .$$

It follows that

$$\mu(\{u \in [0,1];\ \| \max_{1\le i\le 2^n} |y_i(u)| \|_Z \ge 2B_s MQ/\eta\}) \le \eta/2 \ ;$$

i.e., that

$$\mu(\{u \in [0,1];\ \| \max_{1\le i\le 2^n} |y_i(u)| \|_Z < 2B_s MQ/\eta\}) > 1 - \eta/2 \ .$$

In conclusion, there exists $u \in [0,1]$ or, equivalently, a choice of signs $\{\epsilon_{k,j}\}_{j=1,k\in\sigma_m}^{2^k}$ so that if we set

$$\tilde{h}_{n,i} = \sum_{k\in\sigma_m} \alpha_k \sum_{h_{k,j}\subset E_{n,i}} \epsilon_{k,j} h_{k,j} \ ;\ i = 1,2,\ldots,2^n$$

then

$$\| \max_{1\le i\le 2^n} |\chi_{D_m} T\tilde{h}_{n,i}| \|_Y \ge \| \sum_{i=1}^{2^n} \chi_{D_m\cap G_i} T\tilde{h}_{n,i} \|_Y \ge \eta\sqrt{\delta/2}$$

and

$$\| \max_{1\le i\le 2^n} |\chi_{D_m} T\tilde{h}_{n,i}| \|_Z < 2B_s MQ/\eta \ .$$

The above construction of the functions $\{\tilde{h}_{n,i}\}_{i=1}^{2^n}$ was performed under the assumption that n is fixed by using the subset of integers σ_m with $m = m(n)$. Obviously, this construction can be made for every $n \ge 0$ so that the sets $\{\sigma_{m(n)}\}_{n=0}^{\infty}$ are pairwise disjoint. In this way we obtain that the corresponding functions $\{\tilde{h}_{n,i}\}_{n=0,i=1}^{\infty \ \ 2^n}$ form a gaussian Haar system in X which, by the s-concavity of X and 6.2, is an unconditional basic sequence in X equivalent to the usual Haar basis $\{h_{n,i}\}_{n=0,i=1}^{\infty \ \ 2^n}$ of X. Hence, it is possible to define a new isomorphism $\tilde{T}$ from X into Y by putting

$$\tilde{T}h_{n,i} = T\tilde{h}_{n,i}$$

for all $1 \le i \le 2^n$ and $n = 0,1,2,\ldots$. The operator $\tilde{T}$ satisfies $\| \max_{1\le i\le 2^n} |\chi_{D_m} \tilde{T}h_{n,i}| \|_Y \ge \eta\sqrt{\delta/2}$ and $\| \max_{1\le i\le 2^n} |\chi_{D_m} \tilde{T}h_{n,i}| \|_Z < 2B_s MQ/\eta$, for all $n \ge 0$; thus, by 6.6 (ii), there exists a constant $R < \infty$ so that if we set $H_n = \{t \in [0,1] \ ; \ \max_{1\le i\le 2^n} |\tilde{T}h_{n,i}| > R\}$ and $\tilde{H}_n = [0,1] \sim H_n$, then, for each $n \ge 0$, we have

$$\|\chi_{D_m \cap H_n} \max_{1\le i\le 2^n} |\tilde{T}h_{n,i}| \|_Y \le \eta\sqrt{\delta/2}\ /2 \ .$$

Thus,

$$\|\chi_{\tilde{H}_n} \max_{1\le i\le 2^n} |\tilde{T}h_{n,i}|\|_Y \ge \|\chi_{D_m \cap \tilde{H}_n} \max_{1\le i\le 2^n} |\tilde{T}h_{n,i}|\|_Y \ge \eta\sqrt{\delta/2}\ /2 \ ,$$

for all $n \ge 0$. This gives (i) in 6.8.

Case II: $\|\tilde{\Lambda}\|_{Y_{1/2}} = 0$. In this case $\inf \Lambda_n$ in Y is equal to 0 , and, thus, by [58] and the fact that Y contains no isomorphic copies of c_0 , we get that

$$\lim_n \|\Lambda_n\|_{Y_{1/2}} = 0 \quad .$$

We first assume that $\Lambda([0,1)) \neq 0$. Such a situation occurs, for example, if $\Lambda(E) = \mu(E)\Lambda$ for all $E \in \mathcal{S}$, where Λ is a fixed element of Y . We shall actually show that on some σ-field B of Lebesgue measurable sets the measure $\Lambda(\cdot)$ is indeed of this type. This σ-field B will be generated by a tree $\{F_{k,i}\}_{k=0,i=1}^{\infty\ \ 2^k}$ which is constructed in such a manner that $\Lambda(F_{k+1,2i})$ and $\Lambda(F_{k+1,2i-1})$ are almost equal to $\Lambda(F_{k,i})/2$ for every $1 \leq i \leq 2^k$ and $k \geq 0$.

Put $\gamma = \|\Lambda([0,1))\|_{Y_{1/2}} > 0$ and $F_{0,1} = [0,1)$. In order to construct the sets $F_{1,1}$ and $F_{1,2}$ we choose an integer n so that

$$\|\Lambda_n\|_{Y_{1/2}} = \|\max_{1 \leq i \leq 2^n} \Lambda(E_{n,i})\|_{Y_{1/2}}$$

is very small with respect to γ . (The precise requirement will be described in the sequel.) Then, by 6.7, one can find a subset $\sigma \subset \{1,2,\ldots,2^n\}$ consisting of exactly 2^{n-1} integers such that

$$\|\sum_{i\in\sigma} \Lambda(E_{n,i}) - \sum_{i\notin\sigma} \Lambda(E_{n,i})\|_{Y_{1/2}}$$

is very small compared with $\gamma = \|\sum_{i=1}^{2^n} \Lambda(E_{n,i})\|_{Y_{1/2}}$. Thus, if we put $F_{1,1} = \bigcup_{i\in\sigma} E_{n,i}$ and $F_{1,2} = \bigcup_{i\notin\sigma} E_{n,i}$ then $\|\Lambda(F_{1,1}) - \Lambda(F_{1,2})\|_{Y_{1/2}}$ will be very small which means that both $\Lambda(F_{1,1})$ and $\Lambda(F_{1,2})$ are almost equal to $(\Lambda(F_{1,1}) + \Lambda(F_{1,2}))/2 = \Lambda(F_{0,1})/2$. This procedure can be continued inductively in order to produce a tree $\{F_{k,i}\}_{i=1,k=0}^{2^k\ \ \infty}$ of subsets in $\mathcal{S}$ such that

$$\sum_{k=0}^{\infty} \sum_{i=1}^{2^k} \|\Lambda(F_{k,i})/\mu(F_{k,i}) - \Lambda(F_{k+1,2i})/\mu(F_{k+1,2i})\|_{Y_{1/2}}^{1/2} \leq \gamma^{1/2}/8 \ .$$

By (*), one can find a strictly increasing sequence $\{m(k,i)\}_{k=0,\, i=1}^{\infty \quad 2^k}$ for which $w_{m(k,i)}(F_{k,i})$ is well-defined whenever $1 \leq i \leq 2^k$, $k \geq 0$,

$$\sum_{k=0}^{\infty} \sum_{i=1}^{2^k} \|\chi_{D_{m(k,i)}} w^2_{m(k,i)}(F_{k,i})/\mu(F_{k,i}) -$$

$$- \chi_{D_{m(k+1,2i)}} w^2_{m(k+1,2i)}(F_{k+1,2i})/\mu(F_{k+1,2i})\|_{Y_{1/2}}^{1/2} \leq \gamma^{1/2}/4$$

and

$$\|\chi_{D_{m(0,1)}} w^2_{m(0,1)} - \Lambda([0,1))\|_{Y_{1/2}} \leq \gamma/16 \ .$$

Set

$$\Phi = |\chi_{D_{m(0,1)}} w^2_{m(0,1)} - \Lambda([0,1))| +$$

$$+ \sum_{k=0}^{\infty} \sum_{i=1}^{2^k} |\chi_{D_{m(k,i)}} w^2_{m(k,i)}(F_{k,i})/\mu(F_{k,i}) -$$

$$- \chi_{D_{m(k+1,2i)}} w^2_{m(k+1,2i)}(F_{k+1,2i})/\mu(F_{k+1,2i})|$$

and notice that

$$\|\Phi\|_{Y_{1/2}} \leq \gamma/4$$

since $Y_{1/2}$ is a $\frac{1}{2}$-normed space. Moreover, since we clearly have

$$|\Lambda([0,1)) - \chi_{D_{m(k,i)}} w^2_{m(k,i)}(F_{k,i})/\mu(F_{k,i})| \leq \Phi \quad ,$$

it follows that

$$\Phi_1 \leq \chi_{D_{m(k,i)}} w^2_{m(k,i)}(F_{k,i})/\mu(F_{k,i}) \leq \Phi_2 \quad ,$$

for every $1 \le i \le 2^k$ and $k \ge 0$, where $\Phi_1 = (\Lambda([0,1)) - \Phi)_+$ and $\Phi_2 = \Phi + \Lambda([0,1))$. Note that $\|\Phi_1\|_{Y_{1/2}} > 0$ since

$$\|\Phi\|_{Y_{1/2}} \le \|\Lambda([0,1))\|_{Y_{1/2}}/4 .$$

Let us construct now a gaussian Haar system relative to the tree $\{F_{k,i}\}_{k=0,i=1}^{\infty \;\; 2^k}$ by putting

$$\tilde{h}_{k,i} = \sum_{n\in\sigma_{m(k,i)}} \alpha_n \sum_{h_{n,j}\subset F_{k,i}} h_{n,j} \; ; \; i = 1,2,\dots,2^k \; , \; k = 0,1,2,\dots$$

By the remark following 6.2, it follows that $\{\tilde{h}_{k,i}\}_{k=0,i=1}^{\infty \;\; 2^k}$ is an unconditional basic sequence in X which is equivalent to the ordinary Haar basis $\{h_{n,i}\}_{n=0,i=1}^{\infty \;\; 2^n}$ of X. By using (*) from the Introduction we conclude that there exists a constant $D_1 < \infty$ so that, for every choice of scalars $\{a_{k,i}\}_{k=0,i=1}^{\infty \;\; 2^k}$ we have

$$D_1^{-1}\Big\|\sum_{k=0}^{\infty}\sum_{i=1}^{2^k} a_{k,i}h_{k,i}\Big\|_X \le \Big\|\Big(\sum_{k=0}^{\infty}\sum_{i=1}^{2^k} a_{k,i}^2 \sum_{n\in\sigma_{m(k,i)}} \alpha_n^2 \sum_{h_{n,j}\subset F_{k,i}} |Th_{n,j}|^2\Big)^{1/2}\Big\|_Y \le$$

$$\le D_1\Big\|\sum_{k=0}^{\infty}\sum_{i=1}^{2^k} a_{k,i}h_{k,i}\Big\|_X \; ,$$

or, equivalently,

$$D_1^{-1}\Big\|\sum_{k=0}^{\infty}\sum_{i=1}^{2^k} a_{k,i}h_{k,i}\Big\|_X \le \Big\|\Big(\sum_{k=0}^{\infty}\sum_{i=1}^{2^k} a_{k,i}^2 w_{m(k,i)}^2(F_{k,i})\Big)^{1/2}\Big\|_Y \le$$

$$\le D_1\Big\|\sum_{k=0}^{\infty}\sum_{i=1}^{2^k} a_{k,i}h_{k,i}\Big\|_X \; .$$

Put

$$g_{k,i} = \chi_{C_{m(k,i)}} w_{m(k,i)}(F_{k,i}) \; ; \; i = 1,2,\dots,2^k, \; k = 0,1,2,\dots$$

and notice that these functions are mutually disjoint since $\{m(k,i)\}_{k=0,i=1}^{\infty \; 2^k}$ is a strictly increasing sequence of integers. Therefore,

$$\sum_{k=0}^{\infty} \sum_{i=1}^{2^k} a_{k,i} g_{k,i} = \left(\sum_{k=0}^{\infty} \sum_{i=1}^{2^k} a_{k,i}^2 \chi_{C_{m(k,i)}} w_{m(k,i)}^2 (F_{k,i}) \right)^{1/2} ,$$

for every choice of $\{a_{k,i}\}_{k=0,i=1}^{\infty \; 2^k}$. On the other hand,

$$\Phi_1 \sum_{k=0}^{\infty} \sum_{i=1}^{2^k} a_{k,i}^2 \mu(F_{k,i}) \leq \sum_{k=0}^{\infty} \sum_{i=1}^{2^k} a_{k,i}^2 \chi_{D_{m(k,i)}} w_{m(k,i)}^2 (F_{k,i}) \leq$$

$$\leq \Phi_2 \sum_{k=0}^{\infty} \sum_{i=1}^{2^k} a_{k,i}^2 \mu(F_{k,i})$$

and

$$\left\| \sum_{k=0}^{\infty} \sum_{i=1}^{2^k} a_{k,i} h_{k,i} \right\|_2 = \left(\sum_{k=0}^{\infty} \sum_{i=1}^{2^k} a_{k,i}^2 \mu(F_{k,i}) \right)^{1/2} ,$$

for every choice of $\{a_{k,i}\}_{k=0,i=1}^{\infty \; 2^k}$. Combining these facts with the identity $w_m = \chi_{C_m} w_m + \chi_{D_m} w_m$ for all m, we get that there is a constant $D < \infty$ so that

$$D^{-1} \left\| \sum_{k=0}^{\infty} \sum_{i=1}^{2^k} a_{k,i} h_{k,i} \right\|_X \leq \max\left\{ \left\| \sum_{k=0}^{\infty} \sum_{i=1}^{2^k} a_{k,i} g_{k,i} \right\|_Y , \left\| \sum_{k=0}^{\infty} \sum_{i=1}^{2^k} a_{k,i} h_{k,i} \right\|_2 \right\} \leq$$

$$\leq D \left\| \sum_{k=0}^{\infty} \sum_{i=1}^{2^k} a_{k,i} h_{k,i} \right\|_X .$$

This, of course, completes the proof of Case II under the additional assumption that $\|\Lambda([0,1))\|_{Y_{1/2}} > 0$ since it leads to the alternative (ii) of 6.8 with $w = 1$.

If $\|\Lambda([0,1))\|_{Y_{1/2}} = 0$, then, by (*), we get that

$$\lim_{m \to \infty} \|\chi_{D_m} w_m\|_Y = 0 .$$

Thus, by passing to a subsequence if necessary, we can assume with no loss of generality that $\{w_m\}_{m=1}^{\infty}$ is equivalent to the disjointly supported sequence $\{\chi_{C_m} w_m\}_{m=1}^{\infty}$. Then, by defining a gaussian Haar system and functions $\{g_{k,i}\}_{k=0,i=1}^{\infty\ \ 2^k}$ as in the case when $\|\Lambda([0,1))\|_{Y_{1/2}} > 0$, we can easily check that a series $\sum_{k=0}^{\infty}\sum_{i=1}^{2^k} a_{k,i}h_{k,i}$ converges in X if and only if $\sum_{k=0}^{\infty}\sum_{i=1}^{2^k} a_{k,i}g_{k,i}$ converges in Y. This leads to alternative (ii) of 6.8 with $w = 0$.

□

The arguments used in 6.8 essentially prove the following proposition, a simple case of which (6.11) will be applied in section 7.

Proposition 6.9: Let X *be a r.i. function space on* $[0,1]$ *and* Y *a good lattice of functions on* $[0,1]$. *Assume that the Haar system* $\{h_{n,i}\}_{n=0,i=1}^{\infty\ \ 2^n}$ *is an unconditional basis for* X, *and that* Y *does not contain uniformly isomorphic copies of* ℓ_∞^n. *If* X *embeds into* Y, *one can choose an embedding* U *of the form* $U_1 + U_2$, *where the vectors* $\{U_1 h_{n,i}\}_{n=0,i=1}^{\infty\ \ 2^n}$ *are disjoint in* Y *and where the sequence*

$$s_n = \Big(\sum_{i=1}^{2^n} (U_2 h_{n,i})^2\Big)^{1/2}$$

is equi-integrable in Y'.

Proof: Let T be an embedding from X into Y. Following the proof and the notations of 6.8, there exist an additive measure Λ on $\mathcal{S} = \bigcup_{k=0}^{\infty} \mathcal{S}_k$ which takes values in the space of measurable functions on $[0,1]$, a sequence $\{\sigma_m\}_{m=1}^{\infty}$ of mutually disjoint finite subsets of the integers and positive reals $\{\alpha_n\}_{n=1}^{\infty}$ with $\sum_{n\in\sigma_m} \alpha_n^2 = 1$, for all m such that the functions

$$w_m(E) = \Big(\sum_{n\in\sigma_m} \alpha_n^2 v_n^2(E)\Big)^{1/2}, \quad E \in \mathcal{S}_k, \ k \le \min \sigma_m, \ m = 1,2,\ldots$$

have the property that the sequence $w_m^2(E)$ (which is eventually well-defined) converges μ-a.e. to $\Lambda(E)$.

Furthermore, passing to a subsequence (still denoted by $\{w_m\}_{m=1}^\infty$ for simplicity) we can find measurable subsets $\{A_m\}_{m=1}^\infty$ of $[0,1]$ so that

$$w_m = \chi_{A_m} w_m + \chi_{B_m} w_m \text{ , where } B_m = \sim A_m$$

for all m, such that the sets $\{A_m\}_{m=1}^\infty$ are pairwise disjoint and the sequence $\{\chi_{B_m} w_m\}_{m=1}^\infty$ is equi-integrable in Y.

By 6.6 we may construct a lattice of functions Z on $[0,1]$ so that the sequence $\{\chi_{B_m} w_m\}_{m=1}^\infty$ is equi-integrable in Z and the unit ball B_Z of Z is equi-integrable in Y. Since Y does not contain uniformly isomorphic copies of ℓ_∞^n, we may assume Y and Z to be s-concave for some $s < \infty$, with s-concavity constant $\leq M$. It follows that Z is a good lattice on $[0,1]$. Since for every $E \in \mathcal{E}$, the sequence $\{\chi_{B_m} w_m(E)\}_{m=1}^\infty$ is equi-integrable in Z and converges μ-a.e. to $\sqrt{\Lambda(E)}$, it follows that $\sqrt{\Lambda(0,1)}$ belongs to Z, and then $\{|\chi_{B_m} w_m^2(E) - \Lambda(E)|^{1/2}\}_{m=1}^\infty$ is equi-integrable in Z and converges μ-a.e. to 0, which yields

$$\lim_{m\to\infty} \|\Lambda(E) - \chi_{B_m} w_m^2(E)\|_{Z_{1/2}} = 0 .$$

Let us denote by $\{E_{k,i}\}_{k=0,i=1}^{\infty \quad 2^k}$ the fundamental tree of sets in $[0,1]$. We may find a strictly increasing sequence $\{m(k,i)\}_{k=0,i=1}^{\infty \quad 2^k}$ for which $w_{m(k,i)}(E_{k,i})$ is well-defined and such that

$$\|\chi_{B_{m(k,i)}} w_{m(k,i)}^2(E_{k,i}) - \Lambda(E_{k,i})\|_{Z_{1/2}}^{1/2} \leq 2^{-k} ,$$

for every $i = 1,\dots,2^k$ and $k = 0,1,\dots$. Thus,

$$\|\sum_{i=1}^{2^k} \chi_{B_{m(k,i)}} w_{m(k,i)}^2(E_{k,i}) - \Lambda(0,1)\|_{Z_{1/2}} \leq 1 .$$

This shows that the sequence $u_k = (\sum_{i=1}^{2^k} \chi_{B_{m(k,i)}} w^2_{m(k,i)}(E_{k,i}))^{1/2}$,

$K = 0,1,\ldots$ is bounded in Z .

We need now the following lemma, the proof of which is postponed until after the proof of 6.9.

Lemma 6.10: Let Z *be a s-concave lattice, with s-concavity constant* M *, and* $\{z_\alpha\}_{\alpha\in A}$ *a finite family of elements of* Z *. If* $A = \bigcup_{k=1}^{K} A_k$ *is a finite partition of* A *, we may select signs* $\{\epsilon_\alpha\}_{\alpha\in A}$ *such that*

$$\|\{\sum_{k=1}^{K} |\sum_{\alpha\in A_k} \epsilon_\alpha z_\alpha|^2\}^{1/2}\|_Z \le \sqrt{2}\, B_s M \|(\sum_{a\in A} z_\alpha^2)^{1/2}\|_Z .$$

We continue the proof of 6.9. If we recall that

$$u_k = (\sum_{i=1}^{2^k} \chi_{B_{m(k,i)}} \sum_{n\in\sigma_{m(k,i)}} \alpha_n^2 v_n^2(E_{k,i}))^{1/2} ,$$

$$= (\sum_{i=1}^{2^k} \chi_{B_{m(k,i)}} \sum_{n\in\sigma_{m(k,i)}} \alpha_n^2 \sum_{h_{n,\ell}\subset E_{k,i}} |Th_{n,\ell}|^2)^{1/2}$$

we may find according to 6.10 signs $\{\epsilon_{n,\ell}\}_{n=0,\ell=1}^{\infty\ \ 2^n}$ such that

$$s_k = (\sum_{i=1}^{2^k} | \sum_{n\in\sigma_{m(k,i)}} \alpha_n \chi_{B_{m(k,i)}} \sum_{h_{n,\ell}\subset E_{k,i}} \epsilon_{n,\ell} Th_{n,\ell}|^2)^{1/2} ;$$

$k = 0,1,\ldots,$ is bounded in Z and therefore equi-integrable in Y . Define a gaussian Haar system $\tilde{h}_{k,i}$ by

$$\tilde{h}_{k,i} = \sum_{n\in\sigma_{m(k,i)}} \alpha_n \sum_{h_{n,\ell}\subset E_{k,i}} \epsilon_{n,\ell} h_{n,\ell};\ i = 1,\ldots,2^k,\ k = 0,1,\ldots$$

We can define a new embedding U of X into Y by

$$Uh_{k,i} = T\tilde{h}_{k,i}\ ;\ i = 1,\ldots,2^k\ ,\ k = 0,1,\ldots\ .$$

Define now U_1 by $U_1 h_{k,i} = \chi_{A_{m(k,i)}} \cdot Uh_{k,i}$; $i = 1,\ldots,2^k$, $k = 0,1,\ldots$.

As announced, the images $\{U_1 h_{k,i}\}_{i=1,k=0}^{2^k \ \infty}$ are pairwise disjoint in Y . The fact that U_1 is bounded follows from a "diagonal principle" that we have already used several times:

$$\|U_1(\sum_{k,i} \lambda_{k,i} h_{k,i})\|_Y \leq \|\max_{k,i} |\lambda_{k,i} Uh_{k,i}|\ \|_Y$$

$$\leq \int_0^1 \|\sum_{k,i} r_{k,i}(u)\lambda_{k,i} Uh_{k,i}\|_Y du \sim \|U(\sum_{k,i} \lambda_{k,i} h_{k,i})\|_Y \sim$$

$$\sim \|\sum_{k,i} \lambda_{k,i} h_{k,i}\|_X$$

since $\{Uh_{k,i}\}_{i=1,k=0}^{2^k \ \infty}$ is unconditional in Y and U is an isomorphic embedding. If we set $U_2 = U - U_1$, U_2 is bounded and

$$s_k = (\sum_{i=1}^{2^k} |U_2 h_{k,i}|^2)^{1/2} = (\sum_{i=1}^{2^k} |\chi_{B_{m(k,i)}} \tilde{T}h_{k,i}|^2)^{1/2}$$

is equi-integrable in Y according to our construction. □

Proof of 6.10: The notations being as in the statement of 6.10, we have:

$$\int_0^1 \|(\sum_{k=1}^{K} |\sum_{\alpha\in A_k} r_\alpha(u) z_\alpha|^2)^{1/2}\|_Z du$$

$$\leq \sqrt{2} \int_0^1 \| \int_0^1 |\sum_{k=1}^{K} r_k(v) \sum_{\alpha\in A_k} r_\alpha(u) a_\alpha| dv\|_Z du$$

$$\leq \sqrt{2} \int_0^1 \int_0^1 \|\sum_{k=1}^{K} r_k(v) \sum_{\alpha\in A_k} r_\alpha(u) z_\alpha\|_Z du dv = \sqrt{2} \int_0^1 \|\sum_{a\in A} r_\alpha(u) z_\alpha\|_Z du$$

$$\leq \sqrt{2}\, M\|(\int_0^1 |\sum_{\alpha\in A} r_\alpha(u) z_\alpha|^s du)^{1/s}\|_Z \leq \sqrt{2}\, B_s M\|(\sum_{\alpha\in A} a_\alpha^2)^{1/2}\|_Z \ .$$

□

We prove now an easier version of 6.9, to be used in section 7.

Proposition 6.11: Let X *be a r.i. function space on* [0,1] *and* Y *a good lattice of functions on* [0,1]. *Assume that the Haar system* $\{h_{n,i}\}_{n=0,i=1}^{\infty\ \ 2^n}$ *is an unconditional basis for* X *and that* Y *is* p*-convex for some* $p > 2$ *and does not contain uniformly isomorphic copies of* ℓ_∞^n . *If* T *is an isomorphism from* X *into* Y , *then there is another isomorphism* U *such that the sequence* $s_k = (\sum_{i=1}^{2^k} (Uh_{k,i})^2)^{1/2}$, $k = 0,1,\ldots,$ *is equi-integrable in* Y *and* $UX \subset TX$.

Proof: Set

$$v_n = (\sum_{i=1}^{2^n} (Th_{n,i})^2)^{1/2} .$$

Passing to a subsequence if necessary, we can write $v_n = \chi_{A_n} v_n + \chi_{\sim A_n} v_n$, where the sets $\{A_n\}_{n=1}^\infty$ are pairwise disjoint and the sequence $\{\chi_{\sim A_n} v_n\}_{n=1}^\infty$ is equi-integrable in Y .

The space Y is 2-convex; therefore, by 6.6 we can build a lattice Z , 2-convex and s-concave for some $s < \infty$, such that the unit ball B_Z is equi-integrable in Y and the sequence $\{\chi_{\sim A_n} v_n\}_{n=1}^\infty$ is bounded in Z .

The very simple idea to eliminate the contribution of the sets $\{A_n\}_{n=1}^\infty$ is to form blocks of the form $N^{-1/2}(\sum_{j=1}^N r_j)$. Since Y is p-convex with some constant M and the sets $\{A_j\}_{j=1}^\infty$ are disjoint we get

$$\|\sum_{j=1}^N N^{-1/2}\chi_{A_j} Tr_j\|_Y \le M \cdot N^{-1/2}(\sum_{j=1}^N \|Tr_j\|^p)^{1/p} \le M\|T\|N^{1/p-1/2} ,$$

which is as small as we like for big N . It is then clear that we can construct a gaussian Haar system $\tilde{h}_{k,i}$ by putting

$$\tilde{h}_{k,i} = N_k^{-1/2} \sum_{n\in\sigma_k} \sum_{h_{n,\ell}\subset E_{k,i}} h_{n,\ell}\epsilon_{n,\ell}; \quad i = 1,\ldots,2^k; \ k = 0,1,\ldots,$$

where $N_k = \overline{\overline{\sigma}}_k$, and $\{\sigma_k\}_{k=0}^\infty$ are disjoint finite subsets of N , in such a way that

$$(\%)\qquad \|N_k^{-1/2} \sum_{n\in\sigma_k} \chi_{A_n} \sum_{h_{n,\ell}\subset E_{k,i}} T(\epsilon_{n,\ell} h_{n,\ell})\|_Y \le 4^{-k} .$$

On the other hand, since Z is 2-convex with some constant M_1, we have

$$\|(\sum_{i=1}^{2^k} N_k^{-1} \sum_{n\in\sigma_k} \sum_{h_{n,\ell}\subset E_{k,i}} \underset{\sim}{\chi}_{A_n} (T(\epsilon_{n,\ell} h_{n,\ell}))^2)^{1/2}\|_Z$$

$$= \|(N_k^{-1} \sum_{n\in\sigma_k} \underset{\sim}{\chi}_{A_n} v_n^2)^{1/2}\|_Z \le$$

$$\le M(\sum_{n\in\sigma_k} N_k^{-1} \|\underset{\sim}{\chi}_{A_n} v_n\|_Z^2)^{1/2} \le M_1 \sup_n \|\underset{\sim}{\chi}_{A_n} v_n\|_Z .$$

It follows from lemma 6.10 that we can find signs $\{\epsilon_{n,\ell}\}_{\ell=1,n=0}^{2^n\ \ \infty}$ such that

$$\sup_k \|(\sum_{i=1}^{2^k} |\sum_{n\in\sigma_k} N_k^{-1/2} \sum_{h_{n,\ell}\subset E_{k,i}} \underset{\sim}{\chi}_{A_n} T(\epsilon_{n,\ell} h_{n,\ell})|^2)^{1/2}\|_Z < \infty .$$

Since $\{\tilde{h}_{k,i}\}_{k=0,i=1}^{\infty\ \ 2^k}$ is equivalent to the Haar system $\{h_{k,i}\}_{k=0,i=1}^{\infty\ \ 2^k}$ by 6.2, we can define a new isomorphic embedding U from X into Y by setting

$$Uh_{k,i} = T\tilde{h}_{k,i} .$$

We have

$$(\sum_{i=1}^{2^k} (Uh_{k,i})^2)^{1/2} \le (\sum_{i=1}^{2^k} |\sum_{n\in\sigma_k} N_k^{-1/2} \sum_{h_{n,\ell}\subset E_{k,i}} \underset{\sim}{\chi}_{A_n} T(\epsilon_{n,\ell} h_{n,\ell})|^2)^{1/2}$$

$$+ (\sum_{i=1}^{2^k} |\sum_{n\in\sigma_k} N_k^{-1/2} \sum_{h_{n,\ell}\subset E_{k,i}} \chi_{A_n} T(\epsilon_{n,\ell} h_{n,\ell})|^2)^{1/2} .$$

The first term forms, when k varies, an equi-integrable sequence in Y since it is bounded in Z. The second term clearly forms an equi-integrable sequence in Y, since this sequence is dominated by the element of Y defined by

$$y = \sum_{k=0}^{\infty} \sum_{i=1}^{2^k} \Big| \sum_{n \in \sigma_k} N_k^{-1/2} \sum_{h_{n,\ell} \subset E_{k,i}} \chi_{A_n} T(\epsilon_{n,\ell} h_{n,\ell}) \Big|$$

(this is convergent because of (%).)

□

We complete this section by applying the construction in 6.8 to the problem of uniqueness of r.i. structure for super reflexive r.i. function spaces on $[0,1]$.

Theorem 6.12: Let Y *be a super-reflexive r.i. function space on* $[0,1]$. *If* X *is a r.i. function space on* $[0,1]$ *which embeds into* Y *as a complemented subspace,* X^* *is not* $L_2(0,1)$ *up to an equivalent renorming, and the Haar basis for* X^* *is not equivalent to a disjoint sequence in* Y^*, *then* $X^* = Y^*$, *up to an equivalent renorming.*

Proof: Let T be an isomorphism from X into Y so that TX is a complemented subspace of Y. We follow exactly the notation of the proof of 6.8.

Since TX is complemented in Y, there is a sequence $\{f_{n,i}\}_{n=0,i=1}^{\infty \quad 2^n}$ in Y^* which is, say, M-equivalent to the Haar basis for X^* and which satisfies

$$\int f_{n,i} T h_{m,j} d\mu = \begin{cases} 2^{-n}, & \text{if } (n,i) = (m,j) \\ 0, & \text{otherwise} \end{cases} .$$

We build v'_n, w'_n from the family $\{f_{n,i}\}_{n=0,i=1}^{\infty \quad 2^n}$ in the same way as v_n, w_n are built from $\{Th_{n,i}\}_{n=0,i=1}^{\infty \quad 2^n}$ on page 146. That is, for $E \in \mathcal{S}_k$, $n \geq k \geq 0$,

$$v'_n(E) = \Big(\sum_{h_{n,i} \subset E} |f_{n,i}|^2 \Big)^{1/2}$$

and for $E \in \mathcal{S}_k$, $k \leq \min \sigma_m$, $m = 1,2,\ldots,$

$$w'_m(E) = \Big(\sum_{n\in\sigma_m} \alpha_n^2 v'_n(E)^2\Big)^{1/2} ,$$

where the σ_m's and α_n's are chosen as on page 146 to make $\{w_m^2(F)\}_{m=1}^\infty$ converge μ-a.e. to $\Lambda(F)$ for all $F \in \bigcup_{k=1}^\infty \mathcal{S}_k$. (We could also demand that $\{w'_m(F)\}_{m=1}^\infty$ converge μ-a.e., but this is not needed.) As in the middle of page 146, we can assume that there are disjoint sets $\{C_m\}_{m=1}^\infty$ so that (setting $D_m = [0,1] \sim C_m$), $\{\chi_{D_m} w_m\}_{m=1}^\infty$ is equi-integrable in Y.

By 6.8 we have $C\|f\|_{X^*} \geq \|f\|_{Y^*}$ for some constant C and all $f \in X^*$, which is the same as $C\|f\|_Y \geq \|f\|_X$ for all $f \in Y$. Therefore, if case I of 6.8 applies for our embedding T of X into Y, then $X = Y$ up to an equivalent renorming. So assume (see page 151) that $\|\tilde{\Lambda}\|_{Y_{1/2}} = 0$. We now break the proof into the two subcases considered under case II of 6.8.

Case (i). $C\|f\|_X \geq \|f\|_2$ *for some constant* C *and all* $f \in X$.

Case (ii). Case (i) does not occur.

We treat the easier Case (ii) first. We then have from the proof of 6.8 that $\Lambda[0,1) = 0$ and $\|\chi_{D_m} w_m\|_Y \to 0$ as $m \to 0$ (see the bottom of page 154) and hence we can choose the sequence $\{m(k,i)\}_{k=0,i=1}^{\infty \quad 2^k}$ to satisfy

($) $$\|\chi_{D_{m(k,i)}} w_{m(k,i)}\|_Y \leq \delta_{k,i}$$

for any preassigned sequence $\{\delta_{k,i}\}_{k=0,i=1}^{\infty \quad 2^k}$ of positive numbers. (Since $\Lambda[0,1) = 0$, the construction on pages 151-152 of the tree $\{F_{n,i}\}_{n=0,i=1}^{\infty \quad 2^n}$ and the selection of $\{m(k,i)\}_{k=0,i=1}^{\infty \quad 2^k}$ is not needed either here or for the proof of 6.8; we simply let $\{F_{n,i}\}_{n=0,i=1}^{\infty \quad 2^n}$ be the usual tree and choose $\{m(k,i)\}_{k=0,i=1}^{\infty \quad 2^k}$ to satisfy ($), and use ($) in place of the more complicated looking inequality in lines 4-5 on page 152.)

We want to show that the sequence

$$\{u_{k,i}\}_{k=0,i=1}^{\infty\ \ 2^k} \equiv \{\chi_{C_{m(k,i)}} w'_{m(k,i)}(F_{k,i})\}_{k=0,i=1}^{\infty\ \ 2^k} \quad \text{in} \quad Y^*$$

is equivalent to the Haar basis for X^*, which in view of the disjointness of the C_m's will contradict the hypothesis of 6.12. To see the equivalence, first note that

$$\|\sum_{k,i} a_{k,i}u_{k,i}\|_Y = \|(\sum_{k,i} a_{k,i}^2 \chi_{C_{m(k,i)}} w'_{m(k,i)}(F_{k,i})^2)^{1/2}\|_Y \leq$$

$$\leq \|(\sum_{k,i} a_{k,i}^2 w'_{m(k,i)}(F_{k,i})^2)^{1/2}\|_Y =$$

$$= \|(\sum_{k,i} a_{k,i}^2 \sum_{n\in\sigma_{m(k,i)}} \alpha_n^2 \sum_{h_{n,j}\subset F_{k,i}} |f_{n,j}|^2)^{1/2}\|_Y \approx \|\sum_{k,i} a_{k,i}h_{k,i}\|_{X^*}.$$

The last equivalence follows from (*) of the introduction because

$$\tilde{f}_{k,i} \equiv \sum_{n\in\sigma_{m(k,i)}} \alpha_n \sum_{h_{n,j}\subset F_{k,i}} f_{n,j};\ 1 \leq i \leq 2^k,\ k = 0,1,2,\ldots,$$

is the image under an isomorphism of a gaussian Haar system in X. For future reference in the proof of Case (i), we note that the above reasoning did not use ($).

We now need to check that $\{u_{k,i}\}_{k=0,i=1}^{\infty\ \ 2^k}$ in Y^* dominates the Haar system in X^*. Since $\{g_{k,i} = \chi_{C_{m(k,i)}} w_{m(k,i)}(F_{k,i})\}_{k=0,i=1}^{\infty\ \ 2^k}$ in Y is equivalent to the Haar basis for X (see the comment at the top of page 155), and the C_m's are disjoint, we only need to verify that

$$\int u_{k,i}g_{k,i}\,d\mu \geq \delta 2^{-k}$$

for some constant $\delta > 0$. If $\delta_{k,i}$ is chosen sufficiently small, this follows from the following string of inequalities:

$$2^{-k} = \int \sum_{n\in\sigma_{m(k,i)}} \sum_{h_{n,j}\subset F_{k,i}} \alpha_n^2 f_{n,j} Th_{n,j}\, d\mu \leq$$

$$\leq \int \Big(\sum_{n\in\sigma_{m(k,i)}} \alpha_n^2 \sum_{h_{n,j}\subset F_{k,i}} |f_{n,j}|^2\Big)^{1/2} \Big(\sum_{n\in\sigma_{m(k,i)}} \alpha_n^2 \sum_{h_{n,j}\subset F_{k,i}} |Th_{n,j}|^2\Big)^{1/2} d\mu$$

$$= \int w_{m(k,i)}(F_{k,i}) w'_{m(k,i)}(F_{k,i})\, d\mu = \int g_{k,i} u_{k,i}\, d\mu +$$

$$+ \int \chi_{D_{m(k,i)}} w_{m(k,i)}(F_{k,i}) w'_{m(k,i)}(F_{k,i})\, d\mu$$

$$\leq \int g_{k,i} u_{k,i}\, d\mu + \|\chi_{D_{m(k,i)}} w_{m(k,i)}(F_{k,i})\|_Y \|w'_{m(k,i)}(F_{k,i})\|_{Y^*} \leq$$

$$\leq \int g_{k,i} u_{k,i}\, d\mu + \delta_{k,i} C 2^{-k} \quad ,$$

where C is a constant which depends on the constant of equivalence of $\{f_{k,i}\}_{k=0,i=1}^{\infty\ \ 2^k}$ to the Haar basis for X^* and on the p-concavity constant of Y^*.

We turn now to the proof of case (i). If $\Lambda[0,1) = 0$, the proof of case (ii) applies. Otherwise, we carry out exactly the construction on pages 151-154 and define v'_n, w'_n, $u_{k,i}$ as in the proof of case (ii). As noted in the proof of case (ii), we also have here that $\{u_{k,i}\}_{k=0,i=1}^{\infty\ \ 2^k}$ is dominated by the Haar basis for X^*. We will show that there is a constant $\delta > 0$ (in fact, $\delta = \frac{1}{2}$) so that

$$(\#) \qquad \delta 2^{-k} \leq \int u_{k,i} g_{k,i}\, d\mu$$

for at least $[\delta 2^k]$ values of i, $1 \leq i \leq 2^k$ (say for $i \in \tau_k$), and for all sufficiently large k. This yields that $\{u_{k,i}\}_{k=0,i\in\tau_k}^{\infty}$ dominates the corresponding subset of the Haar basis for X^*. To see this note that even though $\{g_{k,i}\}_{k=0,i=1}^{\infty\ \ 2^k}$ may not be equivalent to the Haar basis for X, it is unconditional (being disjointly supported in Y) and

$$\|(\sum_{k,i} \alpha_{k,i}^2 g_{k,i}^2)^{1/2}\|_Y \le \|(\sum_{k,i} \alpha_{k,i}^2 w_{k(m,i)}^2)^{1/2}\|_Y \approx \|\sum_{k,i} \alpha_{k,i} h_{k,i}\|_X$$

so $\{g_{k,i}\}_{k=0,i=1}^{\infty \;\; 2^k}$ is dominated by the Haar basis for X. This combines with (#) to give that $\{u_{k,i}\}_{k=0,i\in\tau_k}^{\infty}$ dominates the corresponding subset of the Haar basis for X. Once (#) is verified, we conclude the proof by observing that the closed span in X^* of $\{h_{k,i}\}_{k=0,i\in\tau_k}^{\infty}$ is isomorphic to X^*. Indeed, if X^* were $L_p(0,1)$ for some $1 < p < \infty$, this would follow from a theorem of Gamlen and Gaudet [28]; it was pointed out in [1] that the general case follows from this special case by an interpolation argument.

Finally, we come to the proof of (#). Reasoning as in case (ii), we have that $1 \le i \le 2^k$, $k = 0,1,\ldots,$

$$2^{-k} \le \int g_{k,i} u_{k,i} d\mu + \int w'_{m(k,i)}(F_{k,i}) \chi_{D_{m(k,i)}} w_{m(k,i)}(F_{k,i}) d\mu .$$

Suppose that for 2^{k-1} values of i (say, $i \in \gamma_k$), $1 \le i \le 2^k$, and infinitely many k, we had

$$2^{-k-1} \le \int w'_{m(k,i)}(F_{k,i}) \chi_{D_{m(k,i)}} w_{m(k,i)}(F_{k,i}) d\mu .$$

From page 154, lines 1-8 we have that there is a constant C so that for all k and all scalars $\{a_i\}_{i=1}^{2^k}$,

$$\|(\sum_{i=1}^{2^k} a_i^2 \chi_{D_{m(k,i)}} w_{m(k,i)}(F_{k,i})^2)^{1/2}\|_Y \le C\|\sum_{i=1}^{2^k} a_i h_{k,i}\|_2 = C2^{-k/2}(\sum_{i=1}^{2^k} a_i^2)^{1/2} .$$

Thus, if $\sum_{i\in\gamma_k} a_i^2 = 1$, we have that

$$\|(\sum_{i\in\gamma_k} a_i^2 w'_{m(k,i)}(F_{k,i})^2)^{1/2}\|_{Y^*}$$

$$\geq C^{-1}2^{k/2}\int(\sum_{i\in\gamma_k} a_i^2 w'_{m(k,i)}(F_{k,i})^2)^{1/2}(\sum_{i\in\gamma_k} a_i^2 \chi_{D_{m(k,i)}} w_{m(k,i)}(F_{k,i})^2)^{1/2}d\mu)$$

$$\geq C^{-1}2^{k/2}\int \sum_{i\in\gamma_k} a_i^2 w'_{m(k,i)}(F_{k,i})\chi_{D_{m(k,i)}} w_{m(k,i)}(F_{k,i})d\mu$$

$$\geq (2C)^{-1}2^{k/2}2^{-k} = (2C)^{-1}2^{-k/2} = (2C)^{-1}\|\sum_{i\in\gamma_k} a_i h_{k,i}\|_2 \ .$$

Since

$$\|(\sum_{i=1}^{2^k} a_i^2 w'_{m(k,i)}(F_{k,i})^2)^{1/2}\|_{Y^*} \approx \|\sum_{i=1}^{2^k} a_i h_{k,i}\|_{X^*} ,$$

we have that $\{h_{k,i}\}_{i\in\gamma_k}$ in X^* uniformly dominates $\{h_{k,i}\}_{i\in\gamma_k}$ in $L_2(0,1)$ which, in view of the r.i. structure of $L_2(0,1)$ and X^*, means that $C\|f\|_{X^*} \geq \|f\|_2$ for some constant C and all f in X^*. However, in case (i), we already have the reverse inequality $C'\|f\|_2 \geq \|f\|_{X^*}$ for some constant C' and all f in $L_2(0,1)$, so we conclude that $X = L_2(0,1)$, up to an equivalent renorming.

□

Corollary 6.13: If X *is a super-reflexive r.i. function space on* $[0,1]$ *and the Haar basis for* X *is not equivalent to a sequence of disjoint functions in* X*, then* X *has a unique representation as a r.i. function space on* $[0,1]$.

Proof: By 6.12, it is enough to observe that if Y is a r.i. function space on $[0,1]$ which is isomorphic to X, then the Haar basis for Y is not equivalent to a sequence of disjoint vectors in X. This is true because if X embeds into Y, then the Haar basis for X is equivalent to a block basis of the Haar basis for Y (see [45]).

□

7. ORLICZ FUNCTION SPACES

In this section the results of sections 5 and 6 are applied to Orlicz function spaces. The reader is referred to [49] for background on Orlicz spaces. Here we mention only that an Orlicz function, F, is a strictly increasing convex function in $[0,\infty)$ with $F(0) = 0$. We always assume that the Orlicz function satisfies the Δ_2-condition both at 0 and at ∞; i.e., that

$$\sup_{0<s<\infty} F(2s)/F(s) < \infty .$$

The Orlicz space $L_F(0,1)$ (respectively, $L_F(0,\infty)$) is the collection of all measurable functions, f, on $[0,1]$ (respectively, on $[0,\infty)$) for which $\int_0^1 F(\lambda|f(t)|)dt < \infty$ (respectively, $\int_0^\infty F(\lambda|f(t)|)dt < \infty$) for some -or, equivalently, for every $\lambda > 0$. The unit ball of $L_F(0,1)$ (respectively, $L_F(0,\infty)$) is the collection of all such f for which $\int_0^1 F(|f(t)|)dt \leq 1$ (respectively, $\int_0^\infty F(|f(t)|)dt \leq 1$). The convexity of F guarantees that the unit ball of L_F is convex.

In order that $\|\chi_{(0,1)}\|_{L_F} = 1$, we require the normalization $F(1) = 1$. Then $L_F(0,1)$ and $L_F(0,\infty)$ are r.i. spaces in the sense used in the introduction.

We recall that non-negative functions F and G on $[0,\infty)$ are said to be equivalent at ∞ (respectively, at 0) provided that there are $0 < t_0 < \infty$, and $0 < A, B, a, b,$ such that $AG(at) \leq F(t) \leq BG(bt)$ for all $t \geq t_0$, (respectively, $t \leq t_0$). If F and G are equivalent at ∞ (respectively, at 0) and F satisfies the Δ_2-condition at ∞ (respectively, at 0) then so does G, and in the equivalence inequality both a and b can be taken to be one.

Observe that, up to an equivalent renorming, the space $L_F(0,1)$ is determined by the behavior of $F(t)$ for $t \geq 1$; i.e., if an Orlicz function G is equivalent to F at ∞ then $L_G(0,1) = L_F(0,1)$ (with equivalent norms).

The space $L_F(0,\infty)$ is p-convex with constant one if and only if $F(t^{1/p})$ is a convex function. To see this, observe that the unit ball of the p-concavification $(L_F(0,\infty))_{1/p}$ consists of the functions f such that $\int_0^\infty F(|f(u)|^{1/p})du \leq 1$, and so is convex if and only if $F(t^{1/p})$ is convex. In general, it can be checked that $L_F(0,\infty)$ (respectively, $L_F(0,1)$) is p-convex if and only if there is a constant M such that, whenever $0 < s \leq t < \infty$, $F(s)/s^p \leq M \cdot F(t)/t^p$. This is the same as saying that there is a function F_1 equivalent to F (respectively, equivalent at ∞) such that $F_1(t^{1/p})$ is a convex function. A similar argument shows that L_F is q-concave exactly when F is equivalent to a function F_1 such that $F_1(t^{1/q})$ is concave, and it is known that this happens for some $q < \infty$ if and only if F satisfies the Δ_2-condition.

In the analysis in sections 5 and 6 of embeddings of a r.i. function space X on $[0,1]$ into a r.i. function space Y, the complicated case occurs when the Haar basis for X is equivalent to a sequence of disjoint vectors in Y. This pathology cannot occur when Y is an Orlicz space unless X is $L_2(0,1)$. Indeed, a sequence of disjoint vectors in an Orlicz function space spans a modular sequence space which embeds into an Orlicz sequence space (cf. [81]). But it was proved in [47, theorem 3 and remark following] that $L_2(0,1)$ is the only r.i. function space on $[0,1]$ which embeds into an Orlicz sequence space. (Note that the normalized Rademacher functions on any set of positive measure in a r.i. function space is equivalent, up to a constant which depends only on the p-concavity of the r.i. space, to the unit vector basis of ℓ_2. Thus the condition in the remark following

Theorem 3 of [47] is satisfied in the situation considered here). Alternately, one can simply observe that the proof of Theorem 3 in [47] really shows that $L_2(0,1)$ is the only r.i. function space on $[0,1]$ which embeds into the span of a sequence of disjoint functions in an Orlicz function space.

This simple observation yields improved versions of the results of sections 5 and 6 in the context of Orlicz spaces. For example, the main result of section 6 gives

Theorem 7.1: Let F *be an Orlicz function (satisfying the* Δ_2 *-condition) and let* X *be a r.i. function space on* $[0,1]$ *which is not* $L_2(0,1)$ *up to an equivalent renorming.*

(a) If the Haar system is unconditional in X *and if* X *is isomorphic to a subspace of* $L_F(0,\infty)$ *then there is a positive constant* δ *so that* $\|f\|_X \geq \delta \|f\|_{L_F(0,1)}$ *for every* f *in* X.

(b) If X *is isomorphic to a complemented subspace of a reflexive* $L_F(0,\infty)$*-space then* $X = L_F(0,1)$ *, up to an equivalent renorming. In particular, a reflexive space* $L_F(0,1)$ *has a unique representation as a r.i. function space on* $[0,1]$.

(c) If X *is a reflexive Orlicz function space so that* $L_F(0,1)$ *embeds isomorphically into* X *and* X *embeds isomorphically into* $L_F(0,\infty)$ *then* $X = L_F(0,1)$ *, up to an equivalent renorming.*

Proof: 7.1. (a) follows from 6.1 and the discussion preceding the statement of the theorem. By [9], the Haar basis in a reflexive Orlicz space is unconditional. The fact that the Haar basis in X is also unconditional follows from its reproducibility (cf. [45]). Then (b) follows by applying (a) to the embeddings of X into $L_F(0,\infty)$ and of X^* into $L_F(0,\infty)^*$. The proof of (c) uses the fact that the Haar basis is unconditional in X and in $L_F(0,1)$.

□

Remarks: From 5.5 it follows that a non-reflexive space $L_F(0,1)$ which is q-concave for some $q < 2$ has a unique representation as a r.i. function space on $[0,1]$, but the general case does not follow from the results of section 5 and 6. However, the argument for 6.1 does yield 7.1. (a) (hence also 7.1. (c)) in the case when X does not have an unconditional basis. We give a brief sketch of the argument for this in an appendix to this section.

The result 7.1 (c) cannot be improved even in the case of L_p. Indeed, in section 10 it is proved that for $1 < p < 2$, there is a r.i. function space X on $[0,1]$, different from $L_p(0,1)$, so that X embeds into $L_p(0,1)$ and $L_p(0,1)$ is isomorphic to a complemented subspace of X. In particular, a r.i. function space which embeds into an Orlicz function space need not be isomorphic to an Orlicz space. On the other hand, the case L_p, $p > 2$, and the classification formula 2.1 suggest that the situation should be simpler for type 2 Orlicz function spaces. The rest of this section is devoted to classifying the r.i. function spaces which embed into, or which are finitely crudely representable in, a type 2 Orlicz function space. (Recall that a Banach space X is *finitely crudely representable* in a Banach space Y provided that there is a constant K so that every finite dimensional subspace of X is K-isomorphic to a subspace of Y. If every $K > 1$ is a suitable constant, then X is said to be *finitely representable* in Y.)

Before proceeding, we make some preliminary comments. Even when we consider the space $L_F(0,1)$, it is still convenient to work in the framework of r.i. function spaces on $[0,\infty)$. Since the space $L_F(0,1)$ is determined by the behavior of F at ∞, we may replace F by a function G which is equivalent to F at ∞ and is equivalent to x^2 at 0; the reason for doing this is that $L_F(0,1) = L_G(0,1)$ and $L_G(0,\infty)$ is 2-convex if $L_F(0,1)$ is 2-convex.

The following simple lemma concerning r.i. function spaces will be used in the sequel. It is convenient to state the lemma for r.i. function spaces on a general interval $[0,s)$.

Lemma 7.2: Let $\{f_i\}_{i=1}^n$ *be non-negative disjointly supported functions in a r.i. function space* X *on* $[0,s)$, $0<s\leq\infty$. *Suppose that* $\{g_i\}_{i=1}^n$ *is a sequence of functions on* $[0,s)$ *such that for each* $1\leq i\leq n$, g_i *has the same distribution as* f_i . *Then*

$$\|\sum_{i=1}^n f_i\|_X \leq \|\sum_{i=1}^n g_i\|_X .$$

Proof: Since the functions which have bounded support are assumed to be dense in a r.i. function space on $[0,\infty)$, we may assume without loss of generality that $s<\infty$. For $1\leq i\leq n$, let τ_i be a measure preserving transformation from $\operatorname{supp} f_i$ onto $\operatorname{supp} g_i$ so that $f_i = g_i(\tau_i)$. Assume for the moment that we can identify X^* with a r.i. function space on $[0,s)$ (this occurs, e.g., if X is q-concave for some $q<\infty$ and we use the lemma only in this situation). Pick a norm one positive element H in X^* supported on $\bigcup_{i=1}^n \operatorname{supp} f_i$ so that

$$H(\sum_{i=1}^n f_i) = \sum_{i=1}^n \int_0^s H(t)f_i(t)dt = \|\sum_{i=1}^n f_i\|_X .$$

Define $G\in X^*$ by $G(t) = \max H(\tau_i^{-1}(t))$ for $t\in\bigcup_{i=1}^n \operatorname{supp} g_i$ (where the max is taken over all i's such that $t\in\operatorname{supp} g_i$), and $G(t)=0$ for $t\notin\bigcup_{i=1}^n \operatorname{supp} g_i$.

Since the decreasing rearrangement of H dominates that of G ,

$$\|G\|_{X^*} \leq \|H\|_{X^*} = 1$$

and

$$G(\sum_{i=1}^n g_i) = \sum_{i=1}^n \int_0^s G(t)g_i(t)dt \geq \sum_{i=1}^n \int_0^s H(t)f_i(t)dt .$$

In the general case, since $L_\infty(0,1)$ is dense in X , there is a norm determining subspace of X^* which can be identified with a r.i. function space on $[0,s)$. The proof goes as above, except that, given $\varepsilon > 0$, we pick the norm one positive functional H out of this norm determining subspace of X^* to satisfy $\|\sum_{i=1}^n f_i\|_X \leq H(\sum_{i=1}^n f_i) - \varepsilon$.

□

Corollary 7.3: Given $M < \infty$ *and* $K < \infty$ *, there is a constant* $C = C(M,K)$ *so that if* X *is a r.i. function space on some interval* $[0,s)$, $0 < s \leq \infty$, *with 2-convexity constant* $\leq M$, *then the following is true. If* $\{x_i\}_{i=1}^n$ *is a K-unconditional sequence in* X , *and* $\{y_i\}_{i=1}^n$ *are disjointly supported vectors in* X *such that* y_i *has the same distribution as* x_i *for* $1 \leq i \leq n$, *then for all choices* $\{a_i\}_{i=1}^n$ *of scalars,*

$$\|\sum_{i=1}^n a_i y_i\|_X \leq C\|\sum_{i=1}^n a_i x_i\|_X .$$

Proof: As in 7.2, we can assume that $s < \infty$. We may also assume that the 2-convexity constant of X is one, and thus the "2-concavification", $X_{1/2}$, of X is a (normed) r.i. function space on $[0,1]$.

Since $\{x_i\}_{i=1}^n$ is K-unconditional we have that

$$K\|\sum_{i=1}^n a_i x_i\|_X \geq \int_0^1 \|\sum_{i=1}^n a_i r_i(u) x_i\|_X du \geq \|\int_0^1 |\sum_{i=1}^n a_i r_i(u) x_i| du\|_X \geq$$

$$\geq A_1 \|(\sum_{I=1}^n |a_i x_i|^2)^{1/2}\|_X$$

for every choice $\{a_i\}_{i=1}^n$ of scalars.

Applying 7.2 to the positive functions $|a_i y_i|^2$ in $X_{1/2}$, we have for all scalars $\{a_i\}_{i=1}^n$, that

$$\|\sum_{i=1}^n a_i y_i\|_X = \|\sum_{i=1}^n |a_i y_i|^2\|_{X_{1/2}}^{1/2} \leq \|\sum_{i=1}^n |a_i x_i|^2\|_{X_{1/2}}^{1/2} =$$

$$= \|(\sum_{i=1}^n |a_i x_i|^2)^{1/2}\|_X \leq K A_1^{-1} \|\sum_{i=1}^n a_i x_i\|_X .$$

□

Remark: It was essentially observed by Dacunha-Castelle and Schreiber [16] that, in the case of an Orlicz space L_F, 7.3 is a simple consequence of the convexity of $F(\sqrt{t})$.

We associate several sets with an Orlicz function F: for an interval I, we put

$$\mathcal{S}(F,I) = \{F(st)/F(s);\ s \in I\}$$

and let $\mathcal{C}(F,I)$ denote the convex hull of $\mathcal{S}(F,I)$. When $I = (0,\infty)$ we write, for simplicity, $\mathcal{S}(F)$ and $\mathcal{C}(F)$ instead of $\mathcal{S}(F,(0,\infty))$ and $\mathcal{C}(F,(0,\infty))$.

We denote by $\overline{D}$ the closure of the set D (where D is one of the above sets of functions) in $C[0,\infty)$ when this space is given the topology of uniform convergence on compact intervals. It is well known and easy to prove (see, e.g., [47]) that $\mathcal{C}(F)$ is a relatively compact subset of $C[0,\infty)$.

Before stating the next theorem, we indicate why the sets just introduced (or, at least, $\mathcal{C}(F)$) play an important role in the study of Orlicz spaces. Let $0 < s \leq \infty$ and suppose that $\{x_i\}_{i=1}^n$ is a sequence in $L_F(0,s)$ which has symmetry constant K. An argument very close to the argument used in 1.2 yields that $\{x_i\}_{i=1}^n$ is K-equivalent to a symmetrically exchangeable sequence $\{y_i\}_{i=1}^n$ in $L_F(0,s)$, which means here that

$$\mu(y_1 \in E_1;\dots y_n \in E_n) = \mu(\varepsilon_1 y_{\pi(1)} \in E_1;\dots;\varepsilon_n y_{\pi(n)} \in E_n),$$

(where the two members are possibly ∞) for every sequence $\{E_i\}_{i=1}^n$ of measurable subsets of the real line, every choice of signs $\{\varepsilon_i\}_{i=1}^n$ and every permutation π of $\{1,\dots,n\}$. The actual construction of $\{y_i\}_{i=1}^n$ may be achieved in the following way: select a partition of $[0,s)$ into $2^n \cdot n!$ disjoint measurable subsets $\{A_{\varepsilon,\pi}\}$ (where $\varepsilon \in \{-1,+1\}^n$ and π is an arbitrary permutation of $\{1,\dots,n\}$) such that $\mu(A_{\varepsilon,\pi}) = s/2^n \cdot n!$, for every ε and π. Select then for every ε and π, a measurable mapping $\varphi_{\epsilon,\pi}$ from $A_{\epsilon,\pi}$ onto $[0,s)$ such that $\mu(\varphi_{\epsilon,\pi}^{-1}(B)) = \mu(B)/2^n \cdot n!$, for every integrable set $B \subset [0,s)$. Finally define

$$y_i = \sum_{\epsilon,\pi} \epsilon_i x_{\pi(i)}(\varphi_{\epsilon,\pi}) \cdot \chi_{A_{\epsilon,\pi}} .$$

Note that if the x_i's have disjoint supports, then so do the y_i's. When the y_i's have disjoint supports, it is easy to compute $\|\sum_{i=1}^n a_i y_i\|_{L_F(0,s)}$; indeed, $\|\sum_{i=1}^n a_i y_i\|_{L_F(0,s)} = 1$ if and only if

$$1 = \int_0^s F(\sum_{i=1}^n |a_i y_i(t)|)dt = \sum_{i=1}^n \int_0^s F(|a_i y_1(t)|)dt$$

$$= \sum_{i=1}^n \int_0^\infty F(|a_i|u)/F(u)d\nu(u) ,$$

for a suitable measure ν on $(0,\infty)$. If we normalize so that $\|y_i\|_{L_F(0,s)} = 1$, then ν is a probability measure, and the function G defined by $G(t) = \int_0^\infty F(tu)/F(u)d\nu(u)$ is in $\overline{\mathcal{C}}(F)$. Of course, $\{y_i\}_{i=1}^n$ is isometrically equivalent to the first n unit vectors in the Orlicz sequence space ℓ_G.

If the x_i's are not disjoint, but the space $L_F(0,\infty)$ is of type 2, we can still use this kind of reasoning. Let $\{\tilde{y}_i\}_{i=1}^n$ represent a sequence of functions on $(0,\infty)$ with disjoint supports each of which has the same distribution as y_1. Since $\{y_i\}_{i=1}^n$ is symmetrically exchangeable, the classification formula 2.1 simplifies to give:

$$D^{-1}\|\sum_{i=1}^n a_i y_i\|_{L_F} \le \max\{\|\max_{1\le i\le n} |a_i y_i|\|_{L_F} , \frac{\|\sum_{i=1}^n y_i\|_{L_F}}{\sqrt{n}} (\sum_{i=1}^n |a_i|^2)^{1/2}\}$$

$$\le \max\{\|\sum_{i=1}^n a_i \tilde{y}_i \|_{L_F} , \frac{\|\sum_{i=1}^n y_i\|_{L_F}}{\sqrt{n}} (\sum_{i=1}^n |a_i|^2)^{1/2}\} ,$$

for some constant D which depends only on F. On the other hand, by the right hand side inequality in 2.1 and 7.3, we have

$$\max\{\|\sum_{i=1}^n a_i\tilde{y}_i\|_{L_F}, \frac{\|\sum_{i=1}^n y_i\|_{L_F}}{\sqrt{n}}(\sum_{i=1}^n |a_i|^2)^{1/2}\} \leq D_1\|\sum_{i=1}^n a_iy_i\|_{L_F},$$

for some constant $D_1 < \infty$. Assuming that $\|y_i\|_{L_F} = 1$ and, as before, letting G be the function in $\overline{\mathcal{C}}(F)$ defined by

$$G(t) = \int_0^\infty F(t|y_1(u)|)du ,$$

we have that

$$\|\sum_{i=1}^n a_ix_i\|_{L_F} \approx \max\{\|\sum_{i=1}^n a_ie_i\|_{\ell_G}, \frac{\|\sum_{i=1}^n x_i\|_{L_F}}{\sqrt{n}}(\sum_{i=1}^n a_i^2)^{1/2}\},$$

where $\{e_i\}_{i=1}^\infty$ is the unit vector basis in ℓ_G and the equivalence "$\approx$" depends only on F and on the symmetry constant K of $\{x_i\}_{i=1}^n$.

This discussion shows that every space spanned by a symmetric sequence in a type 2 Orlicz function space is an Orlicz sequence space. More precisely, it is a weighted diagonal of $\ell_G^n \oplus \ell_2^n$ for some $G \in \overline{\mathcal{C}}(F)$. It is also true that a r.i. function space on $[0,1]$ or on $[0,\infty)$ which is finitely crudely representable in a type 2 Orlicz function space $L_F(0,\infty)$ is an Orlicz function space L_G, for some function G. The possible values of G are computed in 7.5 and 7.6. We begin with a lemma.

If N is an integer and X a r.i. function space on $[0,N]$, $X(n)$ will denote the span in X of the sequence $\{x_{n,i}\}_{i=1}^{N2^n}$, where $x_{n,i} = \chi_{[(i-1)2^{-n}, i2^{-n})}$, as before.

Lemma 7.4: Let F *be an Orlicz function such that* $F(\sqrt{t})$ *is equivalent to a convex function. For every* $K \geq 1$ *there is a constant* $D = D(K)$ *such that if* X *is a r.i. function space on some interval* $[0,N]$, n *an arbitrary integer,* T *a K-isomorphism from* X(n) *into* $L_F(0,\infty)$, *then there are* $G \in \overline{\mathcal{C}(F)}$ *and* $0 < w \leq D$ *such that*

$$D^{-1}\|f\|_X \leq \max\{w\|f\|_{L_G} , \ \rho\|f\|_{L^2}\} \leq D\|f\|_X ,$$

for every $f \in X(n)$, *where* $\rho = \|\chi_{[0,N]}\|_X/\sqrt{N}$. *More precisely, if* $z_i = Tx_{n,i}$; $i = 1,\dots,N2^n$, w *is defined by*

$$\sum_{i=1}^{N2^n} \int_0^\infty F(w^{-1}|z_i(u)|)\frac{du}{N} = 1 ,$$

and

$$G(t) = \int_0^\infty (F(tu)/F(u))\,d\nu(u) ,$$

where ν *is the probability measure on* $(0,\infty)$ *defined by*

$$\nu([a,b]) = \sum_{i=1}^{N2^n} \int_{\{wa \leq |z_i| \leq wb\}} F(w^{-1}|z_i(u)|)\frac{du}{N} ,$$

for $0 < a \leq b < \infty$. *The function* G *is also given by*

$$G(t) = \sum_{i=1}^{N2^n} \int_0^\infty F(tw^{-1}|z_i(u)|)\frac{du}{N} ,$$

or

$$G(t) = \sum_{i=1}^{N2^n} \int_0^\infty F(tw^{-1}|y_i(u)|)\frac{du}{N} ,$$

where $\{y_i\}_{i=1}^{N2^n}$ *is the symmetrically exchangeable sequence constructed from* $\{z_i\}_{i=1}^{N2^n}$ *(by the procedure described above).*

Proof: The proof of the lemma is based on the discussion preceding its statement. If $\{y_i\}_{i=1}^{N2^n}$ is the symmetrically exchangeable sequence K-equivalent to $\{z_i\}_{i=1}^{N2^n}$ obtained by the procedure described above and $\{\tilde{y}_i\}_{i=1}^{N2^n}$ a disjointly supported sequence of vectors, each of which has the same distribution as y_1, we can write, for a function $f = \sum_{i=1}^{N2^n} a_i x_{n,i}$,

$$\|f\|_X \approx \max\{\|\sum_{i=1}^{N2^n} a_i\tilde{y}_i\|_{L_F}, \frac{\|\sum_{i=1}^{N2^n} z_i\|}{\sqrt{N}} \cdot \left(\frac{\sum_{i=1}^{N2^n} a_i^2}{2^n}\right)^{1/2}\} \approx$$

$$\approx \max\{\|\sum_{i=1}^{N2^n} a_i\tilde{y}_i\|_{L_F}, \rho\|f\|_{L^2}\} .$$

Define w by the relation:

$$\sum_{i=1}^{N2^n} \int_0^\infty F(w^{-1}|z_i(u)|)\frac{du}{N} = 1 .$$

To prove that w is bounded by a constant depending only upon K, it is enough to show that the expression $\sum_{i=1}^{N2^n} \int_0^\infty F(|z_i(u)|)\frac{du}{N}$ is bounded. But if we consider disjoint vectors $\{\tilde{z}_i\}_{i=1}^{N2^n}$ with $\tilde{z}_i$ distributed as z_i; $1 \leq i \leq N2^n$, we know according to 7.3 that

$$\|\sum_{i=m\cdot 2^n+1}^{(m+1)2^n} \tilde{z}_i\|_{L_F} \leq D(K)\|\sum_{i=m\cdot 2^n+1}^{(m+1)2^n} z_i\|_{L_F} \leq K\cdot D(K) ,$$

for every $0 \leq m < N$. It follows easily that

$$\sum_{i=m2^n+1}^{(m+1)2^n} \int_0^\infty F(|z_i(u)|du ; \quad 0 \leq m < N$$

is uniformly bounded and this proves our claim. Now, set

$$\nu([a,b]) = \sum_{i=1}^{N2^n} \int_{\{wa \leq |z_i| \leq wb\}} F(w^{-1}|z_i(u)|)\frac{du}{N} .$$

The choice of w and the fact that $\nu(\{0\}) = 0$ make ν a probability measure on $(0,\infty)$. Define the Orlicz function G by

$$G(t) = \int_0^\infty (F(tu)/F(u))d\nu(u) .$$

We have $G(1) = 1$ since ν is a probability measure and

$$G(t) = \sum_{i=1}^{N2^n} \int_0^\infty F(tw^{-1}|z_i(u)|)\frac{du}{N} .$$

The reason for the last equality is as follows: if we introduce for every $i = 1,\dots,N2^n$, the finite measures $d\lambda_i(u) = F(w^{-1}|z_i(u)|)\frac{du}{N}$ and ν_i is defined by $\nu_i([a,b]) = \int_{wa\le|z_i|\le wb} F(w^{-1}|z_i(u)|)\frac{du}{N} = \lambda_i(a \le w^{-1}|z_i| \le b)$, we see that ν_i is the image of λ_i under the measurable mapping $w^{-1}|z_i|$ and, therefore,

$$\int_0^\infty (F(tv)/F(v))d\nu_i(v) = \int_0^\infty (F(tw^{-1}|z_i(u)|)/F(w^{-1}|z_i(u)|)d\lambda_i(u)$$

$$= \int_0^\infty F(tw^{-1}|z_i(u)|)\frac{du}{N} .$$

The desired equality then follows by summing over $1 \le i \le N2^n$.

Suppose, without loss of generality, that $\{a_i\}_{i=1}^{N2^n}$ are chosen to satisfy

$$\|\sum_{i=1}^{N2^n} a_i\tilde{y}_i\|_{L_F} = w .$$

Then

$$1 = \int_0^\infty F(|\sum_{i=1}^{N2^n} w^{-1}a_i\tilde{y}_i(u)|du = \sum_{i=1}^{N2^n} \int_0^\infty F(w^{-1}|a_iy_1(u)|)du .$$

Therefore, by the construction of y_1 we have

$$\int_0^\infty F(w^{-1}|a_iy_1(u)|)du = \sum_{j=1}^{N2^n} \frac{1}{N2^n} \int_0^\infty F(w^{-1}|a_iz_j(v)|)dv ,$$

for every $1 \le i \le N\cdot 2^n$, which implies that

$$1 = 2^{-n} \sum_{i=1}^{N2^n} \sum_{j=1}^{N2^n} \int_0^\infty F(w^{-1}|a_i z_j(v)|)\frac{dv}{N} = \sum_{i=1}^{N2^n} 2^{-n}G(|a_i|) .$$

This shows finally that $\|\sum_{i=1}^{N2^n} a_i\tilde{y}_i\|_{L_F} = w\|f\|_{L_G(0,N)}$ (recall that $f = \sum_{i=1}^{N2^n} a_i x_{n,i}$) . □

Remark: If the vectors $\{z_i\}_{i=1}^{N2^n}$ are already disjoint in $L_F(0,\infty)$, it is easy to check that

$$\inf\{\|\sum_{i\in I} z_i\|_{L_F} ; \bar{\bar{I}} = 2^n\} \le w \le \sup\{\|\sum_{i\in I} z_i\|_{L_F} , \bar{\bar{I}} = 2^n\} ,$$

which means that w is bounded and bounded away from zero by a constant depending only upon K . Since the functions in $\overline{\mathcal{C}}(F)$ uniformly dominate t^2 , the result in this case is simply

$$D_1^{-1}\|f\|_X \le \|f\|_{L_G} \le D_1\|f\|_X ,$$

where D_1 is another constant depending only upon K , G is as in 7.4 and $w \ge D_1^{-1}$.

Theorem 7.5: Let X *be a r.i. space on* $[0,1]$ *which is not* $L_2(0,1)$ *, even up to an equivalent renorming.*

(1) X *is finitely crudely representable in a 2-convex Orlicz function space* $L_F(0,\infty)$ *if and only if* X *is* $L_G(0,1)$ *, up to an equivalent renorming, for some* G *in* $\overline{\mathcal{C}}(F)$.

(2) X *is finitely crudely representable in a 2-convex Orlicz function space* $L_F(0,1)$ *if and only if* X *is* $L_G(0,1)$ *, up to an equivalent renorming, for some* G *in* $\overline{\mathcal{C}}(F,[1,\infty))$.

Proof: Necessity: We know from Lemma 7.4 that there is a constant D such that for every n , there are $0 \le w_n \le D$ and a function $G_n \in \overline{\mathcal{C}}(F)$ so that

$$D^{-1}\|f\|_X \le \max\{w_n\|f\|_{L_{G_n}} , \|f\|_{L^2}\} \le D\|f\|_X ,$$

for every $f \in X(n)$. Since X is not $L_2(0,1)$, we have $w = \liminf w_n > 0$. Let G be the limit of some subsequence $\{G_{n_k}\}_{k=1}^{\infty}$ (such that $w = \lim_k w_{n_k}$, for convenience). If f is a dyadic step function then

$$\|f\|_{L_G} = \lim_{k\to\infty} \|f\|_{L_{G_{n_k}}}$$

which implies

$$D^{-1}\|f\|_X \le \max\{w\|f\|_{L_G} , \|f\|_{L_2}\} \le D\|f\|_X ,$$

for every dyadic step function f , and hence, for every $f \in X$. Of course, this says that X is $L_G(0,1)$ since the norm in $L_G(0,1)$ dominates that in $L_2(0,1)$. This proves the necessity of (1) but for (2) we have to be more precise. We may assume in this case that $F(t) = t^2$, for $0 \le t \le 1$. We know that, for every n , the function G_n is of the form

$$G_n(t) = \int_0^{\infty} (F(tu)/F(u))d\nu_n(u) ,$$

for some probability measure ν_n on $(0,\infty)$. Denoting by λ_n the restriction of ν_n to $(0,1]$, we may assume that the sequence $\{n_k\}_{k=1}^{\infty}$ above has been chosen so that $\{\lambda_{n_k}\}_{k=1}^{\infty}$ is w^*-convergent to some measure λ on $[0,1]$. Since for every fixed t , the function $\psi(u) = F(tu)/F(u)$ is extended to a continuous function on $[0,1]$ by assigning the value t^2 at 0 , we get that the limit function G admits the decomposition

$$G(t) = \lambda(\{0\})\cdot t^2 + \int \chi_{[0,1]}(u)(F(tu)/F(u))d\lambda(u) + c\cdot G_1(t) ,$$

where $G_1(t)$ is an Orlicz function in $\overline{\mathcal{C}}(F,[1,\infty))$ and c is a suitable constant. Since X is not $L_2(0,1)$, even up to an equivalent renorming, we have $\lambda(\{0\}) < 1$. Since $L_G(0,1)$ is 2-convex the function G is equivalent to the Orlicz function $\tilde{G}$ defined by

$$\tilde{G}_0(t) = \int \chi_{(0,1]}(u)(F(tu)/F(u))d\lambda(u) + c \cdot G_1(t)$$

and

$$\tilde{G}(t) = \tilde{G}_0(t)/\tilde{G}_0(1) .$$

If $\lambda((0,1]) = 0$ then the proof is completed. Otherwise, we set

$$H(t) = \lambda((0,1])^{-1}\int\chi_{(0,1]}(u)(F(tu)/F(u))d\lambda(u) ,$$

and observe that, in order to complete the proof, it suffices to show that H is equivalent to F .

Now $F(t)/t^2$ is equivalent to an increasing function, since $F(\sqrt{t})$ is equivalent to a convex function and $F(u) = u^2$ for $0 \leq u \leq 1$, so, for $0 < u < 1$, $F(tu)/F(u) \leq KF(t)$, for some constant K , and hence $H(t) \leq$ $\leq KF(t)$. On the other hand, choose $0 < s < 1$ so that $\lambda(s,1) = \delta > 0$ and observe that the Δ_2-condition on F gives a constant K' so that

$$F(t) \leq K'F(tu)/F(u) \quad \text{for} \quad s \leq u \leq 1 \quad \text{and} \quad t \geq 0 .$$

Hence,

$$F(t) \leq \delta\, K'\, H(t) .$$

The sufficiency is quite simple. Let G be in $\overline{\mathcal{C}}(F)$ (G in $\overline{\mathcal{C}}(F,[1,\infty))$ for case (2)) and let $\{x_{n,i}\}_{i=1}^{2^n}$ have their usual meaning. For fixed n and arbitrary $\varepsilon > 0$, we can find G_n in $\mathcal{C}(F)$ (G_n in $\mathcal{C}(F,[1,\infty))$ in case (2)) so that $\{x_{n,i}\}_{i=1}^{2^n}$ is, in $L_G(0,1)$, $1+\varepsilon$-equivalent to $\{x_{n,i}\}_{i=1}^{2^n}$ in $L_{G_n}(0,1)$. Write

$$G_n(t) = \sum_{j=1}^{k} \lambda_j F(ts_j)/F(s_j)$$

for appropriate $0 < s_1 < \ldots < s_k$ (and $1 \le s_1$, in case (2)), where $\sum_{j=1}^{k} \lambda_j = 1$; $\lambda_j > 0$. It is easily checked that $L_{G_n}(0,1)$ is isometric to a sublattice of $L_F(0,\infty)$ in such a way that the image of $\chi_{(0,1)}$ is a function $\sum_{j=1}^{k} s_j \chi_{A_j}$, where the A_j's are pairwise disjoint intervals such that $\mu(A_j) = \lambda_j/F(s_j)$. Of course, in case (2), $s_j \ge 1$, so $F(s_j) \ge 1$ and hence $\sum_{j=1}^{k} \mu(A_j) \le 1$, whence the embeddings of $L_{G_n}(0,1)$ goes into $L_F(0,1)$. We omit the details, because a more general construction and result is given by 7.7 (a) below. □

Remarks: 1. The proof of case (2) yields that a r.i. function space on $[0,1]$ which is finitely crudely representable in a 2-convex space $L_F(0,\infty)$, where $F(t)$ is equivalent to t^2 at 0, is also finitely crudely representable in $L_F(0,1)$. In 7.6 we show that $L_F(0,\infty)$ is itself finitely crudely representable in $L_F(0,1)$, and in section 8 it is shown that in this case $L_F(0,1)$ and $L_F(0,\infty)$ are actually isomorphic.

2. Say that a Banach lattice Y is *lattice finitely crudely representable* in a Banach lattice X provided that there is a constant K so that every finite sequence of disjoint vectors in Y is K-equivalent to a sequence of disjoint vectors in X . The proof of 7.5 shows that if Y is a r.i. function space on $[0,1]$, Y is not $L_2(0,1)$, even up to an equivalent renorming, and Y is finitely crudely representable in a 2-convex Orlicz function space on $[0,1]$ or on $[0,\infty)$, then Y is lattice finitely crudely representable in the Orlicz space.

3. There is an intermediate step in the proof of 7.5 which gives a "uniformity" result in the following sense: assume F is given so that $F(\sqrt{t})$ is equivalent to a convex function. For every $K \ge 1$, there is a constant $D = D(K)$ such that, for every r.i. function space X on $[0,1]$

which is K-finitely representable in $L_F(0,\infty)$, there is a constant $0 \leq w \leq D$ and a function $G \in \overline{\mathcal{C}}(F)$ so that

$$D^{-1}\|f\|_X \leq \max\{w\|f\|_{L_G}, \|f\|_{L^2}\} \leq D\|f\|_X ,$$

for every $f \in X$.

Theorem 7.6: Let X *be a r.i. function space on* $[0,\infty)$.

(1) X *is finitely crudely representable in a 2-convex Orlicz function space* $L_F(0,\infty)$ *if and only if* X *is* $L_G(0,\infty)$ *, up to an equivalent renorming, where*

(a) $G(t)$ *is equivalent to* t^2 *(both at* 0 *and at* ∞ *), or*

(b) G *is equivalent to a function in* $\overline{\mathcal{C}}(F)$ *, or*

(c) $G(t)$ *is equivalent to* $\max(H(t),t^2)$ *for some* H *in* $\overline{\mathcal{C}}(F)$ *(so that* X *is* $L_H(0,\infty) \cap L_2(0,\infty)$*).*

(2) X *is finitely crudely representable in a 2-convex Orlicz function space* $L_F(0,1)$ *if and only if* X *is* $L_G(0,\infty)$*, up to an equivalent renorming, where*

(a) $G(t)$ *is equivalent to* t^2 *, or*

(b) G *is equivalent to a function in* $\bigcap_{1\leq s<\infty} \overline{\mathcal{C}}(F,[s,\infty))$ *, or*

(c) $G(t)$ *is equivalent to* $\max(H(t),t^2)$ *for some* H *in* $\overline{\mathcal{C}}(F,[1,\infty))$ *(so that* X *is* $L_H(0,\infty) \cap L_2(0,\infty)$*)* .

Proof: We prove the "if" parts for (1) and (2) simultaneously. Case (a) is obvious, because $L_G(0,\infty)$ is $L_2(0,\infty)$. Case (b) follows from 7.5. To see this, for each $n = 1,2,\ldots$ represent $L_G(0,n)$ as an Orlicz function space $L_{G_n}(0,1)$ on $[0,1]$ via the isometry U_n , where

$$U_nf(t) = \|\chi_{(0,n)}\|_{L_G(0,n)}f(tn) \qquad (f \in L_G(0,n))$$

and G_n is defined by

$$G_n(t) = G(\|\chi_{(0,n)}\|^{-1}_{L_G(0,n)}t)/G(\|\chi_{(0,n)}\|^{-1}_{L_G(0,n)})$$

$$= nG(\|\chi_{(0,n)}\|^{-1}_{L_G(0,n)}t)$$

since this choice of G_n makes the equality

$$\int_0^n G(|f(t)|)dt = \int_0^1 G_n(|U_nf(t)|)dt$$

valid for every measurable function f on $[0,n]$. Obviously, G_n is in $\overline{\mathcal{C}}(F)$ when G is in $\overline{\mathcal{C}}(F)$, and G_n is in $\overline{\mathcal{C}}(F,[1,\infty))$ when G is in $\bigcap_{1\le\lambda<\infty} \overline{\mathcal{C}}(F,[\lambda,\infty))$. Thus, by the proof of 7.5, $L_{G_n}(0,1)$ is finitely representable in $L_F(0,\infty)$ ($L_F(0,1)$ in case (2)) . Since $\bigcup_{n=1}^{\infty} L_G(0,n)$ is dense in $L_G(0,\infty)$, we have that $L_G(0,\infty)$ is finitely representable in $L_F(0,\infty)$ ($L_F(0,1)$, in case (2)).

Part (c) of (1) follows from the earlier cases. $L_H(0,\infty)$ and $L_2(0,\infty)$ are both finitely representable in $L_F(0,\infty)$ and $L_G(0,\infty) = L_H(0,\infty) \cap L_2(0,\infty)$ embeds isomorphically into $L_H(0,\infty) \oplus L_2(0,\infty)$ as the "diagonal" $\{(f,f): f \in L_G(0,\infty)\}$, hence $L_G(0,\infty)$ is finitely crudely representable in $L_F(0,\infty) \oplus L_F(0,\infty)$, which is isomorphic to $L_F(0,\infty)$.

Part (c) of (2) requires a little work. Since $L_G(0,1)$ is finitely representable in $L_F(0,1)$ by 7.5, it is sufficient to show that $L_G(0,\infty)$ is finitely crudely representable in $L_G(0,1)$. We assume, without loss of generality, that $G(t) = t^2$ for $0 \le t \le 1$ and that $G(t) \ge t^2$ for $t > 1$.

Let $\{f_{n,i}\}_{i=1}^{\infty}$ be symmetric, independent random variables on $[0,1]$ such that $|f_{n,i}| = \chi_{C_{n,i}}$ for some set $C_{n,i}$ of measure 2^{-n} . It is enough to show that, for each fixed n and arbitrarily large m , the sequence $\{x_{n,i}\}_{i=1}^{m}$ (with $x_{n,i} = \chi_{[(i-1)2^{-n},i2^{-n})}$) in $L_G(0,\infty)$ is K-equivalent to the sequence $\{f_{n,i}\}_{i=1}^{m}$ in $L_F(0,1)$, where K is a constant independent of n and m .

As in the discussion preceding 7.4, we have by the classification formula 2.1 and by 7.3 that there is a constant M so that for all n,m, and scalars $\{a_i\}_{i=1}^m$,

$$M^{-1}\|\sum_{i=1}^m a_i f_{n,i}\|_{L_G(0,1)} \le$$

$$\le \max\{\|\sum_{i=1}^m a_i x_{n,i}\|_{L_G(0,\infty)}, (\sum_{i=1}^m |a_i|^2)^{1/2} \cdot \frac{\|\sum_{i=1}^m f_{n,i}\|_{L_G(0,1)}}{\sqrt{m}}\}$$

$$\le M\|\sum_{i=1}^m a_i f_{n,i}\|_{L_G(0,1)} .$$

Thus we only need to check that if m is large then

$$(\sum_{i=1}^m |a_i|^2)^{1/2} \cdot \frac{\|\sum_{i=1}^m f_{n,i}\|_{L_G(0,1)}}{\sqrt{m}} \le C\|\sum_{i=1}^m a_i x_{n,i}\|_{L_G(0,\infty)} ,$$

for some constant C which is independent of n. Choose $p < \infty$ such that $L_G(0,1)$ is p-concave and let K be the norm of the injection from $L_p(0,1)$ into $L_G(0,1)$ (cf. 2.7). By the central limit theorem we have that

$$\limsup_{m\to\infty} \frac{\|\sum_{i=1}^m f_{n,i}\|_{L_G(0,1)}}{\sqrt{m}} \le K \lim_{m\to\infty} \frac{\|\sum_{i=1}^m f_{n,i}\|_{L_p(0,1)}}{\sqrt{m}} =$$

$$= K\|f_{n,1}\|_{L_2(0,1)}\|f\|_{L_p(0,1)}$$

where f is a random variable with gaussian distribution whose $L_2(0,1)$ norm is one; the important things for us are that $\|f\|_{L_p(0,1)}$ is finite and $\|f_{n,1}\|_{L_2(0,1)} = 2^{-n/2}$. To complete the proof, it is sufficient to verify the inequality

$$2^{-n/2}(\sum_{i=1}^m |a_i|^2)^{1/2} \le \|\sum_{i=1}^m a_i x_{n,i}\|_{L_G(0,\infty)} .$$

This is clear, because $\|\sum_{i=1}^{m} a_i x_{n,i}\|_{L_G(0,\infty)} = 1$ if and only if

$$1 = \int_0^\infty G(\sum_{i=1}^{m} |a_i x_{n,i}(t)|)dt = \sum_{i=1}^{m} 2^{-n} G(|a_i|) \geq 2^{-n} \sum_{i=1}^{m} |a_i|^2 .$$

(We thank Gilles Pisier for suggesting the use of the central limit theorem in the above argument.)

We turn now to the "if" part.

It follows from Lemma 7.4 and the same discussion as in 7.5 that there is a constant D such that for every integer n, there are a constant $0 \leq w_n \leq D$ and a function $G_n \in \overline{\mathcal{C}}(F)$ so that

$$D^{-1}\|f\|_X \leq \max\{w_n\|f\|_{L_{G_n}}, \rho_n\|f\|_{L_2}\} \leq D\|f\|_X ,$$

for every f supported on $[0,n]$, with $\rho_n = \|\chi_{(0,n)}\|_X/\sqrt{n}$. (Notice that $\{\rho_n\}_{n=1}^\infty$ is bounded since X is of type 2.) If we choose a subsequence $\{n_k\}_{k=1}^\infty$ such that $\lim_{k\to\infty} w_{n_k} = w$, $\lim_{k\to\infty} \rho_{n_k} = \rho$ and $\lim_{k\to\infty} G_{n_k} = G \in \overline{\mathcal{C}}(F)$, we get

$$D^{-1}\|f\|_X \leq \max\{w\|f\|_{L_G}, \rho\|f\|_{L_2}\} \leq D\|f\|_X ,$$

for every f with bounded support, and hence for every $f \in X$. Now, w and ρ cannot be both 0. The possibility $w = 0$, $\rho > 0$ yields 1(a) or 2(a). If w and ρ are > 0, then 1(c) occurs in the $L_F(0,\infty)$ case. If w and ρ are > 0 and X is finitely crudely representable in $L_F(0,1)$, we already know that X is the space $L_{\tilde{G}}(0,\infty)$, where $\tilde{G}(t)$ is equivalent to t^2 at 0 and to $G(t)$ at ∞; for example, $\tilde{G}(t) = \max\{t^2; G(t)\}$. Since $L_{\tilde{G}}(0,1) = L_G(0,1)$ is finitely crudely representable in $L_F(0,1)$, we know from 7.5 that G is equivalent at ∞ to a function $H \in \overline{\mathcal{C}}(F,[1,\infty))$, and thus $\tilde{G}$ is equivalent to $\max\{t^2,H(t)\}$, which gives 2(c). Now $w > 0$, $\rho = 0$ gives immediately 1(b) in the $L_F(0,\infty)$ case, but 2(b) requires more work. Since $\rho = 0$ we know that X is isomorphic to $L_G(0,\infty)$, and we may assume that $G(t)$ is not equivalent to t^2 at 0 (otherwise, this is case 2(c) again). For any fixed n, the infinite sequence $\{x_{n,i}\}_{i=1}^{\infty}$ in $L_G(0,\infty)$ is equivalent, up to a constant which depends upon n, to the unit vector basis in the Orlicz sequence space ℓ_G, which is not ℓ_2 in this case. So, by 10.5, for each m and n, $\{x_{n,i}\}_{i=1}^{m}$ in $L_G(0,\infty)$ is K_1-equivalent to a sequence of disjoint vectors in $L_F(0,1)$, where K_1 is the constant of the finite crudely representability of $L_G(0,\infty)$ in $L_F(0,1)$. This yields that for each $\varepsilon > 0$, $L_G(0,\infty)$ is lattice finitely representable, up to the constant K_1, in $L_F(0,\varepsilon)$. Indeed, if n and k are fixed, m is chosen so that $m\varepsilon > k(1+\varepsilon)$, and T is a lattice isomorphism of constant K_1 from $\operatorname{span}\{x_{n,i}\}_{i=1}^{m\cdot 2^n}$ in $L_G(0,\infty)$ into $L_F(0,1)$, then $\mu(\operatorname{supp} Tx_{n,i}) \le \varepsilon 2^{-n}k^{-1}$ for at least $k2^n$ values of i, so that $\{x_{n,i}\}_{i=1}^{k2^n}$ is lattice finitely representable in $L_F(0,\varepsilon)$ up to the constant K_1.

Let $s > 1$ and ε_0 be given, and N, n be integers. Let us choose $\varepsilon > 0$ so that $\|\chi_{(0,\varepsilon)}\|_{L_F} \le \varepsilon_0(sN2^n)^{-1}$, and let T be a K_1-isomorphism from the span of $\{x_{n,i}\}_{i=1}^{N2^n}$ in $L_G(0,\infty)$ into $L_F(0,\varepsilon)$ such that the images $z_i = Tx_{n,i}$; $i = 1,\dots,N2^n$ are disjoint. Since $\sum_{i=1}^{N2^n} \|z_i\chi_{\{|z_i|\le s\}}\|_{L_F} \le \varepsilon_0$ we may assume (by a standard perturbation argument) that for each $i = 1,\dots,N2^n$, $|z_i|$ takes on no value in $(0,s)$. It follows from 7.4 and the remark thereafter that there is a constant D such that for every N and n, we have for any f in $\operatorname{span}\{x_{n,i}\}_{i=1}^{N2^n}$ that

$$D^{-1}\|f\|_X \le \|f\|_{L_{G_{N,n}}} \le D\|f\|_X$$

with

$$G_{N,n}(t) = \int_0^\infty F(tu)/F(u)\,d\nu_{N,n}(u) ,$$

where $\nu_{N,n}$ is the probability measure on $(0,\infty)$ defined by the formula

$$\nu_{N,n}([a,b]) = \sum_{i=1}^{N2^n} \int_{\{a \le w_{N,n}^{-1}|z_i| \le b\}} F(tw_{N,n}^{-1}|z_i(u)|)\frac{du}{N}$$

for $0 < a \le b < \infty$ and for some $D^{-1} \le w_{N,n} \le D$. The measure $\nu_{N,n}$ is supported by $[D^{-1}s,\infty)$, since the $|z_i|$'s take on no value in $(0,s)$, and hence $G_{N,n} \in \overline{\mathcal{C}}(F,[D^{-1}s,\infty))$. Since N and n are arbitrary, a limit argument as before gives $G \in \overline{\mathcal{C}}(F,[D^{-1}s,\infty))$, and since s is arbitrary, the proof is over. □

Before investigating the classification of the r.i. spaces which embed into an Orlicz function space, we outline a simple method for producing sublattices of a r.i. space which are themselves isometric to r.i. spaces on $[0,1]$. (This construction was hinted at in the proof of 7.5.) Let us begin with the following consideration. If X is a r.i. function space on $[0,\infty)$ (respectively, on $[0,1]$) and $(\Omega,\mathcal{A},\lambda)$ is a measure space (respectively, a probability space) we define a norm on the integrable simple functions on $(\Omega,\mathcal{A},\lambda)$ by putting

$$\|f\|_{X(\Omega,\mathcal{A},\lambda)} = \|f^*\|_X ,$$

where f^* denotes the decreasing rearrangement of f. We denote by $X(\Omega,\mathcal{A},\lambda)$ the completion of the integrable simple functions under this norm. If the measure λ has no atoms and is infinite (respectively, has no atoms) the space $X(\Omega,\mathcal{A},\lambda)$ contains a sublattice which is lattice isometric to X. If, furthermore, the σ-algebra $\mathcal{A}$ is countably generated then it is possible to construct a measure equivalence φ between $(\Omega,\mathcal{A},\lambda)$ and $[0,\infty)$ (respectively, $[0,1]$). This means that, up to sets of measure zero, φ is invertible and $\lambda(E) = \mu(\varphi(E))$, for every $E \in \mathcal{A}$. In this case the mapping $f \to f(\varphi)$ is a linear isometry and a lattice isomorphism from X onto $X(\Omega,\mathcal{A},\lambda)$. This is, for example, the case when $\Omega = [0,\infty) \times [0,\infty)$ and X

is a r.i. space on $[0,\infty)$ or when $\Omega = [0,1] \times [0,1]$ and X is on $[0,1]$. For some constructions like the one presented here it is convenient to replace X by $X([0,\infty)^2)$ (respectively $X([0,1]^2)$).

Let X be a r.i. function space on $[0,\infty)$ or on $[0,1]$. It is possible to associate with every positive function g of norm one in X a new r.i. function space X_g which is a sublattice of X. Let s denote either 1 or ∞, according to whether X is on $[0,1]$ or on $[0,\infty)$, and put, for every integrable simple function f on $[0,s)$,

$$\|f\|_{X_g} = \|f \otimes g\|_{X([0,s)^2)} ,$$

where $(f \otimes g)(t,u) = f(t)g(u)$. The space X_g is then defined as the completion of the integrable simple functions under this norm. It is obvious that this formula defines a r.i. norm, and that the embedding T from X_g into $X([0,s)^2)$ given by $Tf = f \otimes g$ is a lattice isomorphism. It is also clear that the norm in X_g always dominates a positive constant times the norm in X. In the case where X is an Orlicz function space $L_F(0,s)$ it is easy to identify X_g. Indeed, if f is a simple function $\|f\|_{X_g} = \|f \otimes g\|_{L_F} = 1$ is equivalent to

$$1 = \int_0^s\int_0^s F(f(t)g(u))dtdu = \int_0^s G(f(t))dt$$

if we define G by

$$G(t) = \int_0^s F(tg(u))du .$$

(We have $G(1) = 1$ because $\|g\|_{L_F} = 1$ by assumption.) This shows that X_g is isometric to $L_G(0,s)$. As in 7.4, we may also write

$$G(t) = \int_0^\infty (F(tu)/F(u))d\nu(u) ,$$

where ν is the probability measure on $(0,\infty)$ defined by

$$\nu([a,b]) = \int_{\{a\leq g\leq b\}} F(g(u))du \quad \text{for every} \quad 0 \leq a \leq b < \infty .$$

It is clear that ν is the image of the probability measure $F(g(u))du$ under the measurable mapping g. It follows that

$$\int \chi_{[a,\infty)}(1/F(v))d\nu(v) = \int \chi_{\{g\geq a\}}(u) \cdot \frac{1}{F(g(u))} \cdot F(g(u))du, \text{ for every } a > 0$$

or

$$\mu(\{g \geq a\}) = \int (F(v))^{-1}\chi_{[a,\infty)}(v)d\nu(v) .$$

In particular, in the case of Orlicz function spaces on $[0,1]$,

$$\mu(\{g > 0\}) = \int (F(v))^{-1}d\nu(v) \leq 1 .$$

Conversely, if ν is a probability measure on $(0,\infty)$ (respectively, such that $\int (F(u))^{-1}d\nu(u) \leq 1$) and if $G(t) = \int F(tu)(F(u))^{-1}d\nu(u)$, we want to show that $L_G(0,s)$ is isometric to $(L_F(0,s))_g$ for some $g \geq 0$ of norm one in $L_F(0,s)$.

To this end, define g by its distribution:

$$\mu\{g \geq a\} = \int (F(u))^{-1}\chi_{[a,\infty)}(u)d\nu(u) \quad \text{for every} \quad a > 0 .$$

(In the case $\int (F(u))^{-1}d\nu(u) \leq 1$, we have $\mu(\{g > 0\}) \leq 1$, which means that we can construct g on $[0,1]$. In every case $\int F(g(u))du = \int d\nu(u) = 1$, so that g has norm one.) With this definition of g, it is clear that $G(t) = \int F(tg(u))du$, and the preceding discussion shows that $L_G(0,s) = (L_F(0,s))_g$.

These considerations yield the first part of

Theorem 7.7: (a) Let F *be an Orlicz function and let* ν *be a probability measure supported on* $(0,\infty)$ *(respectively,* $\int(F(u))^{-1}d\nu(u) \leq 1$), *and define* $G(t) = \int_0^\infty F(ts)/F(s)d\nu(s)$. *Then* $L_G(0,1)$ *is lattice isomorphic to a sublattice of* $L_F(0,\infty)$ *(respectively,* $L_F(0,1)$*)* .

(b) Let F *be an Orlicz function and assume that* $L_F(0,\infty)$ *(respectively,* $L_F(0,1)$*) is p-convex for some* $p > 2$. *If* Y *is a r.i. space on* $[0,1]$ *which embeds into* $L_F(0,\infty)$ *(respectively,* $L_F(0,1)$*)* , *and* Y *is not* $L_2(0,1)$, *up to an equivalent renorming, then* Y *is* $L_G(0,1)$, *up to an equivalent renorming, for some* G *of the form* $G(t) = \int_0^\infty F(st)/F(s)d\nu(s)$, *where* ν *is a probability measure supported on* $[0,\infty)$ *(repectively, on* $[1,\infty)$, *so that* $\int F(u))^{-1}d\nu(u) \le 1$ *)*.

Proof of (b): We know from 7.4 that

$$(\%) \qquad D^{-1}\|f\|_Y \le \max\{w_n\|f\|_{L_{G_n}} , \|f\|_{L_2}\} \le D\|f\|_Y ,$$

for every $f = \sum_{i=1}^{2^n} a_i x_{n,i}$, where $G_n(t) = \int_0^\infty F(st)/F(s)d\nu_n(s)$ for appropriate probability measures ν_n . However, we must make sure that the sequence $\{\nu_n\}_{n=1}^\infty$ is tight on the locally compact space $(0,\infty)$, so that when we pass to the limit G of $\{G_n\}_{n=1}^\infty$, G has the desired representation. (Recall that a family $\mathcal{M}$ of finite positive measures on a locally compact space $\mathcal{J}$ is tight provided they are of uniformly bounded total mass, and for every $\epsilon > 0$, there exists a compact subset K of $\mathcal{J}$ such that $\nu(\mathcal{J} \sim K) \le \epsilon$, for every $\nu \in \mathcal{M}$.)

Before proceeding to the proof of 7.7 (b), we make a definition. Say that a bounded subset $\mathcal{M}$ of an Orlicz space $L_F(0,\infty)$ is *tight* provided that, for each $\epsilon > 0$ there are $a > 0$ and $b < \infty$ such that $\sup_{f\in\mathcal{M}} \int_{|f|\le a} F(|f(t)|)dt < \epsilon$ and $\sup_{f\in\mathcal{M}} \int_{|f|\ge b} F(|f(t)|)dt < \epsilon$. It is clear that an equi-integrable set in $L_F(0,\infty)$ is tight; the converse is true if $\mathcal{M} \subseteq L_F(0,1)$, but is false in general, since the tight family $\{\chi_{(i,i+1)}\}_{i=0}^\infty$ is not equi-integrable. (However, it is true that $\mathcal{M}$ is tight in $L_F(0,\infty)$ if and only if the collection of decreasing rearrangements of the functions in $\mathcal{M}$ is equi-integrable in $L_F(0,\infty)$, but this observation is not useful for us.)

We turn now to the proof of 7.7 (b). By 6.11, since L_F is p-convex for some $p > 2$, there is an embedding T of Y into $L_F(0,\infty)$ (respectively, $L_F(0,1)$) such that the sequence $\{v_n\}_{n=1}^{\infty}$ defined by $v_n = (\sum_{i=1}^{2^n} |Th_{n,i}|^2)^{1/2}$ is equi-integrable and hence tight.

Let $\{y_{n,i}\}_{i=1}^{2^n}$ be the symmetrization of $\{Th_{n,i}\}_{i=1}^{2^n}$ produced by the construction in 2.1, which was recalled before 7.4. Thus $\{y_{n,i}\}_{i=1}^{2^n}$ is symmetrically exchangeable and $v'_n = (\sum_{i=1}^{2^n} |y_{n,i}|^2)^{1/2}$ has, by construction, the same distribution as v_n so that $\{v'_n\}_{n=1}^{\infty}$ is also tight in $L_F(0,\infty)$. Let $\{y'_{n,i}\}_{i=1}^{2^n}$ be disjoint functions on $[0,\infty)$, each of which has the same distribution as $y_{n,1}$, and let $r'_n = \sum_{i=1}^{2^n} y'_{n,i}$ be the first Rademacher element over $\{y'_{n,i}\}_{i=1}^{2^n}$.

Claim: $\{r'_n\}_{n=1}^{\infty}$ *is tight in* $L_F(0,\infty)$.

We postpone the proof of the claim for a while and complete the proof of 7.7 (b). Since the sequence $\{h_{n,i}\}_{i=1}^{2^n}$ is isometrically equivalent to $\{x_{n,i}\}_{i=1}^{2^n}$, we may apply 7.4 with $z_i = Th_{n,i}$. The function G_n introduced in (%) is thus given by the formula

$$G_n(t) = \sum_{i=1}^{2^n} \int_0^{\infty} F(tw_n^{-1}|Th_{n,i}(u)|du ,$$

where w_n is chosen to make $G_n(1) = 1$. By 7.4 we have also

$$G_n(t) = 2^n \int_0^{\infty} F(tw_n^{-1}|y_{n,1}(u)|)du = \int_0^{\infty} F(tw_n^{-1}|r'_n(u)|)du .$$

If we write $G_n(t) = \int_0^{\infty} (F(tu)/F(u))d\nu_n(u)$, we have

$$\nu_n([a,b]) = \int_{\{a\le w_n^{-1}|r_n'|\le b\}} F(w_n^{-1}|r_n'(u)|)du \ .$$

Clearly, the sequence $\{\nu_n\}_{n=1}^{\infty}$ is tight on the locally compact space $(0,\infty)$ if and only if the sequence $\{w_n^{-1}r_n'\}_{n=1}^{\infty}$ is tight in $L_F(0,\infty)$. Since Y is not $L_2(0,1)$, even up to an equivalent renorming, we know that $\liminf_{n\to\infty} w_n > 0$. By 7.4 $\limsup_{n\to\infty} w_n < \infty$, so the tightness of $\{w_n^{-1}r_n'\}_{n=1}^{\infty}$ follows then from the tightness of $\{r_n'\}_{n=1}^{\infty}$. Thus we can select a sequence $\{n_k\}_{k=1}^{\infty}$ of integers so that $\{G_{n_k}\}_{k=1}^{\infty}$ is pointwise convergent to some function G and $\{\nu_{n_k}\}_{k=1}^{\infty}$ is narrowly convergent to some probability measure ν on $(0,\infty)$; i.e., $\int h(t)d\nu_{n_k}(t) \to \int h(t)d\nu(t)$, for each bounded, continuous function h on $(0,\infty)$. Since $u \to F(tu)/F(u)$ is bounded (by Δ_2-condition) and continuous for each fixed t, $G(t) = \int_0^{\infty} (F(tu)/F(u))d\nu(u)$. Passing to the limit in the formula (%), we deduce that Y is $L_G(0,1)$ up to an equivalent renorming and this completes the proof of 7.7 (b) in the $L_F(0,\infty)$ case.

In the $L_F(0,1)$ case, we use the fact that $F(t)$ can be taken to be equivalent to t^p on $[0,1]$ (where $L_F(0,1)$ is p-convex, $p > 2$) and, since $F(t^{1/p})$ is equivalent to a convex function, that $F(t)/t^{1/p}$ is equivalent to an increasing function. Thus $F(st)/F(s) \le K\,F(t)$ for some constant K and all $0 < s < 1$, so $\int_0^1 (F(tu)/F(u))d\nu(u) \le K\,\nu(0,1)F(t)$. We know already from 7.1 (a) that $\|f\|_{L_F(0,1)} \le C\|f\|_Y$ for some constant C, hence if we replace the mass of ν on $(0,1)$ by a point mass at 1, call the resulting measure $\tilde{\nu}$, and define $\tilde{G}$ by

$$\tilde{G}(t) = \int_1^{\infty} (F(tu)/F(u))d\nu(u) \ ,$$

Then $L_G(0,1) = L_{\tilde{G}}(0,1)$.

It remains to prove the claim. For this we need a version of 6.6. We use the following simple lemma.

Lemma 7.8: Suppose that H *is an Orlicz function such that* $H(t)/t \to 0$, *and* $\mathcal{M}$ *is a tight subset of* $L_H(0,\infty)$. *Then there is an Orlicz function* J *such that* $H(t)/J(t) \to 0$ *as* $t \to \infty$ *and as* $t \to 0$, *and* $\mathcal{M}$ *is bounded in* $L_J(0,\infty)$.

(The point of the conditions on J is that they imply that bounded subsets of $L_J(0,\infty)$ are tight in $L_H(0,\infty)$.)

Proof: We can assume, without loss of generality, that H is continuously differentiable, so that H' is increasing by the convexity of H , and $H'(0) = 0$ by the condition on H .

The tightness condition yields that there is an increasing sequence $\{a_n\}_{n=-\infty}^{\infty}$ of positive numbers so that $a_n \to \infty$ as $n \to \infty$, $a_n \to 0$ as $n \to -\infty$, $a_0 = 1$, and

$$\sup_{f\in\mathcal{M}} \int_{a_n}^{\infty} H'(t)\ \mu(\{|f| \geq t\})dt < 4^{-n} \qquad (n \geq 1) ,$$

$$\sup_{f\in\mathcal{M}} \int_{0}^{a_n} H'(t)\ \mu(\{|f| \geq t\})dt < 4^{n} \qquad (n \leq -1) .$$

Further, we can assume that $a_{n+1} \geq 2a_n$ for each n , and (since $H'(t) \to 0$ as $t \to 0$) that $H'(a_n) < 4^{-1}H'(a_{n+1})$ for $n < 0$.

Define a function $\tilde{g}$ on $(0,\infty)$ by letting $\tilde{g}(t) = 2^{|n|}$ on the interval $[a_n, a_{n+1})$. Note that $\{\tilde{g}(a_n)H'(a_n)\}_{n=-\infty}^{\infty}$ is an increasing sequence which goes to 0 as $n \to -\infty$, so that there is a continuous function g on $(0,\infty)$ such that $\tilde{g} \leq g \leq 2\tilde{g}$ and $g(t)H'(t)$ is increasing on $(0,\infty)$.

We can now define J by

$$J(t) = \int_0^t g(s)H'(s)ds \Big/ \int_0^1 g(s)H'(s)ds \ ;$$

then J is an increasing convex function which satisfies $H(t)/J(t) \to 0$ as $t \to \infty$ and as $t \to 0$. The reader can check that, since F and g both satisfy the Δ_2-condition, J does also, so that J is an Orlicz function. Finally, for $f \in \mathcal{M}$,

$$\int_0^\infty J(|f(t)|)dt = \int_0^\infty g(t)H'(t)\mu(\{|f| \geq t\})dt \leq$$

$$\leq 2 \sum_{n=-\infty}^{\infty} 2^{|n|} \int_{a_n}^{a_{n+1}} H'(t)\mu(\{|f| \geq t\})dt \leq 4 + 4\int_0^\infty H(|f(t)|)dt ,$$

so that $\mathcal{M}$ is bounded in $L_J(0,\infty)$.

□

Now we turn to the proof of the claim. Let us again mention that in the $L_F(0,1)$ case, we can assume that $F(t)$ is equivalent at 0 to t^p for an appropriate $p > 2$, so that $L_F(0,\infty)$ is p-convex $(p > 2)$ in both cases. Hence, by replacing F by an equivalent Orlicz function, we can assume that $H(t) = F(\sqrt{t})$ is convex, and $H(t)/t \to 0$ as $t \to 0$. Further, since $\{v_n'\}_{n=1}^\infty$ is tight in $L_F(0,\infty)$, $M = \{(v_n')^2\}_{n=1}^\infty$ is tight in $L_H(0,\infty)$, so that we can apply 7.7 to get an Orlicz function J . Letting $G(t) = J(t^2)$, we have from 7.7 that $L_G(0,\infty)$ is a 2-convex Orlicz space which contains $\{v_n'\}_{n=1}^\infty$ as a bounded subset; further, $F(t)/G(t) \to 0$ as $t \to \infty$ and as $t \to 0$, so that bounded subsets of $L_G(0,\infty)$ are tight in $L_F(0,1)$.

Since $\{y_{n,i}\}_{i=1}^{2^n}$ is symmetrically exchangeable,

$$\|\sum_{i=1}^{2^n} y_{n,i}\|_{L_G(0,\infty)} = \int_0^1 \|\sum_{i=1}^{2^n} r_i(t)y_{n,i}\|_{L_G(0,\infty)}dt ,$$

but, by (*) from the Introduction, this Rademacher average of $\{y_{n,i}\}_{i=1}^{2^n}$ is smaller than $K\|(\sum_{i=1}^{2^n} |y_{n,i}|^2)^{1/2}\|_{L_G(0,\infty)} = K\|v_n'\|_{L_G(0,\infty)}$ for some constant K independent of n . That is, $\{\|\sum_{i=1}^{2^n} y_{n,i}\|_{L_G(0,\infty)}\}_{n=1}^\infty$ is bounded; hence, by 7.3, $\{\|\sum_{i=1}^{2^n} y_{n,i}'\|_{L_G(0,\infty)}\}_{n=1}^\infty$ is also bounded, whence $\{\sum_{i=1}^{2^n} y_{n,i}'\}_{n=1}^\infty$ is tight in $L_F(0,\infty)$, as desired.

□

Remark: It follows in particular that a r.i. space on $[0,1]$ different from $L_2(0,1)$ which embeds into a p-convex $(p > 2)$ Orlicz function space

$L_F(0,1)$ (or $L_F(0,\infty)$) is lattice isomorphic to a sublattice of $L_F(0,1)$ (or $L_F(0,\infty)$).

From 7.7 it follows that there are Orlicz spaces other than $L_p(0,1)$ which have only two isomorphic types of r.i. subspaces (namely, $L_2(0,1)$ and themselves). Let us say that an Orlicz function F is *submultiplicative* at ∞ provided that there are a positive constant K and a real number λ such that $F(st) \leq K\,F(s)F(t)$ whenever $\lambda \leq \min(s,t)$. (By adjusting K , λ can always be taken to be one.) Of course, for any p , there are p-convex Orlicz spaces $L_F(0,1)$ which are not isomorphic to any $L_r(0,1)$ space, for which F is submultiplicative at ∞ ; e.g., take $F(t) = t^p \max(1,\log t)$.

Corollary 7.9: Suppose that $L_F(0,1)$ *is p-convex for some* $p > 2$. *Then* F *is submultiplicative at* ∞ *if and only if* $L_2(0,1)$ *and* $L_F(0,1)$ *are the only r.i. spaces, up to equivalent renormings, which embed into* $L_F(0,1)$.

Proof: Suppose that $F(st) \leq K\,F(s)F(t)$ for $1 \leq \min(s,t)$, and Y is a r.i. space on $[0,1]$ which embeds into $L_F(0,1)$. If Y is not $L_2(0,1)$, then by 7.7, Y is $L_G(0,1)$, where $G(t) = \int_1^\infty F(st)/F(s)d\nu(s)$ for an appropriate probability measure ν on $[1,\infty)$, and thus $G(t) \leq K\,F(t)$ for $1 \leq t < \infty$.

We can deduce from 7.1(a) that there is a constant K' so that $F(t) \leq K'G(t)$ for $1 \leq t < \infty$. Actually, this inequality is also obvious from the form of G , since if $\nu(1,t_0) = \delta > 0$, then

$$G(t) \geq \int_1^{t_0} F(ts)/F(s)d\nu(s) \geq (\delta/F(t_0))F(t) .$$

Conversely, if F is not submultiplicative at ∞ , then there are sequences $t_i \uparrow \infty$ and $1 \leq s_i \uparrow \infty$, $i = 1,2,\ldots$, so that $F(t_is_i) \geq$ $\leq 4^iF(t_i)F(s_i)$. Letting $G(t) = \sum_{i=1}^{\infty} 2^{-i}F(ts_i)/F(s_i)$, we have from 7.7 that $L_G(0,1)$ is isometric to a sublattice of $L_F(0,1)$, and it is clear that $G(t)$

is not equivalent to $F(t)$ at ∞, so that $L_F(0,1) \neq L_G(0,1)$. Since also $F(t) \leq \lambda G(t)$ $(1 \leq t < \infty)$ for some constant λ, $L_G(0,1)$ is not $L_2(0,1)$.

□

Appendix to Section 7.

Here we sketch a proof of our assertion that 7.1(a) is true when the Haar basis for X is conditional. It is convenient to represent X as a r.i. space $Y = X([0,1]^2)$ on the square $[0,1] \times [0,1]$ and let $T: Y \to L_F(0,\infty)$ be an isomorphism. For $n = 0,1,\dots$ and $1 \le i \le 2^n$, let $y_{n,i}(t,s) = h_{n,i}(t)\, r_{n,i}(s)$, where $\{r_{n,i}\}_{n=0,i=1}^{\infty\ \ 2^n}$ is an independent sequence of Rademacher functions, and $\{h_{n,i}\}_{n=0,i=1}^{\infty\ \ 2^n}$ is the Haar system in X. Then $\{y_{n,i}\}_{n=0,i=1}^{\infty\ \ 2^n}$ is a 1-unconditional sequence in Y and, for each $n = 0,1,2,\dots,$ $\{y_{n,i}\}_{i=1}^{2^n}$ in Y is isometrically equivalent to $\{x_{n,i}\}_{i=1}^{2^n}$ in X.

We simply repeat the argument for 6.8, but replace $h_{n,i}$ in 6.8 by $y_{n,i}$. That is, we let

$$v_n(E) = \Big(\sum_{h_{n,i} \subseteq E} |Ty_{n,i}|^2\Big)^{1/2}$$

and break into the two cases in 6.8. One should notice that the proof of 6.2 yields that any "gaussian" system $\{z_{n,i}\}_{n=0,i=1}^{\infty\ \ 2^n}$ over $\{y_{n,i}\}_{n=0,i=1}^{\infty\ \ 2^n}$ is equivalent to $\{y_{n,i}\}_{n=0,i=1}^{\infty\ \ 2^n}$, so that, in particular, $\{z_{n,i}\}_{i=1}^{2^n}$ in Y is uniformly equivalent to $\{x_{n,i}\}_{i=1}^{2^n}$ in X. Thus, if case 1 of 6.8 occurs (i.e., having selected disjoint finite set $\{\sigma_m\}_{m=1}^{\infty}$ of integers and scalars $\{\alpha_n\}_{n\in\sigma_m}$ so that $\sum_{n\in\sigma_m} \alpha_n^2 = 1$ for $m = 1,2,\dots$ and $\{\sum_{n\in\sigma_m} \alpha_n^2 v_n^2(E)\}_{m=1}^{\infty}$ converges μ-a.e. for each dyadic interval E, we have that the limit of $\{\sum_{n\in\sigma_m} \alpha_n^2 v_n^2(0,1)\}_{m=1}^{\infty}$ is not 0), then we get the dominance inequality

$$\|f\|_{L_F} \le C\|f\|_X, \qquad (f \in X).$$

On the other hand, if case 2 occurs, it follows that $\{y_{n,i}\}_{n=0,i=1}^{\infty\ \ 2^n}$ is equivalent to a diagonal sum of a sequence of disjoint vectors in L_F with an orthogonal sequence in ℓ_2, so that $\{y_{n,i}\}_{n=0,i=1}^{\infty\ \ 2^n}$ is equivalent to a sequence of disjoint vectors in some Orlicz function space. In this case, the argument for Theorem 3 in [47] yields that X is $L_2(0,1)$, up to an equivalent re-norming. To see this, observe that if $E_{n,i} = [(i-1)2^{-n}, i2^{-n})$ is a dyadic interval, and for $m \geq n$ we set $r_m(n,i) = \sum_{h_{m,j} \subseteq E_{n,i}} y_{m,j}$, then

$\{\|x_{n,i}\|_X^{-1} r_m(n,i)\}_{m=1}^{\infty}$ is K-equivalent to the unit vector basis for ℓ_2, for some constant K which depends only on X. The proof of theorem 3 in [47] yields that for fixed n and large m, $\{\|x_{n,i}\|_X^{-1} r_m(n,i)\}_{i=1}^{2^n}$ is K_1-equivalent to the unit vector basis of $\ell_2^{2^n}$ for another constant K_1 which is independent of n and m. But this last sequence in Y is isometrically equivalent to the sequence $\{\|x_{n,i}\|_X^{-1}\ x_{n,i}\}_{i=1}^{2^n}$ in X, so that X is indeed $L_2(0,1)$, up to an equivalent renorming.

8. ISOMORPHISMS BETWEEN R.I. FUNCTION SPACES ON $[0,\infty)$ AND ON $[0,1]$.

We show in this section that every super-reflexive r.i. function space X on $[0,1]$ is isomorphic to a r.i. function space Y on $[0,\infty)$. This gives a strongly negative answer to the conjecture that L_p is the only space having r.i. representations both on $[0,1]$ and on $[0,\infty)$.

We prove actually this result for a class of spaces X larger than that of super-reflexive r.i. function spaces, namely for the class of r.i. function spaces X for which Boyd's indices satisfy $0 < \beta_X \leq \alpha_X < 1$ (the relevant definitions appear later in this section.) This condition is not necessary, since we show e.g. that the space $(L \operatorname{Log} L)(0,1)$ of functions f such that $\int_0^1 |f(t)| \operatorname{Log}^+ |f(t)| dt < \infty$ is isomorphic to a r.i. function space on $[0,\infty)$. However, we will show at the end of this section that some limitations are necessary since there exist r.i. function spaces on $[0,1]$ which are not isomorphic to r.i. function spaces on $[0,\infty)$.

The description of the space $Y(0,\infty)$ constructed from $X(0,1)$ is very simple, and looks like many other formulas in this paper where the interplay between an L_2-norm and another norm were shown. (See for example 1.6, which is a special case of the result in this section.) The space $Y(0,\infty)$ consists of all the measurable functions f on $[0,\infty)$ such that $\mu\{|f| > t\}$ is finite for every $t > 0$, and the decreasing rearrangement f^* of f satisfies

$$f^*\chi_{(0,1)} \in X(0,1) \quad \text{and} \quad f^*\chi_{(1,\infty)} \in L_2(0,\infty) .$$

(We don't try yet to define a norm on Y.)

The introduction of the space $Y(0,\infty)$ provides a possibly more intuitive approach to the construction of Rosenthal's space X_p [70] and to the isometric embedding of L_q in L_r, $1 \leq r < q < 2$. In connection with X_p, we show that to every r.i. function space Y on $[0,\infty)$ is associated a space U_Y with an unconditional basis. The description again is quite simple: choose any sequence $\{A_n\}_{n=1}^{\infty}$ of disjoint measurable subsets of $[0,\infty)$ such that $\lim_{n\to\infty} \mu(A_n) = 0$ and $\sum_{n=1}^{\infty} \mu(A_n) = +\infty$. The space U_Y is isomorphic to the closed linear span of the sequence $\{\chi_{A_n}\}_{n=1}^{\infty}$ in $Y(0,\infty)$ (of course, the problem is to show that up to an isomorphism, this span does not depend on the particular choice of a sequence $\{A_n\}_{n=1}^{\infty}$ with the above properties. The sequence $u_n = \|\chi_{A_n}\|_Y^{-1} \cdot \chi_{A_n}$, $n = 1,2,\ldots$, clearly forms an unconditional basis for the space). The space X_p is simply the space U_Y corresponding to the r.i. function space Y on $[0,\infty)$ constructed from $L_p(0,1)$, as is explained above (the space Y is the non-regular representation of L_p as a r.i. function space on $[0,\infty)$ appearing in 1.7).

For what concerns embeddings of L_q into L_r, we begin with a simple lemma: if a r.i. function space Y on $[0,\infty)$ contains the function $g(t) = t^{-1/q}$, then there exists an isometry from $L_q(0,\infty)$ into $Y(0,\infty)$ (which is also a lattice isomorphism). If X is a r.i. function space on $[0,1]$ containing $g \cdot \chi_{[0,1]}$, and if, in addition, $1 < q < 2$, then it is clear from the description (given above) of the space $Y(0,\infty)$ associated to $X(0,1)$ that g belongs to $Y(0,\infty)$, and hence $L_q(0,\infty)$ embeds into $X(0,1)$ (which is isomorphic to $Y(0,\infty)$.) We should point out that the final embedding of L_q into L_r, $1 \leq r < q < 2$, obtained by composing the lattice embedding of L_q into the space $Y(0,\infty)$ associated to $L_r(0,1)$, and some isomorphism (actually, isometry) from $Y(0,\infty)$ into $L_r(0,1)$ coincides exactly with the so-called q-stable embedding. Therefore, our method does not give a different or new way but it points out an imtermediate step in the construction; namely, the space $Y(0,\infty)$, which seems to be also of some independent interest.

The main tool for the construction of the space $Y(0,\infty)$ will be the Poisson process. This is not surprising, according to our previous discussion, since it is known that Poisson processes are the fundamental pieces from which q-stable processes are built.

Recall that the Poisson process $\{N_t\}_{0\leq t<\infty}$ with parameter θ $(\theta > 0)$ is a random process defined on some probability space $(\Omega,\mathfrak{A},P)$ with the following properties:

(a) For every sequence $0 \leq t_1 \leq t_2 \leq \ldots \leq t_k$, the random variables, $N_{t_1}, N_{t_2}-N_{t_1}, \ldots, N_{t_k}-N_{t_{k-1}}$ are independent.

(b) For every s and $t \geq 0$, the random variables $N_{t+s}-N_s$ takes non-negative integer values, with the distribution

$$P\{N_{t+s}-N_s = n\} = e^{-\theta t}(\theta t)^n/n! \ ; \ n = 0,1,2,\ldots$$

(The distribution of $N_{t+s}-N_s$ does not depend on s. Such a process called a *stationary process*.) We shall use the probabilistic notation $EX = \int X(w)dP(w)$. An easy computation shows that the *characteristic function* (Fourier transform) of N_t is given by

$$E \exp iuN_t = \exp(-\theta t(1-e^{iu})) .$$

For further use in this section and for the convenience of the reader who is not familiar with the Poisson process we shall give a concrete construction of N_t for $0 \leq t \leq 1$. After this is done, the construction of N_t for $1 < t < \infty$ is straightforward: consider independent copies $\{N_{t,n}\}_{0\leq t\leq 1}$; n=1,2,... of the Poisson process restricted to $[0,1]$, and set

$$N_t = \sum_{n=1}^{[t]} N_{1,n} + N_{t-[t],[t]+1} ,$$

where $[t]$ denotes the integer part of t, and $N_{0,0} = 0$.

For the construction of $\{N_t\}_{0\leq t\leq 1}$, let us first consider a sequence $\{X_n\}_{n=1}^{\infty}$ of independent random variables over some probability space $(\Omega',\mathfrak{A}',P')$ such that each variable X_n takes values in $[0,1]$ with the probability

$P'(s \leq X_n \leq t) = t-s$ for $0 \leq s \leq t \leq 1$. (The variables X_n's are said to be *uniformly distributed* on $[0,1]$.) Define now for $k = 0,1,\ldots,$ $t \in [0,1]$ and $\omega \in \Omega'$

$$N_t^{(k)}(\omega) = \sum_{j=1}^{k} \chi_{[0,t]}(X_j(\omega)) .$$

Consider a sequence of disjoint subsets $\{A_k\}_{k=0}^{\infty}$ of $[0,1]$ such that $\mu(A_k) = e^{-\theta} \cdot \theta^k/k!$ (note that $\sum_{k=0}^{\infty} \mu(A_k) = 1$) and define on $(\Omega,\mathfrak{A},P) = (\Omega',\mathfrak{A}',P') \times ([0,1],\mu)$ the process $\{N_t\}_{0 \leq t \leq 1}$ by the formula

$$N_t(\omega,x) = \sum_{k=1}^{\infty} \chi_{A_k}(x)\, N_t^{(k)}(\omega) ,$$

where $0 \leq t \leq 1$, $\omega \in \Omega'$ and $x \in [0,1]$.

Let us first compute the characteristic function of N_t-N_s, when $0 \leq s \leq t \leq 1$. We have

$$E_{\omega,x} \exp iu(N_t-N_s)(\omega,x) = \sum_{k=0}^{\infty} e^{-\theta}(k!)^{-1}\theta^k E_\omega \exp iu(N_t^{(k)}(\omega) - N_s^{(k)}(\omega)) .$$

If we notice that $N_t^{(k)}(\omega) - N_s^{(k)}(\omega)$ is the sum of the k independent variables $\chi_{(s,t]}(X_j)$, $j = 1,\ldots,k$, which all have the same distribution, we have

$$E \exp iu(N_t-N_s) = \sum_{k=0}^{\infty} e^{-\theta}(k!)^{-1}\theta^k (E(\exp iu\, \chi_{[s,t]}(X_1)))^k$$

$$= e^{-\theta} \exp\{\theta\, E \exp iu\, \chi_{(s,t]}(X_1) = e^{-\theta} \exp\{\theta(1-(t-s) + (t-s)e^{iu})\}$$

$$= \exp(-\theta(t-s)(1-e^{iu})) ,$$

which shows that the distribution of N_t-N_s is what we wanted. The same computation shows that for $0 \leq t_1 \leq \ldots \leq t_n \leq 1$

$$E(\exp i\{ \sum_{j=1}^{n} u_j(N_{t_j}-N_{t_{j-1}})\}) = \exp(-\theta \cdot \sum_{j=1}^{n} (t_j-t_{j-1})(1-e^{iu_j})) ,$$

which proves that the increments $N_{t_1}, N_{t_2}-N_{t_1}, \ldots, N_{t_n}-N_{t_{n-1}}$ are independent. This means that the process we constructed possesses the properties stated for the Poisson process, when the time parameter t is restricted to $[0,1]$.

For Banach space purposes it is more convenient to use the symmetrized Poisson process $\{Z_t\}_{0\le t<\infty}$, which can be defined in the following way: consider two independent copies $\{N_t\}_{t\ge 0}$ and $\{\tilde{N}_t\}_{t\ge 0}$ of the Poisson process with parameter θ and set $Z_t = N_t - \tilde{N}_t$. It is immediate that

$$E \exp iuZ_t = (E \exp iuN_t)(E \exp -iu\tilde{N}_t) = \exp\{-2\theta t(1-\cos u)\} .$$

We get a convenient normalization by choosing $2\theta = 1$. Of course, the process $\{Z_t\}_{t\ge 0}$ is a stationary process with independent increments, and each variable Z_t is symmetrically distributed, so that now the sequence $\{Z_{(k+1)/n} - Z_{k/n}\}_{k=0}^{\infty}$ is a symmetrically exchangeable sequence for every integer n. Let us compute the L_2-norm of Z_t. By differentiating twice the characteristic function with respect to u we have

$$E(-Z_t^2 \exp iuZ_t) = (t^2 \sin^2 u - t \cos u)\exp\{-t(1-\cos u)\} ,$$

which gives for $u = 0$,

$$E(Z_t^2) = t .$$

Denote by $\{U_n\}_{n=1}^{\infty}$ the sequence $\{Z_n - Z_{n-1}\}_{n=1}^{\infty}$. This sequence is symmetrically exchangeable, and orthonormal in L_2.

Lemma 8.1: For every $1 \le p < \infty$ *there is a constant* $K_1 = K_1(p)$ *such that*

$$K_1^{-1}\left(\sum_{n=1}^{\infty} a_n^2\right)^{1/2} \le \left\| \sum_{n=1}^{\infty} a_n U_N \right\|_{L_p} \le K_1\left(\sum_{n=1}^{\infty} a_n^2\right)^{1/2} ,$$

for every sequence $\{a_n\}_{n=1}^{\infty}$ *of real numbers.*

Proof: For $1 \leq p \leq 2$, since L_p is of cotype 2 , we have

$$A_1^{-1}(\sum_{n=1}^{\infty} a_n^2)^{1/2} \|U_1\|_{L_p} \leq \|\sum_{n=1}^{\infty} a_n U_n\|_{L_p} \leq \|\sum_{n=1}^{\infty} a_n U_n\|_{L_2} = (\sum_{n=1}^{\infty} a_n^2)^{1/2}$$

and for $p \geq 2$, since L_p is of type 2 ,

$$(\sum_{n=1}^{\infty} a_n^2)^{1/2} = \|\sum_{n=1}^{\infty} a_n U_n\|_{L_2} \leq \|\sum_{n=1}^{\infty} a_n U_n\|_{L_p} \leq B_p \|U_1\|_{L_p} (\sum_{n=1}^{\infty} a_n^2)^{1/2} .$$

□

Denote by $\mathfrak{F}(0,\infty)$ the vector space of step functions f on $[0,\infty)$ such that $\mu\{|f| > 0\}$ is finite. We define a linear mapping T from $\mathfrak{F}(0,\infty)$ to $L_1(\Omega,\mathfrak{A},P)$ by setting, for $0 \leq s < t < \infty$,

$$T\chi_{[s,t]} = Z_t - Z_s ,$$

and extending T to $\mathfrak{F}(0,\infty)$ by linearity. If $[s,t]$ and $[s_1,t_1]$ are disjoint, $T\chi_{[s,t]}$ and $T\chi_{[s_1,t_1]}$ are independent and have each mean zero; therefore, they are orthogonal. Since $E((Z_t-Z_s)^2) = t-s$, it follows that T is an isometry in the L_2-norm. We can thus extend the definition of T to $L_2(0,\infty)$. In particular, $T\chi_A$ makes sense for every integrable subset A of $[0,\infty)$.

The next lemma is routine and we just sketch briefly its proof.

Lemma 8.2: Let $\{A_i\}_{i=1}^k$ *and* $\{B_i\}_{i=1}^k$ *be two finite sequences of pairwise disjoint integrable subsets of* $[0,\infty)$ *such that* $\mu(A_i) = \mu(B_i)$, $i = 1,2,\ldots,k$. *Then the sequences* $\{T\chi_{A_i}\}_{i=1}^k$ *and* $\{T\chi_{B_i}\}_{i=1}^k$ *have the same distribution.*

Proof: The lemma is clear when the A_i's and B_i's are finite unions of dyadic intervals. The proof of the general case goes by approximation. Given $\epsilon > 0$, one can find compact sets $\{K_i\}_{i=1}^k$ such that $K_i \subset A_i$ and $\mu(A_i-K_i) \leq \epsilon$, for $i = 1,\ldots,k$. One can then select disjoint open sets $\{V_i\}_{i=1}^k$ such that $V_i \supset K_i$, $\mu(V_i-K_i) \leq \epsilon$ and each V_i is a finite union of dyadic open intervals. In the same way, choose disjoint open sets $\{W_i\}_{i=1}^k$

which are finite unions of dyadic open intervals, such that $\mu(W_i \Delta B_i) \leq 2\varepsilon$. By removing a piece of measure $\leq 2\varepsilon$ in V_i or W_i, one can assume that $\mu(V_i) = \mu(W_i)$, $\mu(V_i \Delta A_i) \leq 4\varepsilon$, $\mu(W_i \Delta B_i) \leq 4\varepsilon$ for $i = 1,2,\ldots,k$. Now, the sequences $\{T\chi_{V_i}\}_{i=1}^k$ and $\{T\chi_{W_i}\}_{i=1}^k$ have the same distribution and the result follows by letting ε tend to zero.

□

The preceding lemma shows that if φ is a measure preserving transformation from $[0,\infty)$ onto itself and if f is an integrable simple function on $[0,\infty)$, the variables Tf and $T(f(\varphi))$ have the same distribution. Also, if ε is a measurable function on $[0,\infty)$ taking values ± 1, an approximation argument even simpler than the argument in 8.2 shows that Tf and $T(\varepsilon f)$ have the same distribution.

Definition 8.3: Let X *be a r.i. function space on* $[0,1]$ *such that* $Z_t \in X$ *for every* $0 \leq t < \infty$. *For every simple integrable function* f *on* $[0,\infty)$ *set*

$$\|f\| = \|Tf\|_{X(\Omega,\mathfrak{A},P)} .$$

The space $Y_X(0,\infty)$ *(or* $Y(0,\infty)$, *in short) is defined as the completion of* $\mathfrak{F}(0,\infty)$ *under the above norm.*

It follows from 8.2 and the discussion after it that $Y(0,\infty)$ is a r.i. function space on $[0,\infty)$. The linear mapping T extends to an isometry from $Y(0,\infty)$ into $X(\Omega,\mathfrak{A},P)$.

We want to identify the space $Y(0,\infty)$ under a relatively mild assumption on X, involving the indices of X introduced by Boyd in [9]. If X is a r.i. function space on $[0,1]$, define the dilation mapping D_s, $0 < s < \infty$, by the formula

$$(D_s f)(t) = f(st) , \text{ for } t \in [0,1] \text{ and } f \in X .$$

In order to make this definition meaningful, the function f is extended to $[0,\infty]$ by $f(u) = 0$ for $u > 1$. Define now the indices

$$\alpha_X = \inf_{0<s<1} (-\mathrm{Log}\|D_s\|/\mathrm{Log}\ s);\ \beta_X = \sup_{1<s<\infty} (-\mathrm{Log}\|D_s\|/\mathrm{Log}\ s) .$$

It follows from this definition that for every $\epsilon > 0$, there is a constant $C(\epsilon)$ such that

$$s^{-\alpha_X} \leq \|D_s\| \leq C(\epsilon)\ s^{-\alpha_X-\epsilon}$$

for every $0 < s \leq 1$ and

$$s^{-\beta_X} \leq \|D_s\| \leq C(\epsilon)\ s^{-\beta_X+\epsilon}$$

for every $s \geq 1$. It is known and also easily checked that $0 \leq \beta_X \leq \alpha_X \leq 1$. There is an alternative description of the dilation D_s for $s > 1$ which is useful here. If we consider (see the discussion preceding 7.7) X as the space $X([0,1]^2)$ and if $f \in X(0,1)$, the distribution of $D_s f$ coincides with that of $g(u,t) = \chi_{[0,1/s]}(u) \otimes f(t)$, which is an element of $X([0,1]^2)$. By averaging over $v \in [0,1]$ we get, for any $1 \leq p < \infty$,

$$\int |g(u-v,t)|^p dv = |f(t)|^p/s .$$

If we set $g_v(u,t) = g(u-v,t)$, we get when the r.i. function space X is q-concave with constant $\leq M$,

$$\|D_s f\|_X = \|g\|_{X([0,1]^2)} = (\int \|g_v\|^q_{X([0,1]^2)} dv)^{1/q}$$

$$\leq M\|(\int |g_v|^q dv)^{1/q}\|_{X([0,1]^2)} = Ms^{-1/q}\|f\|_X$$

and when X is p-convex with constant $\leq M$,

$$\|D_s f\|_X = \|g\|_{X([0,1]^2)} = (\int \|g_v\|^p_{X([0,1]^2)} dv)^{1/p}$$

$$\geq M^{-1}\|(\int |g_v|^p dv)^{1/p}\|_{X([0,1]^2)} = M^{-1}s^{-1/p}\|f\|_X .$$

It follows in the q-concave case that

$$\beta_X = \sup_{s>1}(-\mathrm{Log}\|D_s\|/\mathrm{Log}\ s) \geq \sup_{s>1}(-\mathrm{Log}\ M/\mathrm{Log}\ s + 1/q) = 1/q\ .$$

In the p-convex case observe first that for $s > 1$

$$D_s(D_{1/s}h) = \chi_{[0,1/s]}h\ ,$$

so that

$$\|D_{1/s}f\|_X \leq M\ s^{1/p}\|D_s(D_{1/s}f)\|_X \leq M\ s^{1/p}\|f\|_X$$

and thus

$$\alpha_X = \inf_{s>1}(-\mathrm{Log}\|D_{1/s}\|/\mathrm{Log}\ 1/s) \leq \inf_{s>1}(-\mathrm{Log}\ M/\mathrm{Log}\ 1/s + 1/p) = 1/p\ .$$

If we recall that a Banach lattice is super-reflexive if and only if it is p-convex and q-concave for some $p > 1$ and $q < \infty$, we see that Boyd's indices of a super-reflexive r.i. function space X satisfy $0 < \beta_X \leq \alpha_X < 1$. The converse however is not true, and there are non-reflexive r.i. function spaces X on $[0,1]$ for which $\beta_X = \alpha_X = 1/2$.

We shall use later in this section a weaker version of Boyd's interpolation theorem [9]: Let X be a r.i. function space on $[0,1]$, p,q such that $0 < 1/q < \beta_X \leq \alpha_X < 1/p \leq 1$, and let $(\Omega,\mathfrak{A},P)$ be a probability space. A linear transformation L, which is bounded from $L_p(\Omega,\mathfrak{A},P)$ to itself and from $L_q(\Omega,\mathfrak{A},P)$ to itself defines a bounded operator from $X(\Omega,\mathfrak{A},P)$ into itself.

Theorem 8.4: Assume that $X(0,1)$ *is a r.i. function space on* $(0,1)$ *such that* $\beta_X > 0$. *The space* $Y_X(0,\infty) = Y(0,\infty)$ *defined in 8.3 consists of all the measurable functions* f *on* $[0,\infty)$ *such that*

$$f^*\chi_{[0,1]} \in X \ \text{ and } \ f^*\chi_{[1,\infty)} \in L_2(0,\infty)\ ,$$

where f^* *denotes the decreasing rearrangement of* f *, and the norm of a function* f *in* $Y(0,\infty)$ *is equivalent to the expression (which is not necessarily a norm)*

$$\|f^*\chi_{[0,1]}\|_X + \|f^*\chi_{[1,\infty)}\|_{L_2(0,\infty)} .$$

Proof: Let us check the assertion concerning the norm. We shall first prove that there is a constant K such that

$$K^{-1}\|f\|_X \le \|f\|_{Y(0,\infty)} = \|Tf\|_{X(\Omega,\mathfrak{A},P)} \le K\|f\|_X ,$$

for every function $f \in X(0,1)$. If we define a linear mapping T_1 from $\mathfrak{F}(0,\infty)$ to $L_1(\Omega,\mathfrak{A},P)$ by $T_1\chi_{(s,t)} = N_t-N_s$, where $\{N_t\}_{0\le t<\infty}$ is the Poisson process with parameter $\theta = 1/2$, we see that Tf has the same distribution as the difference between T_1f and an independent copy $\widetilde{T_1f}$ of T_1f . This yields

$$\|Tf\|_{X(\Omega,\mathfrak{A},P)} \le 2\|T_1f\|_{X(\Omega,\mathfrak{A},P)} ,$$

and

$$P(|Tf| \ge t) \ge P(|T_1f| \ge t \text{ and } \widetilde{T_1f} = 0)$$

$$\ge P(|T_1f| \ge t \text{ and } N_1 = \widetilde{T_1\chi}_{[0,1]} = 0) = e^{-1/2}P(|T_1f| \ge t) .$$

Thus

$$\|T_1f\|_{X(\Omega,\mathfrak{A},P)} \le e^{1/2}\|Tf\|_{X(\Omega,\mathfrak{A},P)} .$$

It is then enough to check that

$$2K^{-1}\|f\|_X \le \|T_1f\|_{X(\Omega,\mathfrak{A},P)} \le (K/2)\|f\|_X ,$$

for every $f \in X(0,1)$. We shall make use of the special representation of the Poisson process $\{N_t\}_{0\le t\le 1}$ discussed above. Recall that

$$N_t(\omega,x) = \sum_{k=0}^{\infty} \chi_{A_k}(x)\cdot N_t^{(k)}(\omega) = \sum_{k=0}^{\infty} \chi_{A_k}(x)\Big(\sum_{j=1}^{k} \chi_{[0,t]}(X_j(\omega))\Big) ,$$

for $\omega \in \Omega'$ and $x \in [0,1]$, where $\{A_k\}_{k=0}^{\infty}$ is a sequence of disjoint subsets of $[0,1]$ with measure $\mu(A_k) = e^{-1/2}/2^k\, k!$. It is clear by linearity that for every $f \in \mathfrak{J}(0,1)$,

$$T_1 f(\omega,x) = \sum_{k=0}^{\infty} \chi_{A_k}(x)\Big(\sum_{j=1}^{k} f(X_j(\omega))\Big) .$$

For every $k \geq 1$ the function of two variables $\chi_{A_k}(x)\big(\sum_{j=1}^{k} f(X_j(\omega))\big)$ has the same distribution as $D_s\big(\sum_{j=1}^{k} f(X_j)\big)$ with $s = e^{1/2}2^k k!$. Since we have $\beta_X > 0$, there are an $\epsilon > 0$ and a C such that $\|D_s\| \leq C\, s^{-\epsilon}$ for every $s \geq 1$. We have therefore,

$$\Big\|\chi_{A_k} \otimes \Big(\sum_{j=1}^{k} f(X_j)\Big)\Big\|_{X(\Omega,\mathfrak{A},P)} \leq C(e^{1/2}2^k k!)^{-\epsilon}\Big\|\sum_{j=1}^{k} f(X_j)\Big\|_{X(\Omega',\mathfrak{A}',P')}$$

$$\leq Ck(e^{1/2}2^k k!)^{-\epsilon}\|f(X_1)\|_{X(\Omega',\mathfrak{A}',P')} = Ck(e^{1/2}2^k k!)^{-\epsilon}\|f\|_X .$$

Since the series $\sum_{k=0}^{\infty} k(e^{1/2}2^k k!)^{-\epsilon}$ is convergent we get that

$$\|T_1 f\|_{X(\Omega,\mathfrak{A},P)} \leq K/2\|f\|_X$$

for every $f \in X(0,1)$. The other side is very simple because

$$\|T_1 f\|_{X(\Omega,\mathfrak{A},P)} \geq \|\chi_{A_1} \otimes f(X_1)\|_{X(\Omega,\mathfrak{A},P)} \geq \frac{e^{-1/2}}{2}\|f\|_X .$$

In order to complete the proof, we will use the fact that the norm of $X(\Omega,\mathfrak{A},P)$ is, up to a constant, less than the norm of $L_q(\Omega,\mathfrak{A},P)$ for some $q < \infty$. (Precisely, the relation $1/q \leq \beta_X$ yields that $\|f\|_X \leq C(\epsilon)\|f\|_{L_{q+\epsilon}}$, for every $\epsilon > 0$.) It follows that the sequence $\{U_n\}_{n=1}^{\infty}$ in $X(\Omega,\mathfrak{A},P)$ is equivalent to the unit vector basis in ℓ_2 since, according to 8.1, we have for all scalars $\{a_n\}_{n=1}^{\infty}$,

$$K_1(1)^{-1}\Big(\sum_{n=1}^{\infty} a_n^2\Big)^{1/2} \leq \Big\|\sum_{n=1}^{\infty} a_n U_n\Big\|_{L_1} \leq \Big\|\sum_{n=1}^{\infty} a_n U_n\Big\|_X \leq$$

$$\leq C\Big\|\sum_{n=1}^{\infty} a_n U_n\Big\|_{L_q} \leq CK_1(q)\Big(\sum_{n=1}^{\infty} a_n^2\Big)^{1/2} .$$

Let us denote $\max\{K_1(1), CK_1(q)\}$ by $K' = K'(X)$ and consider first a simple function f such that $\|f^*\chi_{[0,1]}\|_X + \|f^*\chi_{[1,\infty)}\|_{L_2} \leq 1$. We know already that

$$\|f^*\chi_{[0,1]}\|_{Y(0,1)} = \|T(f^*\chi_{[0,1]})\|_{X(\Omega,\mathfrak{A},P)} \leq K\|f^*\chi_{[0,1]}\|_X .$$

On the other hand, if we observe that $\|f^*\chi_{[0,1]}\|_X \leq 1$ yields $f^*(1) \leq 1$, we get

$$\|f^*\chi_{[1,\infty)}\|_{Y(0,\infty)} \leq \Big\|\sum_{n=1}^{\infty} f^*(n)\chi_{(n,n+1)}\Big\|_{Y(0,\infty)} = \Big\|\sum_{n=1}^{\infty} f^*(n)U_n\Big\|_{X(\Omega,\mathfrak{A},P)}$$

$$\leq K'\Big(\sum_{n=1}^{\infty} f^*(n))^2\Big)^{1/2} = K'\Big\|\sum_{n=1}^{\infty} f^*(n)\chi_{(n-1,n)}\Big\|_{L_2(0,\infty)} \leq$$

$$\leq K'(1 + \|f^*\chi_{[1,\infty)}\|_{L_2(0,\infty)}) .$$

To prove the other directicn, we can assume that f is a non-negative decreasing integrable simple function. We already know that

$$\|f\chi_{[0,1]}\|_{X(0,1)} \leq K\|f\chi_{[0,1]}\|_{Y(0,\infty)} ,$$

and we also have

$$\|f\chi_{[1,\infty)}\|_{L_2(0,\infty)} \leq \Big\|\sum_{n=1}^{\infty} f(n)\chi_{[n,n+1]}\Big\|_{L_2(0,\infty)} \leq K'\Big\|\sum_{n=1}^{\infty} f(n)U_n\Big\|_{X(\Omega,\mathfrak{A},P)}$$

$$= K'\Big\|\sum_{n=1}^{\infty} f(n)\chi_{[n-1,n]}\Big\|_{Y(0,\infty)} \leq K'\|f\|_{Y(0,\infty)} .$$

This completes the verification that $\|f^*\chi_{[0,1]}\|_X + \|f^*\chi_{[1,\infty)}\|_{L_2(0,\infty)}$ defines an expression equivalent to the norm of f in $Y(0,\infty)$. It is now clear that a function f on $[0,\infty)$ such that $f^*\chi_{[0,1]} \in X$ and $f^*\chi_{[1,\infty)} \in L_2(0,\infty)$ belongs to the closure of the simple functions under the norm of $Y(0,\infty)$, and the converse statement is also easy.

□

Notice that in the case of L_p, $2 \leq p < \infty$, the result of 8.4 becomes

$$\|f\|_{Y_p(0,\infty)} \sim \max\{\|f\|_{L_p(0,\infty)}, \|f\|_{L_2(0,\infty)}\} ,$$

where $Y_p(0,\infty)$ is the space associated to L_p by 8.3.

Denote by $\mathcal{Z}$ the vector space of measurable functions generated by the set $\{Z_t;\ 0 \leq t < \infty\}$, and let $\mathcal{Z}_p$ denote the closure of $\mathcal{Z}$ in $L_p(\Omega,\mathfrak{A},P)$. Denote by Q the orthogonal projection from $L_2(\Omega,\mathfrak{A},P)$ onto the closed subspace $\mathcal{Z}_2$.

Lemma 8.5: For every $1 < p < \infty$ *there exists a constant* $K_2 = K_2(p)$ *such that*

$$\|Qf\|_{L_p} \leq K_2\|f\|_{L_p} ,$$

for every simple function f *on* $(\Omega,\mathfrak{A},P)$. *Therefore,* Q *extends to a bounded projection from* $L_p(\Omega,\mathfrak{A},P)$ *onto* Z_p.

Proof: Since Q is self-adjoint, it is enough to consider the case $2 \leq p < \infty$. Denote by $\mathcal{Z}_{n,p}$ the closed linear span of the sequence $\{Z_{k2^{-n}};\ k = 1,2,\ldots\}$ in $L_p(\Omega,\mathfrak{A},P)$. The space $\mathcal{Z}_{n,p}$ is spanned by the symmetrically exchangeable sequence $\{Z_{k2^{-n}} - Z_{(k-1)2^{-n}}\}_{k=1}^{\infty}$, and $\mathcal{Z}_p$ is the closure of the union of the increasing sequence $\{\mathcal{Z}_{n,p}\}_{k=1}^{\infty}$. (Notice that $t \to Z_t$ is continuous from $[0,\infty)$ to $L_p(\Omega,\mathfrak{A},P)$, so that $\mathcal{Z}_p$ coincides with the closed linear span of $\{Z_{k2^{-n}};\ k,n = 0,1,\ldots\}$.) Denote by Q_n the orthogonal projection from $L_2(\Omega,\mathfrak{A},P)$ onto $\mathcal{Z}_{n,2}$. We want to prove that for $2 \leq p < \infty$, the Q_n's are bounded, with a bound independent of n, from

$L_p(\Omega,\mathfrak{A},P)$ to $\mathcal{Z}_{n,p}$. Once this is done, it follows that Q is a bounded projection from $L_p(\Omega,\mathfrak{A},P)$ onto $\mathcal{Z}_p$. Indeed, since $\mathcal{Z}_2$ is the closure of $\bigcup_{n=1}^{\infty} \mathcal{Z}_{n,2}$, we know that for every simple function f the sequence $\{Q_n f\}_{n=1}$ converges to Qf in the L_2-norm. But actually, $\{Q_n f\}_{n=1}^{\infty}$ converges to Qf in every L_q-norm, $2 \le q < \infty$, since it is bounded in $L_{2q}(\Omega,\mathfrak{A},P)$ and convergent in $L_2(\Omega,\mathfrak{A},P)$. It remains to prove the uniform boundedness of the sequence $\{Q_n\}_{n=1}^{\infty}$. Set $V_k = Z_{k2^{-n}} - Z_{(k-1)2^{-n}}$, $k = 1,2,\ldots$. It follows from 8.4 that $\|V_k\|_{L_p} = \|Z_{2^{-n}}\|_{L_p}$ is K-equivalent to $2^{-n/p}$. We have

$$Q_n f = \sum_{k=1}^{\infty} < V_k, f > V_k/\|V_k\|_{L_2}^2 = 2^n \sum_{k=1}^{\infty} < V_k, f > V_k$$

$$= T\Big(\sum_{k=1}^{\infty} 2^n < V_k, f > x_{n,k}\Big) ,$$

where, as usual, $x_{n,k} = \chi_{[(k-1)2^{-n}, k2^{-n})}$. If we set $g = \sum_{k=1}^{\infty} 2^n < V_k, f > x_{n,k}$, we know from 8.4 that

$$\|Q_n f\|_{L_p} \le K \max\{\|g\|_{L_p(0,\infty)} , \|g\|_{L_2(0,\infty)}\} .$$

Since we have

$$\|g\|_{L_2(0,\infty)} = \|Q_n f\|_{L_2(\Omega,\mathfrak{A},P)} \le \|f\|_{L_2(\Omega,\mathfrak{A},P)} \le \|f\|_{L_p(\Omega,\mathfrak{A},P)} ,$$

we need only to estimate $\|g\|_{L_p(0,\infty)}$. If we define q by $1/p + 1/q = 1$, the type q property of L_q gives

$$\Big\|\sum_{k=1}^{\infty} \alpha_k V_k\Big\|_{L_q} \le \|V_1\|_q \cdot \Big(\sum_{k=1}^{\infty} |\alpha_k|^q\Big)^{1/q} \le K2^{-n/q}\Big(\sum_{k=1}^{\infty} |\alpha_k|^q\Big)^{1/q} .$$

It follows by duality that

$$\Big(\sum_{k=1}^{\infty} |< V_k, f >|^p\Big)^{1/p} \le K2^{-n/q}\|f\|_{L_p} ,$$

and therefore,

$$\|g\|_{L_p(0,\infty)} \leq K\|f\|_{L_p} .$$

□

If X is a r.i. function space on $[0,1]$ satisfying $0 < \beta_X < \alpha_X < 1$, it follows from 8.5 and Boyd's interpolation theorem that Q defines a bounded projection from $X(\Omega,\mathfrak{A},P)$ onto the closure $\mathcal{Z}_X$ of $\mathcal{Z}$ in $X(\Omega,\mathfrak{A},P)$. Theorem 8.4 shows that the restriction to $[0,1]$ of the r.i. function space $Y(0,\infty)$ associated to X by 8.3; i.e., $Y(0,1)$, is isomorphic to X. Hence, X is isomorphic to a complemented subspace of $Y(0,\infty)$. On the other hand, $Y(0,\infty)$ is by definition isomorphic to $\mathcal{Z}_X$, which is a complemented subspace of $X(\Omega,\mathfrak{A},P)$, as we said above. We may assume that the σ-algebra $\mathfrak{A}$ is countably generated, and notice that the probability P is necessarily purely non-atomic, which means that we may assume that $X(\Omega,\mathfrak{A},P)$ is isometric to $X(0,1)$. (The first assertion follows from the fact that we can replace $\mathfrak{A}$ by the σ-algebra $\widetilde{\mathfrak{A}}$ generated by the family $\{Z_t;\ t \text{ rational}\}$, which is countably generated, and the second from the fact that $P(Z_t = 0)$ takes all possible values in $(0,1)$ when t varies.) If we notice that $X(0,1)$ and $Y(0,\infty)$ are isomorphic each to their squares $X(0,1) \times X(0,1)$, respectively $Y(0,\infty) \times Y(0,\infty)$, then it follows from the decomposition method that $Y(0,\infty)$ is actually isomorphic to $X(0,1)$. Indeed, writing X for $X(0,1)$ and Y for $Y(0,\infty)$, we know that there are two spaces U and V such that $X \sim Y \times U$ and $Y \sim X \times V$. We have then

$$X \times Y \sim X \times (V \times X) \sim X \times V \sim Y$$

and

$$X \times Y \sim (U \times Y) \times Y \sim U \times Y \sim X .$$

This proves the main result of this section:

Theorem 8.6: Assume that X *is an r.i. function space on* $[0,1]$ *such that* $0 < \beta_X < \alpha_X < 1$ *. The r.i. function space* $Y_X(0,\infty)$ *defined in 8.3 is isomorphic to* X .

Remarks: 1) Another definition of r.i. function spaces used in interpolation theory is that of Luxemburg [50], which is slightly different from ours. One does not assume that $L_\infty(0,1)$ is dense in X but, instead, that $g \in X$ and $|f| \leq |g|$, imply $f \in X$, and that $0 \leq f_n \nearrow f$ almost everywhere with $\lim_n \|f\|_X < \infty$ implies $f \in X$ and $\|f\|_X = \lim_n \|f_n\|_X$.

It is trivial to check that Boyd's interpolation theorem is true with our definition. Let T be an operator which is bounded from $L_p(0,1)$ to $L_p(0,1)$ and from $L_q(0,1)$ to $L_q(0,1)$, and let X be a r.i. function space on $[0,1]$ (in our sense) such that $1/q < \beta_X < \alpha_X < 1/p$. Extend X to a space $\widetilde{X}$ of functions on $[0,1]$ by adding all the measurable functions f on $[0,1]$ such that $|f|$ is the limit a.e. of an increasing sequence $\{f_n\}_{n=1}^\infty$ of positive elements in X with $\sup_n \|f_n\|_X < \infty$. Define the norm of an element f of $\widetilde{X}$ by

$$\|f\|_{\widetilde{X}} = \sup\{\|g\|_X ;\ g \in X,\ 0 \leq g \leq |f|\} .$$

The space $\widetilde{X}$ is a r.i. function space on $[0,1]$ in Luxemburg's sense, and the indices of $\widetilde{X}$ are equal to the indices of X . It follows from Boyd's interpolation theorem that T maps $\widetilde{X}$ into $\widetilde{X}$ and it remains to show that T maps X into X . This is clear since T maps $L_q(0,1)$, and thus $L_\infty(0,1)$, which is dense in X , into $L_q(0,1)$, which is contained in X .

If we want to carry out the construction of $Y_X(0,\infty)$ with Luxemburg's definition of r.i. function spaces, we may define the unit ball of $Y_X(0,\infty)$ as the smallest set C containing $\{f \in \mathfrak{F}(0,\infty);\ \|Tf\|_{X(\Omega,\mathfrak{A},P)} \leq 1\}$ and satisfying all the conditions for a r.i. space in Luxemburg's definition. Then we have of course to justify that T is definable on $Y_X(0,\infty)$ and is still an isometry from $Y_X(0,\infty)$ into $X(\Omega,\mathfrak{A},P)$. We leave the details to the interested reader.

2) If X is an Orlicz function space $L_F(0,1)$, where F satisfies the Δ_2-condition at ∞ (which is equivalent to $0 < \beta_X$ in this case), the space $Y(0,\infty)$ associated to X is also an Orlicz space: let G be a convex function on $[0,\infty)$, equivalent to t^2 at 0 and to F at ∞; for example,

$$G(t) = t^2\chi_{[0,1)}(t) + (2F(t)-1)\chi_{[1,\infty)}(t) .$$

Then it is easily verified that $L_G(0,\infty)$ is equal to the space described in 8.4. This generalizes 7.6 where we showed, under the additional assumption of type 2 for $L_F(0,1)$, that $L_G(0,\infty)$ is finitely crudely representable in $L_F(0,1)$.

3) The condition $0 < \beta_X < \alpha_X < 1$ is not necessary to achieve the construction of $Y(0,\infty)$ and to prove the isomorphism with $X(0,1)$. Actually, the easiest space (except $L_2(\Omega,\mathfrak{A},P)$!) for computing the closed linear span of the Z_t's is the Orlicz function space $L_F(\Omega,\mathfrak{A},P)$, with $F(t) = e^t-1$. (The function F does not satisfy Δ_2 at ∞, and $L_F(\Omega,\mathfrak{A},P)$ is not a r.i. function space in our sense because $L_\infty(\Omega,\mathfrak{A},P)$ is not dense in it. However, it is a r.i. function space in Luxemburg's sense.) If $f = \sum_{i=1}^k a_i\chi_{[s_{i-1},s_i)}$, $0 < s_1 < s_2 < \dots < s_k$, we have for every $\lambda > 0$

$$\begin{aligned} E \exp \lambda Tf &= E \exp(\lambda \cdot \sum_{i=1}^k a_i(Z_{s_i} - Z_{s_{i-1}})) \\ &= \prod_{i=1}^k E \exp(\lambda a_i N_{s_i-s_{i-1}}) \cdot E \exp(-\lambda a_i N_{s_i-s_{i-1}}) \\ &= \exp(\sum_{i=1}^k (s_i-s_{i-1})\{\cosh(\lambda a_i) - 1\}) . \end{aligned}$$

Define a convex function G on $[0,\infty)$ by $G(t) = \cosh t - 1$. The function G is equivalent to t^2 at 0 and to F at ∞. If $\|f\|_{L_G(0,\infty)} = 1$, we have

$$\sum_{i=1}^k (s_i-s_{i-1})\, G(a_i) = 1 ,$$

and, therefore, $E \exp Tf = e$ so that $E(\exp |Tf|-1) \geq E \exp Tf-1 = e-1 \geq 1$, which shows that $\|Tf\|_{L_F(\Omega,\mathfrak{A},P)} \geq 1$. On the other hand, if we notice that $E \exp|Z| \leq 2 E \exp Z$ for any symmetrically distributed variable Z, we get by convexity for $\rho = (2e-1)^{-1} < 1$

$$E(\exp \rho|Tf| -1) \leq \rho E(\exp|Tf| -1)$$

$$\leq \rho(2E \exp Tf - 1) = \rho(2e-1) = 1 .$$

Finally,

$$\|f\|_{L_G(0,\infty)} \leq \|Tf\|_{L_F(\Omega,\mathfrak{A},P)} \leq (2e-1)\|f\|_{L_G(0,\infty)} .$$

It follows from 8.2 that the same relation holds for any simple integrable function f. Furthermore, the computation shows that $E \exp \lambda|Tf|$ is finite for every $\lambda > 0$, which yields that Tf belongs to the closure of $L_\infty(\Omega,\mathfrak{A},P)$ in $L_F(\Omega,\mathfrak{A},P)$, denoted by $L_F^0(\Omega,\mathfrak{A},P)$. Hence, T extends to an isomorphism from $L_G^0(0,\infty)$ (the closure of $\mathfrak{F}(0,\infty)$ in $L_G(0,\infty)$) into $L_F^0(\Omega,\mathfrak{A},P)$. It is also easy to deduce that T actually extends to an isomorphism from $L_G(0,\infty)$ into $L_F(\Omega,\mathfrak{A},P)$. First, T is already defined on $L_G(0,\infty)$ since $L_G(0,\infty)$ is contained in $L_2(0,\infty)$. If f is a norm one element in $L_G(0,\infty)$, $f\chi_{\{|f|\leq n\}}$ belongs to $L_G^0(0,\infty)$ for every n. Since the sequence $\{f\chi_{\{|f|\leq n\}}\}_{n=1}^\infty$ converges to f in $L_2(0,\infty)$, there is a subsequence $\{n_k\}_{k=1}^\infty$ such that $\{T(f\chi_{\{|f|\leq n_k\}})\}_{k=1}^\infty$ converges a.e. to Tf. We get then $E \exp \rho|Tf| \leq 1$ using Fatou's lemma and $\rho = (2e-1)^{-1}$. To prove the other direction, note that $\lim_{n\to\infty} \|f\chi_{\{|f|\leq n\}}\|_{L_G} = 1$ and that $\|Tf\|_{L_F(\Omega,\mathfrak{A},P)} \geq \|T(f\chi_{\{|f|\leq n\}}\|_{L_F(\Omega,\mathfrak{A},P)}$ since Tf is the sum of $T(f\chi_{\{|f|\leq n\}})$ and $T(f\chi_{\{|f|>n\}})$ which are independent and centered. So the relation $\|f\|_{L_G(0,\infty)} \leq \|f\|_{L_F(\Omega,\mathfrak{A},P)}$ extends to $f \in L_G(0,\infty)$.

We want to consider now the dual spaces of $L_F^0(\Omega,\mathfrak{A},P)$ and $L_G^0(0,\infty)$. The dual of $L_F^0(\Omega,\mathfrak{A},P)$ is $L_{F^*}(\Omega,\mathfrak{A},P)$, where $F^*(t)$ is equivalent to tLogt at ∞, while the dual of $L_G^0(0,\infty)$ is $L_{G*}(0,\infty)$, where

$$G^*(t) = \int_0^t \text{Log}(u + \sqrt{u^2+1})du .$$

($\text{Log}(u + \sqrt{u^2+1})$ is the inverse function of sinh, which is the derivative of $G(t)$.) Notice that $F^*(t)$ satisfies the Δ_2-condition at ∞ and that $G^*(t)$ is equivalent to t^2 at 0 and to $F^*(t)$ at ∞. According to 8.4 and remark 2, this shows that $L_{G*}(0,\infty)$ is the space associated to $L_{F*}(0,1)$ by 8.3, or, in other words, the mapping T defined from $\{Z_t\}_{0\leq t<\infty}$ induces an isomorphism from $L_{G*}(0,\infty)$ into $L_{F*}(\Omega,\mathfrak{A},P)$. It follows then almost immediately that the orthogonal projection Q is bounded in the norm of $L_{F*}(\Omega,\mathfrak{A},P)$, and exactly the same decomposition argument as before allows us to conclude that $L_{F*}(0,1) = \text{LLogL}(0,1)$ is isomorphic to the r.i. function space on $[0,\infty)$ $L_{G*}(0,\infty)$.

The proof that Q is bounded in $L_{F*}(\Omega,\mathfrak{A},P)$ goes as follows: assume that $x \in L_2(\Omega,\mathfrak{A},P)$. We can write $Qx = Th$ for some $h \in L_2(\Omega,\mathfrak{A},P)$. Now

$$\|Qx\|_{L_{F*}(\Omega,\mathfrak{A},P)} = \|Th\|_{L_{F*}(\Omega,\mathfrak{A},P)} \leq C\|h\|_{L_{G*}(0,\infty)} = C\sup\{|\langle h,g\rangle|;\|g\|_{L_G(0,\infty)}\leq 1\}$$

$$= C\sup\{|\langle Th,Tg\rangle|;\|g\|_{L_G(0,\infty)}\leq 1\} = C\sup\{|\langle x,Tg\rangle|;\|g\|_{L_G(0,\infty)}\leq 1\}$$

$$\leq C\|x\|_{L_{F*}(\Omega,\mathfrak{A},P)}\cdot\|Tg\|_{L_F(\Omega,\mathfrak{A},P)} \leq (2e-1)C\|x\|_{L_{F*}(\Omega,\mathfrak{A},P)} .$$

We want to discuss now some generalizations of Rosenthal's space X_p. We begin with a lemma:

Lemma 8.7: Let Y *be a r.i. function space on* $[0,\infty)$. *There is a Banach space* U_Y *with unconditional basis such that for every sequence* $\{A_n\}_{n=1}^\infty$ *of disjoint integrable sets* $[0,\infty)$, *such that* $\sum_{n;\mu(A_n)\leq\varepsilon}^{\infty} \mu(A_n) = +\infty$ *for every* $\varepsilon > 0$, *the space* U_Y *is isomorphic to the closed linear span of the sequence* $\{\chi_{A_n}\}_{n=1}^\infty$ *in* Y.

Proof: Let $\{A_n\}_{n=1}^{\infty}$ and $\{B_n\}_{n=1}^{\infty}$ be two sequences of disjoint integrable subsets of $[0,\infty)$ satisfying the above hypothesis, and denote by W and W' the closed linear span in Y of the corresponding sequences of characteristic functions. It is clear that we can pick disjoint subsets of the integers $\{N_k\}_{k=1}^{\infty}$ such that

$$\mu\Big(\bigcup_{n\in N_k} A_n\Big) = \mu(B_k) \quad \text{for} \quad k = 1,2,\ldots$$

(Since there exists a divergent subseries $\sum_{i=1}^{\infty} \mu(A_{n_i})$ such that $\lim_{i\to\infty} \mu(A_{n_i}) = 0$ and since every $x > 0$ can be considered as the sum of some subseries of any divergent series $\sum_n u_n$ with $u_n \geq 0$ and $\lim_{n\to\infty} u_n = 0$.)

If we set $B_k' = \bigcup_{n\in N_k} A_n$, the sequence $\{\chi_{B_k'}\}_{k=1}^{\infty}$ is contained in W (unless $Y(0,1) = L_\infty(0,1)$, up to an equivalent renorming) and is isometrically equivalent to the sequence $\{\chi_{B_k}\}_{k=1}^{\infty}$. Therefore, W' embeds in W . (If $Y(0,1) = L_\infty(0,1)$, instead let N_k be <u>finite</u> and such that

$$(1/2)\mu(B_k) \leq \mu\Big(\bigcup_{n\in N_k} A_n\Big) \leq \mu(B_k) \quad .$$

In this case $\{\chi_{B_k'}\}_{k=1}^{\infty}$ is 2-equivalent to $\{\chi_{A_k}\}_{k=1}^{\infty}$.) Furthermore, the averaging projection

$$Py = \sum_{k=1}^{\infty} \Big(\int_{B_k'} (\mu(B_k'))^{-1} y(t)\,dt\Big)\chi_{B_k'}$$

defines a norm one projection from W onto the span of the sequence $\{\chi_{B_k'}\}_{k=1}^{\infty}$. This shows that W' is isomorphic to a complemented subspace of W , and in the same way W is isomorphic to a complemented subspace of W' . If we know that W and W' are isomorphic, respectively, to $W \times W$ and $W' \times W'$, we could conclude as in 8.6 that W and W' are isomorphic. So, in order to conclude the proof, it is enough to prove that every such space W is isomorphic to $W \times W$.

Let $\{A_n\}_{n=1}^{\infty}$ be a sequence of disjoint subsets of $[0,\infty)$ satisfying the assumption of 8.7, and consider a sequence $\{B_{n,i}\}_{n=1,i=1}^{\infty\ \ \infty}$ of disjoint subsets of $[0,\infty)$ such that $\mu(B_{n,i}) = \mu(A_n)$ for all integers n and i. If we denote by W and W'', respectively, the closed linear span in Y of the sequences $\{\chi_{A_n}\}_{n=1}^{\infty}$ and $\{\chi_{B_{n,i}}\}_{n=1,i=1}^{\infty\ \ \infty}$ it is clear that $W'' \times W''$ and $W \times W''$ are isomorphic to W''. Since W'' is isomorphic to a complemented subspace of W, we can write $W \sim W'' \times Z$ for some Banach space Z and

$$W'' \sim W \times W'' \sim Z \times W'' \times W'' \sim Z \times W'' \sim W .$$

Thus W is isomorphic to $W \times W$.

□

Remark: The space U_Y is isomorphic to a complemented subspace of Y if $Y(0,1)$ is not equal to $L_\infty(0,1)$, even up to an equivalent renorming. In this case, if $\{A_n\}_{n=1}^{\infty}$ is a sequence of disjoint measurable subsets of $[0,\infty)$ satisfying the assumption of 8.7 and W is the closed linear span of $\{\chi_{A_n}\}_{n=1}^{\infty}$, the averaging projection

$$Pf = \sum_{n=1}^{\infty} \mu(A_n)^{-1}\left(\int_{A_n} f(u)du\right)\chi_{A_n}$$

maps Y onto W. It is obvious that the range of P contains W. To show that they are actually equal, it is enough to prove that

$$\lim_{k,N\to\infty} \left\| \sum_{n=k}^{N} \mu(A_n)^{-1}\, \mu(A \cap A_n)\chi_{A_n} \right\|_Y = 0$$

for every integrable set A. (Use the fact that the simple functions with bounded support are dense in Y by our definition of r.i. function spaces.) Since $Y(0,1)$ is not $L_\infty(0,1)$,

$$\lim_{k\to\infty} \|\chi_{A \cap (\bigcup_{n>k} A_n)}\|_Y = 0 ,$$

which completes the argument because

$$\|\sum_{n=k}^{N} \mu(A_n)^{-1} \mu(A \cap A_n)\chi_{A_n}\|_Y \leq \|\chi_A \cap (\bigcup_{n>k} A_n)\|_X .$$

If $Y(0,\infty)$ is the r.i. function space on $[0,\infty)$ associated to $L_p(0,1)$, $2 \leq p < \infty$, the space U_Y is identical to X_p. To see this, define an equivalent norm on $Y(0,\infty)$ by

$$\|f\| = \max\{\|f\|_{L_p(0,\infty)} , \|f\|_{L_2(0,\infty)}\} ,$$

and choose a sequence of disjoint measurable subsets of $[0,\infty)$, say $\{A_n\}_{n=1}^{\infty}$, such that $\alpha_n = \mu(A_n)$ is decreasing to zero when $n \to \infty$, $\alpha_1 \leq 1$ and $\sum_{n=1}^{\infty} \alpha_n = \infty$. If we set $\mu_n = \alpha_n^{-1/p} \chi_{A_n}$, $n = 1,2,\ldots,$ the sequence $\{u_n\}_{n=1}^{\infty}$ is a normalized unconditional basis for $[\chi_{A_n}]_{n=1}^{\infty} \sim U_Y$, and for every sequence $\{\lambda_n\}_{n=1}^{\infty}$ of scalars, we have

$$\|\sum_{n=1}^{\infty} \lambda_n u_n\|_Y = \max\{(\sum_{n=1}^{\infty} (\lambda_n)^p)^{1/p} , (\sum_{n=1}^{\infty} \alpha_n^{1-2/p} \lambda_n^2)^{1/2}\} .$$

If we define $w_n = \alpha_n^{1/2-1/p}$, we find back the usual representation of X_p.

In the general case, if $X(0,1)$ is a r.i. function space such that $0 < \beta_X \leq \alpha_X < 1$ and if $Y(0,\infty)$ is the r.i. function space on $[0,\infty)$ associated to it, the space U_Y is isomorphic to the span in $X(0,1)$ of any sequence of independent symmetrized Poisson variables $\{Z_n\}_{n=1}^{\infty}$ such that $\lim_{n\to\infty} EZ_n^2 = 0$ and $\sum_{n=1}^{\infty} EZ_n^2 = +\infty$. Actually, U_Y is also isomorphic to the span of any sequence $\{T_n\}_{n=1}^{\infty}$ of independent random variables taking only values 0, ± 1 such that $ET_n = 0$ for every n, $\lim_{n\to\infty} ET_n^2 = 0$ and $\sum_{n=1}^{\infty} ET_n^2 = +\infty$. Indeed, if $ET_n^2 = EZ_n^2$ for all n there is a purely formal proof that the sequences $\{T_n\}_{n=1}^{\infty}$ and $\{Z_n\}_{n=1}^{\infty}$ are equivalent, by using simply Boyd's interpolation theorem. Notice first that $\{T_n\}_{n=1}^{\infty}$ and $\{Z_n\}_{n=1}^{\infty}$ are equivalent in $L_p(0,1)$, $1 < p < \infty$. This follows from [70] and the preceding remark. Next, the orthogonal projections Q_1 and Q_2 onto $[T_n]_{n=1}^{\infty}$ and $[Z_n]_{n=1}^{\infty}$, respectively, are continuous in the L_p-norm, according to [70] and a trivial modification of 8.5. (Actually, the proof of 8.5 is taken from [70].) It follows that we can define two operators R_1 and R_2 which are bounded in every L_p, $1 < p < \infty$, in the following way:

$$\text{If } Q_1 x = \sum_{n=1}^{\infty} \alpha_n T_n \text{, then } R_1 x = \sum_{n=1}^{\infty} \alpha_n Z_n$$

$$\text{if } Q_2 x = \sum_{n=1}^{\infty} \alpha_n Z_n \text{, then } R_2 x = \sum_{n=1}^{\infty} \alpha_n T_n \ .$$

By Boyd's interpolation theorem, R_1 and R_2 will be bounded on every r.i. function space $X(0,1)$ such that $0 < \beta_X \leq \alpha_X < 1$, and this clearly yields the equivalence of the sequences $\{T_n\}_{n=1}^{\infty}$ and $\{Z_n\}_{n=1}^{\infty}$ in $X(0,1)$.

The next lemma is the key for understanding embeddings of $L_q(0,\infty)$ into r.i. function spaces:

Lemma 8.8: Let $Y(0,\infty)$ *be a r.i. function space on* $[0,\infty)$ *and assume that the function* $g_0(t) = t^{-1/q}$, $0 < t < \infty$, *belongs to* $Y(0,\infty)$. *If we define* $g = \|g_0\|_Y^{-1} \cdot g_0$, *the space* $Y_g(0,\infty)$ *(discussed in section 7) is equal to* $L_q(0,\infty)$. *(Recall that this space* $Y_g(0,\infty)$ *embeds isometrically as a sublattice of* $Y(0,\infty)$.*)*

Proof: If we identify $Y(0,\infty)$ with the space $Y([0,\infty)^2)$, the norm in $Y_g(0,\infty)$ is defined by

$$\|f\|_{Y_g(0,\infty)} = \|f \otimes g\|_{Y([0,\infty)^2)} \ .$$

Let us consider an integrable simple function $f = \sum_{n=1}^{N} \lambda_n \chi_{A_n}$, where the A_n's are disjoint and $\sum_{n=1}^{N} |\lambda_n|^q \mu(A_n) = 1$. Consider

$$v(s,t) = \|g_0\|_Y \cdot f \otimes g(s,t) = t^{-1/q} \cdot \sum_{n=1}^{N} \lambda_n \chi_{A_n}(s) \ .$$

For every $u > 0$ we have

$$\mu\{(s,t);\ |v(s,t)| > u\} = \sum_{n=1}^{N} \mu\{(s,t);\ |\lambda_n|\chi_{A_n}(s) > t^{1/q}u\}$$

$$= \sum_{n=1}^{N} \mu(A_n)\,\mu\{t;\ |\lambda_n| > t^{1/q}u\} = \sum_{n=1}^{N} \mu(A_n)|\lambda_n|^q/u^q$$

$$= u^{-q} = \mu\{g_o > u\} \quad ,$$

which shows that the distributions of $f \otimes g$ and g are identical.

□

Proposition 8.9: Let X *be a r.i. function space on* [0,1] *such that the function* $g_o(t) = t^{-1/q}$, $0 < t \leq 1$, $1 < q < 2$, *belongs to* X . *The space* X *contains isometrically* $L_q(0,\infty)$.

Proof: Assume first that $0 < \beta_X$ and consider the space $Y(0,\infty)$ isomorphic to X . Since $q < 2$, it is clear from 8.4 that the function $g_o(t) = t^{-1/q}$, $0 < t < \infty$, belongs to $Y(0,\infty)$, and the result follows immediately by 8.8. In the general case, assume we can prove that if $g_o\chi_{(0,1)}$ belongs to $X(0,1)$ then there is a r.i. function space $Z(0,1)$ such that $0 < \beta_Z$,

$$\|f\|_X \leq K\|f\|_Z \quad ,$$

for some K and for every $f \in Z$, and such that $g_o\chi_{(0,1)}$ still belongs to $Z(0,1)$. If we consider the space $Y_Z(0,\infty)$ associated to $Z(0,1)$, we have also

$$\|f\|_{Y(0,\infty)} \leq K\|f\|_{Y_Z(0,\infty)}$$

for every $f \in \mathfrak{F}(0,\infty)$. It is enough by 8.8 to prove that $g_o(t) = t^{-1/q}$, $0 < t < \infty$, belongs to $Y(0,\infty)$, but this is clear since it belongs to $Y_Z(0,\infty)$ by the first part of the argument.

To finish the proof it remains to construct $Z(0,1)$. This is obvious if we work with Luxemburg's definition of r.i. spaces. The space $Z(0,1)$ is then simply the space $L_{q,\infty}(0,1)$ of all measurable functions f such that

$$\sup_{t>0} t^q \mu\{|f| \geq t\} < \infty .$$

This space contains $g_o \chi_{[0,1]}$ and has both indices α and β equal to $1/q$. Our case requires more work. Since $g_o \chi_{[0,1]}$ belongs to X and since $L_\infty(0,1)$ is dense in X , the norm of $g_o \chi_{[0,\varepsilon]}$ tends to zero when $\varepsilon \to 0$. It follows that we can find a function $h \in X$ of the form $h(t) = t^{-1/q} L(t)$, where $L(t)$ is decreasing, non-negative, and $\lim_{t\to 0+} L(t) = +\infty$. Define a r.i. norm on simple functions by

$$\|f\| = \sup_{0<t\leq 1} (th(t))^{-1}\int_0^t f^*(u)du$$

and denote by $Z(0,1)$ the r.i. function space on $[0,1]$ obtained by completing $L_\infty(0,1)$ under the above norm. A simple computation shows that g_o belongs to Z . It is clear that $\|f\|_Z \leq 1$ implies $f^*(t) \leq h(t)$ for $0 < t \leq 1$, so that

$$\|f\|_X \leq \|h\|_X \cdot \|f\|_Z$$

for every $f \in Z$. Another computation shows that $1/q \leq \beta_Z$.

□

We want to point out again for the reader who is familiar with q-stable processes that the final embedding of L_q into X obtained here is given by a q-stable process. Observe that if $g_o(t) = t^{-1/q}$, $0 < t < \infty$, and if T denotes the operator associated to the symmetric Poisson process, the image Tg_o is a q-stable random variable . If $Y(0,\infty)$ is identified with $Y([0,\infty]^2)$, $X_t = T(\chi_{[0,t]} \otimes g_o)$ is a q-stable random process; i.e., a stationary process with independent increments such that $E \exp i u X_t = \exp(-\lambda t|u|^q)$ for some $\lambda > 0$.

We shall show now that there are r.i. function spaces on $[0,1]$ which are not isomorphic to any r.i. function space on $[0,\infty)$. Incidentally, this will show also that "most" of the r.i. function spaces on $[0,\infty)$ are not isomorphic to r.i. function spaces on $[0,1]$.

Lemma 8.10: Let X *be a separable* σ *-complete and* σ *-order continuous Banach lattice, and let* $\{x_n\}_{n=1}^\infty$ *be a K-symmetric sequence in* X . *If the span* $[x_n]_{n=1}^\infty$ *is C-complemented in* X , *the sequence* $\{x_n\}_{n=1}^\infty$ *is equivalent to the unit vector basis of* ℓ_2 *or is* $D = D(K,C)$*-equivalent to a disjointly supported sequence in* X .

Proof: Without loss of generality we can assume that X is a space of measurable functions over some probability space $(\Omega,\mathfrak{A},P)$ such that

$$\|f\|_{L_1(\Omega,\mathfrak{A},P)} \le \|f\|_X \le 2\|f\|_{L_\infty(\Omega,\mathfrak{A},P)} \quad ,$$

for every $f \in L_\infty(\Omega,\mathfrak{A},P)$. As was observed after 6.5, the unit ball of $L_\infty(\Omega,\mathfrak{A},P)$ is equi-integrable in X , which yields that there is a real function g with $\lim_{t\to 0} g(t) = 0$ such that

$$\|\chi_A\|_X \le g(P(A))$$

for every set $A \in \mathfrak{A}$. It follows that $X^* \subset L_1(\Omega,\mathfrak{A},P)$. (If x^* is a positive norm one functional on X , $\lambda(A) = x^*(\chi_A)$ defines an additive measure on $(\Omega,\mathfrak{A})$, but the inequality $\lambda(A) \le g(P(A))$ yields that λ is σ-additive and absolutely continuous with respect to P ; i.e., $\lambda = fP$ for some $f \in L_1(\Omega,\mathfrak{A},P)$.)

Let $\{x_n\}_{n=1}^\infty$ be a K-symmetric sequence in X such that there is some projection Q from X onto $[x_n]_{n=1}^\infty$ with norm $\le C$.

If $\lim_{n\to\infty} \inf \|x_n\|_{L_1(\Omega,\mathfrak{A},P)} = 0$ we can find a subsequence equivalent to a disjointly supported sequence in X (cf. [26]). Otherwise, there is a $\delta > 0$ such that $\|x_n\|_{L_1(\Omega,\mathfrak{A},P)} \ge \delta$ for every integer n and it follows from the cotype 2 property of L_1 that

$$\|\sum_{n=1}^{\infty} a_n x_n\|_X \geq K^{-1}\int_0^1 \|\sum_{n=1}^{\infty} a_n r_n(u)x_n\|_X du$$

$$\geq K^{-1}\int_0^1 \|\sum_{n=1}^{\infty} a_n r_n(u)x_n\|_{L_1} du \geq A_1K^{-1}\,\delta(\sum_{n=1}^{\infty} a_n^2)^{1/2} \quad ,$$

for all scalars $\{a_n\}_{n=1}^{\infty}$.

Let $\{x_n^*\}_{n=1}^{\infty}$ be the biorthogonal system to $\{x_n\}_{n=1}^{\infty}$ in X^* associated with Q. We get by duality

$$\|\sum_{n=1}^{\infty} a_n x_n^*\|_{X^*} \leq KC\sqrt{2}\,\delta^{-1}(\sum_{n=1}^{\infty} a_n^2)^{1/2}$$

for all scalars $\{a_n\}_{n=1}^{\infty}$. Assume now that

$$\delta_1 = \liminf_{n\to\infty}\|x_n^*\|_{L_1(\Omega,\mathfrak{A},P)} > 0 \; .$$

We have, using again that L_1 is of cotype 2 and the fact that $\{x_n^*\}_{n=1}^{\infty}$ is KC-symmetric, that

$$\|\sum_{n=1}^{\infty} a_n x_n^*\|_{X^*} \geq A_1K^{-1}C^{-1}\delta_1(\sum_{n=1}^{\infty} a_n^2)^{1/2} \quad ,$$

for all scalars $\{a_n\}_{n=1}^{\infty}$, and this proves that the sequence $\{x_n^*\}_{n=1}^{\infty}$ is equivalent in X^* to the unit vector basis of ℓ_2. By duality, the sequence $\{x_n\}_{n=1}^{\infty}$ in X is also equivalent to the unit vector basis of ℓ_2.

In the remaining case we may assume by passing to a subsequence that $\lim_{n\to\infty}\|x_n^*\|_{L_1(\Omega,\mathfrak{A},P)} = 0$. Also by passing to a further subsequence we can write (see the discussion after 6.5) the sequence $\{x_n\}_{n=1}^{\infty}$ as $x_n = y_n + z_n$, $n = 1,2,\ldots$, where the sequence $\{z_n\}_{n=1}^{\infty}$ is equi-integrable in X and $y_n = \chi_{A_n} x_n$ for a suitable sequence $\{A_n\}_{n=1}^{\infty}$ of disjoint measurable sets. Observe now that

$$\lim_{k\to\infty}\sup_n |\langle x_k^*, z_n\rangle| = 0 \; .$$

Indeed, for every $\epsilon > 0$ we can decompose $z_n = z'_n + z''_n$, $n = 1,2,\ldots$, with $\sup_n \|z'_n\|_X \leq \epsilon$ and $R = \sup_n \|z''_n\|_\infty < \infty$. It follows that for all integers k and n we have

$$|\langle x^*_k, z_n\rangle| \leq \epsilon\|x^*_k\|_{X^*} + R\|x^*_k\|_{L_1(\Omega,\mathfrak{A},P)}$$

so that

$$\lim_k \sup(\sup_n|\langle x^*_k, z_n\rangle|) \leq \epsilon \sup_k \|x^*_k\|_{X^*}$$

and this proves our claim since ϵ is arbitrary. It follows easily from this fact that we can find a sequence $\{n_k\}_{k=1}^\infty$ of integers such that $\{y_{n_k}, x^*_{n_k}\}_{k=1}^\infty$ is essentially a biorthogonal system, in the sense that

$$\sum_{k=1}^\infty |1 - \langle x^*_{n_k}, y_{n_k}\rangle| + \sum_{j\neq k} |\langle x^*_{n_k}, y_{n_j}\rangle| \leq 1/2K^2C^2 .$$

We have then for all scalars $\{a_k\}_{k=1}^\infty$,

$$\|\sum_{k=1}^\infty a_k x_{n_k}\|_X \leq C \sup\{\sum_{k=1}^\infty a_k b_k;\ \|\sum_{k=1}^\infty b_k x^*_{n_k}\|_{X^*} \leq 1\}$$

$$\leq C \sup\{\langle \sum_{k=1}^\infty b_k x^*_{n_k}, \sum_{k=1}^\infty a_k y_{n_k}\rangle;\ \|\sum_{k=1}^\infty b_k x^*_{n_k}\|_{X^*} \leq 1\} + 1/2C\|\sum_{k=1}^\infty a_k x_{n_k}\|_X .$$

This gives

$$\|\sum_{k=1}^\infty a_k x_{n_k}\|_X \leq 2C\|\sum_{k=1}^\infty a_k y_{n_k}\|_X .$$

On the other hand, by the "diagonal principle" used already several times, for example in the proof of 3.11,

$$\|\sum_{k=1}^\infty a_k y_{n_k}\|_X \leq K\|\sum_{k=1}^\infty a_k x_{n_k}\|_X$$

so that, finally, the symmetric sequence $\{x_n\}_{n=1}^\infty$ is equivalent to the disjointly supported sequence $\{y_{n_k}\}_{k=1}^\infty$.

The next two propositions have essentially the same statement, but the first applies to r.i. function spaces X on $[0,1]$ which are not p-convex for any $p > 1$, while the second requires X to be p-convex for some $p > 1$.

Proposition 8.11: Let X *be a r.i. function space on* $[0,1]$ *and* Y *a r.i. function space on* $[0,\infty)$. *Assume that* X *is p-concave for some* $p < 2$ *and that* Y *is isomorphic to* X. *Then either*

i) Y *is isomorphic to the space* $Y_X(0,\infty)$ *of 8.3 or*

ii) there is a constant D *such that for every integer* n, *the sequence* $\{x_{n,i}\}_{i=1}^{\infty}$ *in* Y *is* D*-equivalent to a disjointly supported sequence in* X.

Proposition 8.12: Let X *be a superreflexive r.i. functions space on* $[0,1]$. *If a r.i. function space* Y *on* $[0,\infty)$ *is isomorphic to* X *then either*

i) the space Y *is the space* $Y_X(0,\infty)$ *of 8.3 or*

ii) there is a constant D *such that for every integer* n, *the sequence* $\{x_{n,i}\}_{i=1}^{\infty}$ *in* Y *is* D*-equivalent to a disjointly supported sequence in* X.

Proof of 8.11 and 8.12: In both cases let T be an isomorphism from Y onto X. There is a constant K such that for every n the sequence $\{Tx_{n,i}\}_{i=1}^{\infty}$ is K-symmetric in X and $[Tx_{n,i}]_{i=1}^{\infty}$ is K-complemented in X. If the sequence $\{Tx_{0,i}\}_{i=1}^{\infty}$ is not equivalent to the unit vector basis of ℓ_2 then the same is true for the sequence $\{Tx_{n,i}\}_{i=1}^{\infty}$ for every integer n. In this case it follows from 8.10 that there is a constant $D = D(K)$ such that for every n the sequence $\{x_{n,i}\}_{i=1}^{\infty}$ in Y is D-equivalent to a disjointly supported sequence in X.

Assume now that the sequence $\{x_{0,i}\}_{i=1}^{\infty}$ is equivalent to the unit vector basis of ℓ_2. Consider first the case when X is p-concave for some $p < 2$. The r.i. function space $Y(0,1)$ is isomorphic to a complemented subspace of X, and hence, by 5.6, we know that $Y(0,1)$ is, up to an equivalent norm, equal to $L_2(0,1)$ or to X. But if $Y(0,1) = L_2(0,1)$, the reasoning used in 8.4 shows that $Y = L_2(0,\infty)$, which is not possible. It follows that $Y(0,1)$ coincides with X and the proof of 8.4 shows that Y is the space $Y_X(0,\infty)$ of 8.3, up to an equivalent renorming.

Assume now that X is super-reflexive and that the sequence $\{x_{0,i}\}_{i=1}^{\infty}$ in Y is equivalent to the unit vector basis of ℓ_2 . Therefore, the indices of $Y(0,1)$ satisfy $0 < \beta_{Y(0,1)} \le \alpha_{Y(0,1)} < 1$, and it follows from 8.4 and 8.6 that Y is equal to $Y_{Y(0,1)}(0,\infty)$ and thus isomorphic to $Y(0,1)$. This yields that X is isomorphic to $Y(0,1)$. If the Haar system in X is not equivalent to a disjointly supported sequence in X , it follows from 6.9 that X is equal to $Y(0,1)$, up to an equivalent norm, and thus Y coincides with $Y_X(0,\infty)$. In the case where the Haar system in X is equivalent to a disjointly supported sequence in X , the result follows from the fact that the K-unconditional sequence $\{Tx_{0,i}\}_{i=1}^{\infty}$ is $D = D(K)$-equivalent to disjoint blocks of the Haar system in X (lemma 8.13 below) and hence equivalent to a disjointly supported sequence in X . □

The next lemma appeared in [73] in the case of L_p .

Lemma 8.13: Let X *be a q-concave r.i. function space on* $[0,1]$ *, with* $q < \infty$ *and q-concavity constant* M *. Every* K *-unconditional sequence in* X *is* $D(K,q,M)$ *-equivalent to a sequence of disjoint blocks of the Haar system in* X .

Proof: Let $\{u_n\}_{n=1}^{\infty}$ be a K-unconditional sequence in X . Identifying X with $X([0,1]^2)$ the usual square function argument shows that $\{u_k \otimes r_{n_k}\}_{k=1}^{\infty}$ is $\sqrt{2}\, B_q M K^2$-equivalent to $\{u_k\}_{k=1}^{\infty}$ for every strictly increasing sequence $\{n_k\}_{k=1}^{\infty}$. The lemma follows easily from the fact that for a fixed k the sequence $\{u_k \otimes r_n\}_{n=1}^{\infty}$ converges weakly to zero in $X([0,1]^2)$. □

An argument similar to that used for 8.11 gives:

Proposition 8.14: For every $1 < p < q < 2$ *the r.i. function space* $Y_{p,q} = L_p(0,\infty) + L_q(0,\infty)$ *is not isomorphic to any r.i. function space on* $[0,1]$.

Proof: Observe that $Y_{p,q}$ is q-concave. Therefore, if $Y_{p,q}$ is isomorphic to some r.i. function space X on $[0,1]$ it follows from 5.6 that X (considered as a complemented subspace of $Y_{p,q}$) is equal to $L_2(0,1)$ but this is not possible, or that $Y_{p,q}(0,1) = L_q(0,1)$ coincides with X. However, ℓ_p embeds into $Y_{p,q}$ but not into $L_q(0,1)$. □

It is possible to investigate further the case ii of 8.11 and 8.12 when X is an Orlicz function space $L_F(0,1)$, where the function F is regularly varying at ∞ in the sense of Karamata, that is to say

$$\lim_{u\to\infty} F(tu)/F(u) = G(t)$$

exists for every $t > 0$. It is known that $G(t)$ is necessarily of the form $G(t) = t^p$ for some $1 \leq p < \infty$. We shall call p the index of F. If p is the index of F, the space $L_F(0,1)$ is $(p-\varepsilon)$-convex and $(p+\varepsilon)$-concave for $\varepsilon > 0$. Let us check, for example, the $(p+\varepsilon)$-concavity. According to the discussion in section 7, we need only to verify that $F(t)/t^{p+\varepsilon}$ is equivalent at ∞ to a decreasing function for every $\varepsilon > 0$, or that

$$F(st)/F(s) \leq K\, t^{p+\varepsilon} \quad ,$$

for every $t \geq 1$ and every $s \geq s(\varepsilon)$.

To see this, let $s(\varepsilon)$ be such that for $s \geq s(\varepsilon)$

$$\frac{F(2s)}{F(s)} \leq 2^{p+\varepsilon} \; .$$

Then

$$\frac{F(2^n s)}{F(s)} \leq 2^{n(p+\varepsilon)}$$

for every $s \geq s(\varepsilon)$ and $n = 0,1,\ldots,$ so that if $2^n \leq t \leq 2^{n+1}$ and $s \geq s(\varepsilon)$, then

$$\frac{F(ts)}{F(s)} \leq 2^{(n+1)(p+\varepsilon)} \leq 2^{p+\varepsilon}\, t^{p+\varepsilon} \quad .$$

Another important property of such spaces $L_F(0,1)$ is that every disjointly supported sequence in $L_F(0,1)$ has, for every $\epsilon > 0$, a subsequence $(1+\epsilon)$-equivalent to the unit vector basis of ℓ_p, where p is the index of F.

Corollary 8.15: Let F *be a regularly varying Orlicz function with index* p, $1 < p < \infty$, *not equivalent at* ∞ *to* t^p. *The space* $Y_X(0,\infty)$ *of 8.3 is the only r.i. function space on* $[0,\infty)$ *isomorphic to* $X = L_F(0,1)$.

Proof: First, the facts recalled before 8.15 and the condition $1 < p < \infty$ yield the super-reflexivity of $X = L_F(0,1)$ (and that the two indices are equal to $1/p$) so that, by 8.6, the space $Y_X(0,\infty)$ is isomorphic to $L_F(0,1)$. Conversely, let Y be a r.i. function space on $[0,\infty)$ isomorphic to $L_F(0,1)$ and assume that for some D and for every n, the sequence $\{x_{n,i}\}_{i=1}^{\infty}$ in Y is D-equivalent to a disjointly supported sequence in $L_F(0,1)$. By the fact recalled before 8.15, this yields that the sequence $\{\|x_{n,1}\|_Y^{-1} \cdot x_{n,i}\}_{i=1}^{\infty}$ is D-equivalent to the unit vector basis of ℓ_p for every n. This immediately implies that Y is, up to an equivalent renorming, equal to $L_p(0,\infty)$. But this in turn yields that $F(t)$ is equivalent at ∞ to t^p, contrary to our assumption.

□

The second part of the proof of 8.15 is valid for $p = 1$: if F is a regularly varying Orlicz function with index 1, such that $\lim_{t\to\infty} F(t)/t = +\infty$ (to ensure that $L_F(0,1)$ is not $L_1(0,1)$), the only r.i. function space on $[0,\infty)$ which can be isomorphic to $X = L_F(0,1)$ is the space $Y_X(0,\infty)$ of 8.3. On the other hand, if $X = L_F(0,1)$ is isomorphic to $Y_X(0,\infty)$, the space ℓ_2 is isomorphic to a complemented subspace of $L_F(0,1)$ (since $[x_{0,i}]_{i=1}^{\infty}$ in $Y_X(0,\infty)$ is isomorphic to ℓ_2 and complemented, by the averaging projection.) We shall see now that ℓ_2 is not complemented in spaces which are too close to L_1, in a sense to be made precise below. This will provide examples of r.i. function spaces on $[0,1]$ which have no representation as r.i. function space on $[0,\infty)$.

Lemma 8.16: Let X *be a good lattice of functions on* [0,1] *(see the discussion preceding 6.3) such that:*

(a) for every $p > 1$, $L_p(0,1)$ *is contained in* X *and if* K_p *denotes the norm of the injection of* $L_p(0,1)$ *into* X *then*

$$\lim_{p\to 1} K_p\sqrt{p-1} = 0$$

(b) The space X *is of cotype* 2 , *and no sequence of disjoint vectors in* X *is equivalent to the unit vector basis of* ℓ_2 .

Then X *does not contain any complemented copy of* ℓ_2 .

Proof: Let Z be a subspace of X isomorphic to ℓ_2 and P a projection from X onto Z . Next we show that the unit ball B_Z of Z is equi-integrable in X . If this were false, we could find a $\delta > 0$, a normalized sequence $\{z_n\}_{n=1}^{\infty}$ in Z which is 2-equivalent to the unit vector basis of ℓ_2 , and a sequence $\{A_n\}_{n=1}^{\infty}$ of disjoint sets such that $\|\chi_{A_n} z_n\|_X \geq \delta$ for every n . But since X is of cotype 2, the disjoint sequence $\{\chi_{A_n} z_n\}_{n=1}^{\infty}$ is already equivalent to the unit vector basis of ℓ_2 , contradicting our assumption. There exists therefore an $\varepsilon > 0$ such that

$$(+) \qquad \|\chi_A z\|_X \leq (1/5\|P\|)\, \|z\|_X ,$$

for every $z \in Z$ and every set A with $\mu(A) \leq \varepsilon$. According to a result of Nikishin [61] there exists a set $\Omega \subset [0,1]$ with $\mu([0,1] \sim \Omega) \leq \varepsilon$ and a constant K such that

$$\int_{\Omega} |z(u)|^{p_o}\, du \leq K^{p_o} \|z\|_X^{p_o}$$

for some $1 < p_o < 2$ and for every $z \in Z$. The property (+) yields that $\|z\|_X \leq 2\|\chi_{\Omega} z\|_X$, for every $z \in Z$ or, in other words, $Z' = \{\chi_{\Omega} z;\ z \in Z\}$ is isomorphic to Z , and thus to ℓ_2 . If we choose p_1 such that $1 < p_1 < p_o$ and define θ by $1/p_1 = 1-\theta + \theta/p_o$, the above property yields

$$\|\chi_\Omega z\|_{p_o} \leq 2K\|\chi_\Omega z\|_X \quad ,$$

for every $z \in Z$ and

$$\|\chi_\Omega z\|_X \leq K_{p_1}\|\chi_\Omega z\|_{p_1} \leq K_{p_1}\|\chi_\Omega z\|_1^{1-\theta}\|\chi_\Omega z\|_{p_o}^{\theta} \leq K_{p_1} \ (2K)^{\theta}\|\chi_\Omega z\|_1^{1-\theta} \ \|\chi_\Omega z\|_X^{\theta} \ .$$

Thus

$$\|\chi_\Omega z\|_X \leq K_{p_1}^{1/(1-\theta)}(2K)^{\theta/(1-\theta)}\|\chi_\Omega z\|_1 \ ,$$

for every $z \in Z$. The preceding inequalities show that there is a constant C such that for $1 \leq p \leq p_o$ the space Z' endowed with the L_p-norm is C-isomorphic to ℓ_2 .

It follows from (+) that the mapping $\chi_\Omega z \to \chi_\Omega P(\chi_\Omega z)$ defines an isomorphism from Z' onto itself. If we set $X(\Omega) = \{f \in X; f = \chi_\Omega f\}$ the mapping

$$Q(f) = U^{-1}(\chi_\Omega P(f))$$

defines a projection from $X(\Omega)$ onto Z' . If we denote now by j_p the injection of $L_p(\Omega)$ into $X(\Omega)$, Qj_p is continuous in the L_p-norm for $1 < p \leq p_o$ since if $Qj_p(f) = \chi_\Omega z$, we have

$$\|\chi_\Omega z\|_p \leq 2K\|\chi_\Omega z\|_X = 2K\|Qj_p(f)\|_X$$
$$\leq 2K\|Q\| \ \|j_p f\|_X \leq 2K\|Q\| \ \|j_p\| \ \|f\|_p \quad .$$

This shows that for every $1 < p \leq p_o$, the space L_p contains a subspace C-isomorphic to ℓ_2 with a projection Q_p of norm less than $C_1\|j_p\|$. The known estimates for the projection constant of ℓ_2 in L_p give

$$\liminf_{p\to 1} \sqrt{p-1} \ \|Q_p\| > 0 \ ,$$

which contradicts our assumption on $\|j_p\|$. □

Proposition 8.17: Let F *be a regularly varying Orlicz function such that*

$$\lim_{t\to\infty} (t\sqrt{\mathrm{Log}t})^{-1}\cdot F(t) = 0 .$$

The space $L_F(0,1)$ *is not isomorphic to any r.i. function space on* $[0,\infty)$.

Proof: According to 8.16 and to the discussion preceding it, it is enough to show that

$$\lim_{p\to 1} \sqrt{p-1}\ \|j_p\| = 0 ,$$

where j_p denotes the injection from $L_p(0,1)$ into $L_F(0,1)$. Define $C_p = \max_{t\geq 1} t^{-p} F(t)$. Let $\varepsilon > 0$ be given. There exists a constant A such that

$$F(t) \leq \varepsilon\, t\sqrt{\mathrm{Log}t} ,$$

for every $t \geq A$. We deduce

$$C_p \leq \sup_{1\leq t\leq A} t^{-p} F(t) + \varepsilon \sup_{t\geq 1} t^{1-p}\sqrt{\mathrm{Log}t} .$$

But

$$\sup_{t\geq 1} t^{1-p}\sqrt{\mathrm{Log}t} = \sup_{u\geq 1} u^{-1}\sqrt{\mathrm{Log}\, u^{1/p-1}} = C/\sqrt{p-1} ,$$

where $C = \sup_{u>1} u^{-1}\sqrt{\mathrm{Log}u}$. This yields immediately that $\lim_{p\to 1}\sqrt{p-1}\, C_p = 0$. Let now f be a function on $[0,1]$ such that $\int_0^1 |f(t)|^p\, dt = 1$ for some $p > 1$. We have, using the convexity of F ,

$$\int_0^1 F((C_p+1)^{-1}|f(t)|)dt \leq (C_p+1)^{-1}\int_0^1 F(|f(t)|)dt$$

$$\leq (C_p+1)^{-1}(1 + \int_{\{|f(t)|>1\}} F(|f(t)|)dt)$$

$$\leq (C_p+1)^{-1}(1 + C_p\int |f(t)|^p dt) \leq 1 .$$

This shows that

$$\limsup_{p\to 1} \sqrt{p-1}\,\|j_p\| \le \limsup_{p\to 1} \sqrt{p-1}\,(C_p+1) = 0 .$$

□

Remark: The simplest examples of Orlicz functions which satisfy the hypothesis of 8.7 are given by

$$F_\alpha(t) = \begin{cases} t & ; 0 \le t \le e \\ t(\log t)^\alpha & ; e < t \end{cases} ,$$

where $0 < \alpha < 1/2$.

In the final part of this section we will present an example showing that case (ii) of 8.11 or 8.12 can really occur for spaces other than L_p. We will construct an Orlicz function F on $[0,\infty)$ which is not equivalent to any function t^p at ∞ or at 0 such that $L_F(0,\infty)$ is linearly and lattice isomorphic to $L_F(0,1)$. To achieve this it is enough to produce a constant K and an increasing sequence $\{\mu_n\}_{n=1}^\infty$ such that $\sum_{n=1}^\infty 1/F(\mu_n) = 1$ and

$$(\cdot) \qquad K^{-1}F(t) \le F(\mu_n t)/F(\mu_n) \le KF(t)$$

for every integer n and every $0 \le t < \infty$.

Indeed, if $\{A_n\}_{n=1}^\infty$ is a sequence of disjoint intervals of $[0,1]$ with measure $\mu(A_n) = 1/F(\mu_n)$, and if φ_n denotes the increasing affine transformation from A_n onto $[n,n+1)$, define an operator T on $\mathfrak{F}(0,\infty)$ by the formula

$$Tf = \sum_{n=1}^\infty \mu_n \chi_{A_n} f(\varphi_n) .$$

We have

$$\int_0^1 F(Tf(t))dt = \sum_{n=1}^\infty \int_{A_n} F(\mu_n f(\varphi_n(t)))dt$$

$$= \sum_{n=1}^\infty \int_n^{n+1} (F(\mu_n f(u))/F(\mu_n))\, du .$$

By the property (•) of F this expression is K-equivalent to

$$\sum_{n=1}^{\infty} \int_{n}^{n+1} F(f(u))du = \int_{0}^{\infty} F(f(u))du .$$

This shows that T extends to an isomorphism from $L_F(0,\infty)$ into $L_F(0,1)$ and clearly this extension is onto.

The condition $\sum_{n=1}^{\infty} 1/F(\mu_n) = 1$ is not a problem, since if we have the property (•) for some increasing sequence $\{\mu_n'\}_{n=1}^{\infty}$ with $\lim_n \mu_n' = +\infty$ we may construct the sequence $\{\mu_n\}_{n=1}^{\infty}$ by repeating or deleting a suitable number of terms in the sequence $\{\mu_n'\}_{n=1}^{\infty}$.

It is more convenient for the construction of F to transform the property (•) into an equivalent additive property, by setting $F(t) = \exp f(\mathrm{Log}\, t)$, where f is a function defined on $(-\infty,+\infty)$. Clearly, the property (•) for F is equivalent to the existence of an increasing sequence $\{\lambda_n\}_{n=1}^{\infty}$ with $\lim_n \lambda_n = +\infty$ such that

$$(\because) \qquad |f(u+\lambda_n) - f(u) - f(\lambda_n)| \leq \mathrm{Log}\, K ,$$

for every integer n and every real number u.

The construction of f is fairly easy. Set

$$f_o(u) = \sum_{k=1}^{\infty} (1 - \cos 2^{-k}\pi u)$$

If we remark that f_o is the real part of $\sum_{k=1}^{\infty} (1-\exp 2^{-k}\pi i\, u)$ we get

$$|(1-\exp 2^{-k}\pi i(u+\lambda_n)) - (1-\exp 2^{-k}\pi i\, \lambda_n) - (1-\exp 2^{-k}\pi i u)|$$

$$= |(\exp 2^{-k}\pi i\, \lambda_n - 1)\, (1-\exp 2^{-k}\pi i u)| \leq |1-\exp 2^{-k}\pi i\, \lambda_n|$$

and

$$|f_o(u+\lambda_n) - f_o(u) - f_o(\lambda_n)| \leq \sum_{k=1}^{\infty} |1-\exp 2^{-k}\pi i\, \lambda_n| .$$

If we define simply $\lambda_n = 2^n$, we get

$$\sum_{k=1}^{\infty} |1-\exp 2^{n-k}\pi i| = \sum_{j=0}^{\infty} |1-\exp 2^{-j}\pi i| = A \quad ,$$

that is,

$$|f_o(u+2^n) - f_o(u) - f_o(2^n)| \leq A \quad .$$

We also have

$$|f_o(s+t) - f_o(s)| \leq \sum_{k=1}^{\infty} |\exp 2^{-k}\pi is - \exp 2^{-k}\pi i(t+s)|$$

$$\leq \sum_{k=1}^{\infty} |1-\exp 2^{-k}\pi it| \; .$$

We want to show that for every $\varepsilon > 0$, there is a constant K_ε such that

$$|f_o(s+t) - f_o(s)| \leq \varepsilon|t| + K_\varepsilon$$

for all real numbers s and t . We have if $|t| \geq 1$

$$|f_o(s+t) - f_o(s)| \leq 2 \log_2|t| + \sum_{\substack{k \\ |t| \leq 2^k}} |1-\exp 2^{-k}\pi it| \quad .$$

If we notice that the second expression is bounded by a constant B independent of t , we get

$$|f_o(s+t) - f_o(s)| \leq 2 \log_2|t| + B$$

which yields the desired property.

We want also to prove that $f_o(t)$ is not bounded when $t \to +\infty$ (or $-\infty$, since $f_o(-t) = f_o(t)$) . If we set $t_n = 4^n + 4^{n-1} + \ldots + 1$, we have for $k = 2j$, $j = 1,2,\ldots,n$

$$\cos(2^{-k}\pi t_n) = \cos(\pi(1 + 1/4 + \ldots + 4^{-j}))$$

and since $1/4 + \ldots + 4^{-j}$ is less than $1/2$, we deduce that $\cos(2^{-k}\pi t_n) \le 0$, and hence, $f_o(t_n) \ge n$.

Let now p be a real number, $1 < p < \infty$, and define a function on $[0,\infty)$ by

$$F(t) = t^p \exp f_o(\mathrm{Log}\,t) .$$

It is clear that F satisfies the property $(\cdot)$ since f_o satisfies $(\because)$. We know that for every $\varepsilon > 0$ there is a constant C_ε such that

$$C_\varepsilon^{-1} t^{p-\varepsilon} \le F(st)/F(s) \le C_\varepsilon t^{p+\varepsilon}$$

for every s and $t > 0$. This yields in particular that $F(t)/t$ is equivalent to an increasing function, which implies that F is equivalent on $[0,\infty)$ to a convex function. Furthermore, the spaces $L_F(0,\infty)$ and $L_F(0,1)$ are $(p-\varepsilon)$-convex and $(p+\varepsilon)$-concave for every $\varepsilon > 0$. Finally, the function F is not equivalent to any function t^q at 0 or at ∞: the only possible function would be t^p, but $F(t)$ is not equivalent to t^p since $f_o(t)$ is not bounded when $t \to \pm\infty$.

Many results from this section are connected to some theorems proved by N. J. Nielsen [84]. For instance, he proved that if F is slowly varying function of index p; $1 \le p < \infty$ so that $L_F(0,1)$ is lattice isomorphic to $L_F(0,\infty)$ then $F(t)$ is equivalent to t^p both at 0 and ∞. (cf. [84] Theorem 2.6). This shows that the function F constructed in the last example of this section cannot be slowly varying. Also, under the same conditions on F if $L_F(0,\infty)$ is isomorphic to a r.i. function space X on $[0,1]$ and $\max(p,1/\beta_{L_F}) < 2$ or $\min(p,1/\alpha_{L_F}) > 1$ then again $F(t)$ must be equivalent to t^p.

9. SUBSPACES OF A R.I. FUNCTION SPACE X ISOMORPHIC TO X

This section is mainly devoted to the proof of the following theorem.

Theorem 9.1: Let X *be a r.i. function space on* [0,1] *such that* X *is s-concave for some* $s < \infty$, *the index* α_X *of* X *(see section 8 for the definition of* α_X *and* β_X *) satisfies* $\alpha_X < 1$ *and the Haar system in* X *is not equivalent to a sequence of disjoint functions in* X . *Then any subspace of* X *which is isomorphic to* X *contains a further subspace which is complemented in* X *and isomorphic to* X . *In particular, the theorem holds for* $X = L_p(0,1)$, $1 < p < \infty$ *and, more generally, for every reflexive Orlicz function space on* [0,1].

It is known (cf. section 8 and [50]) that the $\infty >$ s-concavity of X implies that $\beta_X > 0$ and the condition $0 < \beta_X \leq \alpha_X < 1$ is equivalent to the condition that the Haar system forms an unconditional basis for X . The analogous theorem for $X = L_1(0,1)$ (which is not covered by 9.1) was proved by Enflo and Starbird [22].

The proof of theorem 9.1 in the general form stated depends heavily on the techniques developed in section 6. We give, however, an independent proof for the case $X = L_p(0,1)$, $1 < p < \infty$. Since the proof for $L_p(0,1)$, $1 < p < 2$, is rather complicated and it digresses from the main track of ideas in this section we postpone it to an appendix to this section. We state now a corollary to theorem 9.1. The proof in the case $X = L_p(0,1)$, $1 < p < \infty$ is a straightforward consequence of theorem 9.1 and the decomposition method [63]; this is also the case for reflexive Orlicz function spaces on [0,1] since the Haar basis cannot be equivalent to the unit vector basis of a modular sequence space. However, the general proof requires some work.

Corollary 9.2: Let X *be a r.i. function space satisfying the assumptions of theorem 9.1. If* Y *is a complemented subspace of* X *which contains an isomorphic copy of* X *, then* X *is isomorphic to* Y .

For the proof of 9.2 we need a lemma. Recall that if X is a r.i. function space on $[0,1]$ we denote by $X([0,1]^2)$ the space of all functions on the square with the same distribution as the functions in X and with the obvious norm. For $f \in X$ and $g \in L_\infty(0,1)$ we define $f \otimes g \in X([0,1]^2)$ by

$$f \otimes g(s,t) = f(s) \cdot g(t) .$$

Lemma 9.3: Let X *be a r.i. function space on* $[0,1]$ *with* $0 < \beta_X$. *There exists a constant* K_1 *such that for every sequence* $\{f_n\}_{n=1}^\infty$ *of elements of* X

$$K_1^{-1}\|(\sum_{n=1}^\infty f_n^2)^{1/2}\|_X \le \|\sum_{n=1}^\infty f_n \otimes r_n\|_{X([0,1]^2)} \le K_1\|(\sum_{n=1}^\infty f_n^2)^{1/2}\|_X .$$

Proof: For any sequence $\{f_n\}_{n=1}^\infty$ in X ,

$$\|\sum_{n=1}^\infty f_n \otimes r_n\|_{X([0,1]^2)} = \int_0^1 \|\sum_{n=1}^\infty r_n(u)f_n \otimes r_n\|_{X([0,1]^2)} du$$

$$\ge \|\int_0^1 |\sum_{n=1}^\infty r_n(u)f_n \otimes r_n|du\|_{X([0,1]^2)}$$

$$\ge A_1\|(\sum_{n=1}^\infty f_n^2)^{1/2} \otimes 1\|_{X([0,1]^2)} = A_1\|(\sum_{n=1}^\infty f_n^2)^{1/2}\|_X .$$

If we assume that X is s-concave for some $s < \infty$ we can prove the other side estimate in a similar way; this is enough for the proof of 9.2. However, for further use we prove the lemma under the mere assumption that $\beta_X > 0$. Notice that for any $1 \le q < \infty$, and $0 < u < \infty$

$$\mu(\{t;|\sum_{n=1}^\infty f_n(s)r_n(t)| > u\}) \le B_q^q(\sum_{n=1}^\infty f_n^2(s))^{q/2}u^{-q} .$$

So, in order to finish the proof it is enough to prove

Lemma 9.4: Let X *be a r.i. function space on* $[0,1]$ *with* $0 < \frac{1}{q} < \beta_X$ *. There exists a constant* K_2 *such that whenever* $\|f\|_X = 1$ *and* $g(s,t)$ *satisfies*

$$\mu(\{t; |g(s,t)| > u\}) \le f(s)^q u^{-q} \ ; \quad 0 \le s \le 1 \ , \ 0 < u < \infty$$

then

$$\|g\|_{X([0,1]^2)} \le K_2 \ .$$

Proof: Since

$$\mu(\{t; t^{-1/q} f(s) > u\}) = f(s)^q u^{-q} \ ,$$

it is enough to prove that $\|t^{-1/q} f(s)\|_{X([0,1]^2)} \le K_2$, for some K_2 which depends only on X . Now by the discussion in section 8 concerning the indices, and using "D" to denote the dilation operator, we have

$$\|t^{-1/q} f(s)\|_{X([0,1]^2)} \le \|f(s) \sum_{n=0}^{\infty} 2^{(n+1)/q} \chi_{(2^{-(n+1)},2^{-n}]}(t)\|_{X([0,1]^2)}$$

$$\le \sum_{n=0}^{\infty} 2^{(n+1)/q} \|f(s)\chi_{(0,2^{-n}]}(t)\|_{X([0,1]^2)}$$

$$\le \sum_{n=0}^{\infty} 2^{(n+1)/q} \|D_{2^n}\|_X$$

$$\le \text{const.} \sum_{n=0}^{\infty} 2^{(n+1)/q} \, 2^{-n(\beta_X + 1/q)/2} = K_2 \ .$$

(See Section 8 for the inequality $\|D_s\|_X \le \text{const.} \ s^{-\beta_X + \varepsilon}$.) □

Proof of Corollary 9.2: By theorem 9.1, Y contains a complemented subspace Z which is isomorphic to X. Let

$$P: X \xrightarrow{\text{onto}} Y, \quad Q: Y \xrightarrow{\text{onto}} Z$$

be the given projections and denote by $Y(\ell_2)$ the space of all sequences $\{f_n\}_{n=1}^\infty$ of functions in Y such that the sequence $(\sum_{n=1}^N f_n^2)^{1/2}$, $N = 1,2,\dots,$ converges in X and

$$\|\{f_n\}_{n=1}^\infty\|_{Y(\ell_2)} = \|(\sum_{n=1}^\infty f_n^2)^{1/2}\|_Y .$$

$Z(\ell_2)$ and $X(\ell_2)$ are defined similarly. By the $\infty > s$-concavity of X, we get that the expression $\|(\sum_{n=1}^\infty f_n^2)^{1/2}\|_X$ is equivalent, with constant depending only on s and the s-concavity constant of X, to $\int_0^1 \|\sum_{n=1}^\infty f_n r_n(u)\|_X du$. It follows easily that the operators

$$\tilde{P}: X(\ell_2) \xrightarrow{\text{onto}} Y(\ell_2) \quad \text{and} \quad \tilde{Q}: X(\ell_2) \xrightarrow{\text{onto}} Z(\ell_2)$$

defined by

$$\tilde{P}(f_1,f_2,\dots) = (Pf_1,Pf_2,\dots); \; \tilde{Q}(f_1,f_2,\dots) = (Qf_1,Qf_2,\dots)$$

are bounded projections. A similar construction shows that $X(\ell_2)$ is isomorphic to $Z(\ell_2)$. Thus, $X(\ell_2)$ and $Y(\ell_2)$ are each isomorphic to a complemented subspace of the other, and each is also isomorphic to its square; therefore, $X(\ell_2)$ and $Y(\ell_2)$ are isomorphic.

Now, as is well known and quite easy to prove

$$Pf(s,t) = \sum_{n=1}^\infty \left(\int_0^1 f(s,u) r_n(u)du\right) r_n(t)$$

defines a bounded projection on $L_p([0,1]^2)$ for all $1 < p < \infty$ and thus, by interpolation (see section 8 or [9]), also on $X([0,1]^2)$. By lemma 9.3, the range of this projection is isomorphic to $X(\ell_2)$. Of course, X is a complemented subspace of $X(\ell_2)$ and each of them is isomorphic to its square, so,

$$X \approx X(\ell_2) \approx Y(\ell_2) \approx Y \oplus Y(\ell_2) \approx Y \oplus X$$

and

$$Y \approx X \oplus W \approx X \oplus X \oplus W \approx X \oplus Y$$

for a suitable subspace W of Y. Hence, we conclude that $X \approx Y$.

□

We pass now to some preliminary results needed for the proof of 9.1. We begin with a generalization of an inequality of E. Stein [76] from the setting of $L_p(0,1)$ spaces to that of r.i. function spaces on $[0,1]$.

Proposition 9.5: Let X *be a r.i. function space on* $[0,1]$ *such that* $0 < \beta_X \leq \alpha_X < 1$ *and let* $\{E_n\}_{n=1}^{\infty}$ *be an increasing sequence of conditional expectations on* $[0,1]$. *Then there exists a constant* K_3 *such that*

$$\|(\sum_{n=1}^{\infty} (E_n f_n)^2)^{1/2}\|_X \leq K_3 \|(\sum_{n=1}^{\infty} f_n^2)^{1/2}\|_X$$

for every sequence $\{f_n\}_{n=1}^{\infty} \in X(\ell_2)$.

Proof: The case $X = L_p(0,1)$ is due to Stein [76]; we reproduce his proof for completeness.

For $1 \leq p$, $q < \infty$ let $L_p(\ell_q)$ be the space of all sequence $\{f_n\}_{n=1}^{\infty}$ of measurable functions on $[0,1]$ such that

$$\|\{f_n\}_{n=1}^{\infty}\|_{L_p(\ell_q)} = (\int_0^1 (\sum_{n=1}^{\infty} |f_n|^q)^{p/q} d\mu)^{1/p} < \infty$$

and $L_p(\ell_\infty)$ the space of all such sequences for which

$$\|\{f_n\}_{n=1}^{\infty}\|_{L_p(\ell_\infty)} = (\int_0^1 \sup_n |f_n|^p d\mu)^{1/p} < \infty .$$

Define a linear operator E on the linear space of sequences of bounded measurable functions on $[0,1]$ by

$$E(\{f_n\}_{n=1}^N) = \{E_n f_n\}_{n=1}^N .$$

This operator is clearly of norm one if we consider it as an operator from $L_p(\ell_p)$ to $L_p(\ell_p)$, $1 \le p < \infty$. It is also bounded as an operator from $L_p(\ell_\infty)$ to $L_p(\ell_\infty)$, $1 \le p < \infty$, since by Doob's inequality and the positivity of the E_n's

$$\int_0^1 \sup_n |E_n f_n|^p d\mu \le \int_0^1 (\sup_n E_n(\sup_m |f_m|))^p d\mu$$

$$\le K_p \int_0^1 \sup_m |f_m|^p d\mu$$

for some constant K_p which depends only on p. Interpolating (cf. [5] for a discussion of interpolation theorems in this situation), we get that, if $1 < p < 2$, E is bounded as an operator from $L_p(\ell_2)$ to $L_p(\ell_2)$. Since each of the E_n's is self-adjoint on $L_2(0,1)$ we get that E^* is formally equal to E and thus E is also bounded on $L_q(\ell_2)$ for $q > 2$.

In the general case we look at the projection

$$Pf(s,t) = \sum_{n=1}^{\infty} \left(\int_0^1 f(s,u)r_n(u)du\right)r_n(t) .$$

P is a bounded projection on $L_p([0,1]^2)$ for all $1 < p < \infty$. Define an operator $\tilde{E}$ from

$$L = \{f \in L_p([0,1]^2);\ f = \sum_{n=1}^{\infty} f_n \otimes r_n\}$$

into itself by

$$\tilde{E}\left(\sum_{n=1}^{\infty} f_n \otimes r_n\right) = \sum_{n=1}^{\infty} E_n f_n \otimes r_n .$$

In view of (*) of the introduction, $\tilde{E}$ and thus also $\tilde{E}P$ are bounded operators in $L_p([0,1]^2)$. Interpolating we get that EP is bounded on $X([0,1]^2)$. Now, we restrict our attention to functions of the form $\sum_{n=1}^{\infty} f_n \otimes r_n$ in $X([0,1]^2)$ and use lemma 9.3 to get the result.

□

Next we prove a proposition which gives a condition insuring the existence of a projection onto a reproduction of the Haar system. Let $\{y_i\}_{i=1}^{\infty}$ be an unconditional basis for a r.i. function space. For $y = \sum_{i=1}^{\infty} a_i y_i$, the *square function* of y with respect to $\{y_i\}_{i=1}^{\infty}$ is defined by

$$S(y) = \Big(\sum_{i=1}^{\infty} a_i^2 y_i^2\Big)^{1/2} .$$

In section 6 we defined the notion of a tree of subsets of $[0,1]$. In the present case we need a slightly more general notion. A *C-tree* over a measure space $(\Omega,\mathfrak{F},\lambda)$ is a collection of sets $\{G_{n,i}\}_{i=1,n=0}^{2^n \ \ \infty}$ in $\mathfrak{F}$ such that

$$G_{n,i} = G_{n+1,2i-1} \cup G_{n+1,2i} \ ; \ G_{n,i} \cap G_{n,j} = \emptyset$$

and

$$C^{-1}2^{-n} \le \lambda(G_{n,i}) \le C2^{-n} \ ; \ i,j = 1,\dots,2^n \ , \ i \ne j; \ n = 0,1,\dots$$

The notion of a 1-tree coincides with that of a tree. We recall also that $\{h_{n,i}\}_{i=1,n=0}^{2^n \ \ \infty}$ denotes the Haar system on $[0,1]$ normalized in $L_\infty(0,1)$.

Proposition 9.6: Let X *be a r.i. function space on* $[0,1]$ *such that* $0<\beta_X\le\alpha_X<1$ *and let* $\{y_i\}_{i=1}^{\infty}$ *be an enumeration of the Haar system for* X. *Let* $\{k_{n,i}\}_{i=1,n=0}^{2^n \ \ \infty}$ *be a block basis of* $\{y_i\}_{i=1}^{\infty}$ *and, for some* $C>0$, *let* $\{G_{n,i}\}_{i=1,n=0}^{2^n \ \ \infty}$ *be a C-tree in* $[0,1]$ *such that*

(i) $\{k_{n,i}\}_{i=1,n=0}^{2^n \ \ \infty}$ *is C-dominated by the Haar system; i.e.,* $\|\sum_{n,i} a_{n,i}k_{n,i}\|_X \le C\|\sum_{n,i} a_{n,i}h_{n,i}\|_X$ *for every sequence* $\{a_{n,i}\}_{i=1,n=0}^{2^n \ \ \infty}$ *of scalars.*

(ii) $S(k_{n,i})(t) \le C$ *for all* $t \in G_{n,i}$, $i = 1,\dots,2^n$; $n = 0,1,\dots$

(iii) $\int_{G_{n,i}} S^2(k_{n,i})d\mu \geq C^{-1}2^{-n}$; $i = 1,\ldots,2^n$; $n = 0,1,\ldots$

Then $\{k_{n,i}\}_{i=1,n=0}^{2^n\ \infty}$ *is equivalent to* $\{h_{n,i}\}_{i=1,n=0}^{2^n\ \infty}$ *and the span* $[k_{n,i}]_{i=1,n=0}^{2^n\ \infty}$ *of* $\{k_{n,i}\}_{i=1,n=0}^{2^n\ \infty}$ *is complemented in* X .

For the proof of 9.6 we need a lemma which replaces (*) of the introduction in our situation.

Lemma 9.7: Let X *be a r.i. function space on* $[0,1]$ *with* $0<\beta_X \leq \alpha_X<1$. *Then there exists a constant* K_4 *such that*

$$K_4^{-1}\|\sum_{n,i} a_{n,i}h_{n,i}\|_X \leq \|(\sum_{n,i} a_{n,i}^2 h_{n,i}^2)^{1/2}\|_X \leq K_4\|\sum_{n,i} a_{n,i}h_{n,i}\|_X ,$$

for all sequences $\{a_{n,i}\}_{i=1,n=0}^{2^n\ \infty}$ *of scalars.*

Proof: The right hand side inequality follows from the unconditionality of $\{h_{n,i}\}_{i=1,n=0}^{2^n\ \infty}$ as in the proof of the trivial side of (*). The same inequality holds also in X^* and we can therefore conclude by a standard duality argument.

□

Note that the lemma and its proof remain valid for any orthogonal unconditional basis for X .

Proof of 9.6: We first prove that $\{k_{n,i}\}_{i=1,n=0}^{2^n\ \infty}$ is equivalent to $\{h_{n,i}\}_{i=1,n=0}^{2^n\ \infty}$. Using lemma 9.7 and then proposition 9.5 for E_n being the conditional expectation with respect to the field generated by $\{G_{n,i}\}_{i=1}^{2^n}$, and $f_n = \sum_{i=1}^{2^n} |a_{n,i}|S(k_{n,i})\chi_{G_{n,i}}$, $n = 0,1,2,\ldots$ we get,

$$\|\sum_{n=0}^{\infty}\sum_{i=1}^{2^n} a_{n,i}k_{n,i}\|_X \geq K_4^{-1}\|(\sum_{n=0}^{\infty}\sum_{i=1}^{2^n} a_{n,i}^2 S^2(k_{n,i}))^{1/2}\|_X$$

$$\geq K_4^{-1}\|(\sum_{n=0}^{\infty}[\sum_{i=1}^{2^n}|a_{n,i}|S(k_{n,i})\chi_{G_{n,i}}]^2)^{1/2}\|_X$$

$$\geq K_4^{-1}K_3^{-1}\|(\sum_{n=0}^{\infty}\sum_{i=1}^{2^n}\mu(G_{n,i})^{-2}(\int_{G_{n,i}}|a_{n,i}|S(k_{n,i})d\mu)^2\chi_{G_{n,i}})^{1/2}\|_X .$$

Now (ii) and (iii) imply that $\int_{G_{n,i}} S(k_{n,i})d\mu \geq C^{-2}2^{-n}$ for $1 \leq i \leq 2^n$; $n = 0,1,\ldots$; using this and the fact that $\mu(G_{n,i}) \leq C2^{-n}$ we get that

$$\|\sum_{n=0}^{\infty}\sum_{i=1}^{2^n} a_{n,i}k_{n,i}\|_X \geq K_4^{-1}K_3^{-1}C^{-3}\|(\sum_{n=0}^{\infty}\sum_{i=1}^{2^n} a_{n,i}^2\chi_{G_{n,i}})^{1/2}\|_X .$$

Thus it is enough to prove that

(1) There exists a constant K_5 such that

$$K_5\|(\sum_{n=0}^{\infty}\sum_{i=1}^{2^n} a_{n,i}^2\chi_{G_{n,i}})^{1/2}\|_X \geq \|\sum_{n=0}^{\infty}\sum_{i=1}^{2^n} a_{n,i}h_{n,i}\|_X .$$

Let $\{H_{n,i}\}_{i=1,n=0}^{2^n\ \ \infty}$ be a sequence of subsets of $[0,1]$ such that

$$H_{n,i} = H_{n+1,2i-1} \cup H_{n+1,2i} \ ; \ H_{n,i} \cap H_{n,j} = \emptyset$$

and

$$\mu(H_{n,i}) = C^{-1}2^{-n} \ , \ i,j = 1,\ldots,2^n \ , \ i \neq j \ ; \ n = 0,1,\ldots .$$

Then for each sequence of coefficients $\{a_{n,i}\}_{i=1,n=0}^{2^n\ \ \infty}$, for every positive integer N and every $1 \leq j \leq 2^N$, the function $\sum_{n=0}^{N}\sum_{i=1}^{2^n} a_{n,i}^2\chi_{G_{n,i}}$ takes on the set $G_{N,j}$ the same value as $\sum_{n=0}^{N}\sum_{i=1}^{2^n} a_{n,i}^2\chi_{H_{n,i}}$ takes on $H_{n,j}$. Hence, since $\mu(G_{N,i}) \geq \mu(H_{N,i})$,

$$\|(\sum_{n=0}^{\infty}\sum_{i=1}^{2^n} a_{n,i}^2 x_{G_{n,i}})^{1/2}\|_X \geq \|(\sum_{n=0}^{\infty}\sum_{i=1}^{2^n} a_{n,i}^2 x_{H_{n,i}})^{1/2}\|_X$$

$$\geq C^{-1}\|(\sum_{n=0}^{\infty}\sum_{i=1}^{2^n} a_{n,i}^2 h_{n,i}^2)^{1/2}\|_X$$

$$\geq C^{-1}K_4^{-1}\|\sum_{n=0}^{\infty}\sum_{i=1}^{2^n} a_{n,i}h_{n,i}\|_X$$

by Lemma 9.7. This proves that $\{k_{n,i}\}_{i=1,n=0}^{2^n\ \infty}$ is equivalent to $\{h_{n,i}\}_{i=1,n=0}^{2^n\ \infty}$.

For vectors $x = \sum_{i=1}^{\infty} a_i y_i$ and $y = \sum_{i=1}^{\infty} b_i y_i$ in X we define a "vector valued inner product" by

$$(x,y)(t) = \sum_{i=1}^{\infty} a_i b_i y_i^2(t) \ ; \quad 0 \leq t \leq 1 \ .$$

Note that $S(x)$ is the "vector valued norm" associated with $(.,.)$; i.e., $S^2(x) = (x,x)$. Thus, by the Cauchy-Schwartz inequality $|(x,y)| \leq S(x)S(y)$, which shows that (x,y) is well defined and is in $X_{1/2}$. For every $n = 0,1,\ldots$, let P_n be the natural projection from $X = [y_i]_{i=1}^{\infty}$ onto the support (with respect to $\{y_i\}_{i=1}^{\infty}$) of $k_{n,1}+k_{n,2}+\ldots+k_{n,2^n}$. Define now an operator P from X onto $[k_{n,i}]_{i=1,n=0}^{2^n\ \infty}$ by putting

$$Px = \sum_{n=0}^{\infty}\sum_{i=1}^{2^n} \int_{G_{n,i}} (P_n x, k_{n,i})d\mu\Big[\int_{G_{n,i}} S^2(k_{n,i})d\mu\Big]^{-1}\cdot k_{n,i} ; \ x \in X \ .$$

This operator is clearly a projection onto $[k_{n,i}]_{i=1,n=0}^{2^n\ \infty}$ provided it is bounded.

By (ii), (iii) and the Cauchy-Schwartz inequality we have for all $i = 1,\ldots,2^n$; $n = 0,1,\ldots$, that

$$\int_{G_{n,i}} (P_n x, k_{n,i})d\mu \,[\int_{G_{n,i}} S^2(k_{n,i})d\mu]^{-1} \leq$$

$$\leq \int_{G_{n,i}} S(P_n x)S(k_{n,i})d\mu \; C\, 2^n \leq$$

$$(2) \qquad \leq C^2\, 2^n \int_{G_{n,i}} S(P_n x)d\mu \;\leq$$

$$\leq C^3\, \mu(G_{n,i})^{-1} \int_{G_{n,i}} S(P_n x)d\mu \;.$$

So, by (i) and (1) we get

$$\|Px\|_X \leq C^4 K_5 \|(\sum_{n=0}^{\infty} \sum_{i=1}^{2^n} \mu(G_{n,i})^{-2}[\int_{G_{n,i}} S(P_n x)]^2 \chi_{G_{n,i}})^{1/2}\|_X$$

$$= C^4 K_5 \|(\sum_{n=0}^{\infty} (\sum_{i=1}^{2^n} \mu(G_{n,i})^{-1} \int_{G_{n,i}} S(P_n x)\chi_{G_{n,i}})^2)^{1/2}\|_X \,.$$

Finally, by 9.5 applied for the conditional expectations onto the fields $\mathfrak{F}_n$ generated by $\{G_{n,i}\}_{i=1}^{2^n}$, $n = 0,1,\ldots,$ we get

$$\|Px\|_X \leq C^4 K_5 K_3 \|(\sum_{n=0}^{\infty} S^2(P_n x))^{1/2}\|_X \leq$$

$$\leq C^4 K_5 K_3 \|S(x)\|_X \leq C^4 K_4 K_2 K_3 \|x\|_X \,.$$

□

The next lemma with a somewhat different proof is due to Enflo and Starbird. We recall some notations introduced in section 6: $E_{n,i} = [(i-1)2^{-n}, i2^{-n})$; $i = 1,2,\ldots,2^n$, $n = 0,1,\ldots$. $\mathcal{E}_n$ is the algebra generated by $\{E_{n,i}\}_{i=1}^{2^n}$ and $\mathcal{E} = \bigcup_{n=1}^{\infty} \mathcal{E}_n$.

Lemma 9.8: Let ν *be a measure on* $[0,1]$ *taking values in the set of positive functions in* $L_1(\Omega,\mathcal{F},\lambda)$ *, where* λ *is a finite measure. Assume that*

(i) The semi-variation of ν *is absolutely continuous with respect to the Lebesgue measure* μ *; i.e;*

$$\int \nu(A)d\lambda \to 0 \qquad \text{whenever} \quad \mu(A) \to 0$$

(ii) There exists an $\epsilon > 0$ *such that*

$$\int \max_{1\leq i\leq 2^n} \nu(E_{n,i})d\lambda \geq \epsilon \ , \ n = 0,1,\ldots \ .$$

Then there exist a constant $C > 0$ *, a tree* $\{F_{n,i}\}_{i=1,n=0}^{2^n \ \infty}$ *with* $F_{n,i} \in \mathcal{S}$ *and a C-tree* $\{G_{n,i}\}_{i=1,n=0}^{2^n \ \infty}$ *in* Ω *such that*

$$\int_{G_{n,i}} \nu(F_{n,i})d\lambda \geq C^{-1}2^{-n} \ .$$

Proof: Define

$$M_n(t) = \max_{1\leq i\leq 2^n} \nu(E_{n,i})(t) \ , \qquad t \in \Omega \ .$$

$\{M_n\}_{n=0}^{\infty}$ is a decreasing sequence so we can define

$$M(t) = \lim_{n\to\infty} M_n(t) \ , \qquad t \in \Omega \ .$$

Clearly, $\int M d\lambda \geq \epsilon$. Let $E \in \Omega$ and $\eta > 0$ be such that

$$\int_E M d\lambda \geq \epsilon/2 \text{ and } M(t) \geq \eta \text{ for } t \in E \ .$$

Define a sequence of functions $\varphi_n: E \to [0,1]$ by $\varphi_n(t) = (i-1)2^{-n}$ if $1 \leq i \leq 2^n$ is the first integer such that $\nu(E_{n,i})(t) \geq M(t)$. $\{\varphi_n(t)\}_{n=0}^{\infty}$ is an increasing sequence for each $t \in E$; therefore, we can define

$$\varphi = \lim_{n\to\infty} \varphi_n \ ,$$

We are going to show now that

(a) $\chi_{\varphi^{-1}(A)}(t)M(t) \le \nu(A)(t)$, for every measurable

$A \subset [0,1]$ and λ-almost every t ; and

(b) $\lambda(\varphi^{-1})$ is absolutely continuous with respect to μ .

Notice first that if A is an interval of the form $((i-1)2^{-n}, i2^{-n})$ and $t \in \varphi^{-1}(A)$ then $\varphi(t) \in A$ and therefore $\varphi_k(t), \varphi_k(t)+2^{-k} \in A$ for k large enough. This implies, by the definition of φ_k that

$$\nu(A)(t) \ge \nu(\varphi_k(t), \varphi_k(t) + 2^{-k})(t) \ge M(t) \ .$$

In order to complete the proof of (a) it is thus enough to prove (b). Let A be a measurable set in $[0,1]$ and let $(I_n)_{n=1}^{\infty}$ be a sequence of disjoint open intervals such that $\bigcup_{n=1}^{\infty} I_n \supset A$. If $t \in \varphi^{-1}(A)$ then $t \in I_n$ for some n which implies that

$$\nu(\bigcup_{n=1}^{\infty} I_n)(t) \ge \nu(I_n)(t) \ge M(t) \ .$$

By (i)

$$\int_{\varphi^{-1}(A)} \nu(A)d\lambda \ge \int_{\varphi^{-1}(A)} Md\lambda \ge \eta\lambda(\varphi^{-1}(A)) \ .$$

Therefore, if $\mu(A) \to 0$ then, since $\int \nu(A)d\lambda \to 0$ we have that $\lambda(\varphi^{-1}(A)) \to 0$.

Look at the vector measure

$$m(A) = (\mu(A), \lambda(\varphi^{-1}(A)) \ , \int_{\varphi^{-1}(A)} Md\lambda \ ) \ .$$

Now m is purely non-atomic, since it is absolutely continuous with respect to μ. Thus, we can apply Lyapunov's theorem to find disjoint sets $F'_{1,1}$, $F'_{1,2}$ which have the same m measure. By perturbing these sets a little bit we get dyadic sets $F_{1,1}, F_{1,2} \in \mathcal{B}$ which satisfy

$$\mu(F_{1,1}) = \mu(F_{1,2}) = \frac{1}{2},$$

$$(1-\frac{1}{2})\frac{\lambda(E)}{2} \leq \lambda(\varphi^{-1}(F_{1,1})),\ \lambda(\varphi^{-1}(F_{1,2})) \leq (1+\frac{1}{2})\frac{\lambda(E)}{2}$$

and

$$(1-\frac{1}{2})\frac{1}{2}\int_E M d\lambda \leq \int_{\varphi^{-1}(F_{1,1})} M d\lambda\ ,\ \int_{\varphi^{-1}(F_{1,2})} M d\lambda \leq (1+\frac{1}{2})\frac{1}{2}\int_E M d\lambda .$$

By dividing the sets $F_{1,1}$, $F_{1,2}$ into dyadic sets in a similar manner and continuing the division process we get a dyadic tree $\{F_{n,i}\}_{i=1,n=0}^{2^n\ \ \infty}$ such that for $1 \leq i \leq 2^n$, $n = 0,1,\ldots$,

$$\prod_{i=1}^{\infty}(1-2^{-i})2^{-n}\mu(E) \leq \mu(\varphi^{-1}(F_{n,i})) \leq \prod_{i=1}^{\infty}(1+2^{-i})2^{-n}\mu(E)$$

$$\prod_{i=1}^{\infty}(1-2^{-i})2^{-n}\int_E M d\lambda \leq \int_{\varphi^{-1}(F_{n,i})} M d\lambda \leq \prod_{i=1}^{\infty}(1+2^{-i})2^{-n}\int_E M d\lambda .$$

Define now $G_{n,i} = \varphi^{-1}(F_{n,i})$, $i = 1,\ldots,2^n$; $n = 0,1,\ldots$ and use (a) to get the desired result.

□

We now introduce a concept which will play in the sequel the role of the E-operators in [22]. As in section 6, we use "$\sum_{h_{n,i}\subset E}$" to denote summation over all i for which the support of $h_{n,i}$ is contained in E. "S(f)" will, as above, denote the square function of f relative to the Haar system.

Definition 9.9: Let X *be a r.i. function space on* $[0,1]$. *An operator* $T\colon X\to X$ *is said to have property A if there exists a sequence* $\{h'_{n,i}\}_{i=1,n=0}^{2^n\ \infty}$ *equivalent to the Haar system in* X *such that*

$$(3)\qquad \Lambda(E) = \lim_{n\to\infty} S^2\Big(\sum_{h_{n,i}\subset E} Th'_{n,i}\Big)$$

exists μ *a.e. for every* $E \in \mathcal{S}$, *and*

$$(4)\qquad \inf_n \int_0^1 \max_{1\le i\le 2^n} \Lambda^{1/2}(E_{n,i})\,d\mu > 0 .$$

In the next lemma we mention a stability property of property A .

Lemma 9.10: Let X *be a r.i. function space on* $[0,1]$ *with* $0 < \beta_X \le \alpha_X < 1$ *and let* $T\colon X\to X$ *be an operator having property A* .

(a) If $\{h'_{n,i}\}_{i=1,n=0}^{2^n\ \infty}$ *is a sequence equivalent to* $\{h_{n,i}\}_{i=1,n=0}^{2^n\ \infty}$ *for which (3) and (4) of 9.9 are satisfied, and* $\{g_{n,i}\}_{i=1,n=0}^{2^n\ \infty}$ *is a sequence in* X *satisfying*

$$\sum_{n,i} \|Th'_{n,i}-g_{n,i}\|\ \|Th'_{n,i}\|^{-1} < \infty$$

then the operator $\tilde{T}\colon X\to X$ *defined by*

$$\tilde{T}h_{n,i} = g_{n,i}\ ,\ i = 1,\dots,2^n\ ;\ n = 0,1,\dots$$

has property A; in fact, for each $E \in \mathcal{S}$,

$$\lim_{n\to\infty} S^2\Big(\sum_{h_{n,i}\subset E} g_{n,i}\Big) = \lim_{n\to\infty} S^2\Big(\sum_{h_{n,i}\subset E} Th'_{n,i}\Big) .$$

Furthermore, if $[g_{n,i}]_{i=1,n=0}^{2^n\ \infty}$ *is complemented in* X , *so is* $[Th'_{n,i}]_{i=1,n=0}^{2^n\ \infty}$

(b) There exist a sequence $\{h'_{n,i}\}_{i=1,n=0}^{2^n\ \ \infty}$ *and a block basis* $\{g_{n,i}\}_{i=1,n=0}^{2^n\ \ \infty}$ *of the Haar system which satisfy the conditions in (a).*

Proof: (a) For every $E \in \mathcal{E}$ and every n

$$|S(\sum_{h_{n,i}\subset E} Th'_{n,i}) - S(\sum_{h_{n,i}\subset E} g_{n,i})| \le S(\sum_{h_{n,i}\subset E} (Th'_{n,i}-g_{n,i}))$$

and

$$\sum_{n=1}^{\infty} \int_0^1 S(\sum_{h_{n,i}\subset E} (Th'_{n,i}-g_{n,i}))d\mu \le \sum_{n=1}^{\infty} \|S(\sum_{h_{n,i}\subset E} (Th'_{n,i}-g_{n,i}))\|_X < \infty$$

by Lemma 9.7. This is enough to insure that

$$|S(\sum_{h_{n,i}\subset E} Th'_{n,i}) - S(\sum_{h_{n,i}\subset E} g_{n,i})| \to 0 \quad \mu\text{-a.e.}$$

as $n \to \infty$, and thus $\lim_{n\to\infty} S^2(\sum_{h_{n,i}\subset E} g_{n,i}) = \lim_{n\to\infty} S^2(\sum_{h_{n,i}\subset E} Th'_{n,i})$, which proves that $\tilde{T}$ has property A .

The complementation assertion follows from a well-known stability result. Indeed, by [8], $[Th'_{n,i}]_{i=1,n=N}^{2^n\ \ \infty}$ will be complemented for N large enough if $[g_{n,i}]_{i=1,n=0}^{2^n\ \ \infty}$ is complemented; hence also $[Th'_{n,i}]_{i=1,n=0}^{2^n\ \ \infty}$ will be complemented.

To prove (b) notice first that for every $E \in \mathcal{E}$ $r_n^E = \sum_{h_{n,i}\subset E} h_{n,i}$ converges weakly to zero in X as $n \to \infty$, since $X \supset L_r(0,1)$ for $r > \beta_X^{-1}$ with continuous injection and $r_n^E \to 0$ weakly in $L_r(0,1)$. Suppose now that $\{h'_{n,i}\}_{i=1,n=0}^{2^n\ \ \infty}$ satisfies (3) and (4). We have just seen that

$$\sum_{h_{n,i}\subset E} Th'_{n,i} \to 0 \text{ weakly as } n \to \infty$$

for every $E \in \mathcal{E}$ so, by a standard gliding hump argument, we can find, for every $\epsilon > 0$, a sequence $\{k_{m,j}\}_{j=1,m=0}^{2^m\ \infty}$ of normalized vectors, disjointly supported with respect to the Haar system and a sequence of positive integers $\{n(m,j)\}_{j=1,m=0}^{2^m\ \infty}$ such that by putting $h''_{m,j} = \sum_{h_{n(m,j),i} \subset E_{m,j}} h'_{n(m,j),i}$ we get

$$(5) \qquad \sum_{m,j} \| \|Th''_{m,j}\|^{-1} Th''_{m,j} - k_{m,j}\| < \epsilon \cdot K^{-1}$$

where K is the unconditionality constant of the Haar system in X. Of course, $\{h''_{m,j}\}_{j=1,m=0}^{2^m\ \infty}$ is, by 9.7, equivalent to the Haar system and clearly $\{h''_{n,i}\}_{i=1,n=0}^{2^n\ \infty}$ satisfies (3) and (4) since $\{h'_{n,i}\}_{i=1,n=0}^{2^n\ \infty}$ does. Finally, we set

$$g_{m,i} = \|Th''_{m,i}\| k_{m,i} \ ; \ i = 1,\ldots,2^m \ ; \ m = 0,1,\ldots$$

□

Remark: It is clear that if $\{g_{n,i}\}_{i=1,n=0}^{2^n\ \infty}$ is a block basis of the Haar system and $\Lambda(E) = \lim_{n\to\infty} S^2(\sum_{h_{n,i}\subset E} g_{n,i})$ exists a.e. for every $E \in \mathcal{E}$, then $\Lambda(\cdot)$ is an additive vector valued measure on $\mathcal{E}$. Thus, it follows from the proof of 9.10 that, if Λ is the set function associated to an operator with property A, then Λ defines an additive vector valued measure on $\mathcal{E}$.

The way we are going to use property A becomes clear in the following:

Proposition 9.11: Let X *be a r.i. function space on* $[0,1]$ *with* $0 < \beta_X \le \alpha_X < 1$ *and let* $T: X \to X$ *be an operator having property A. Then* X *contains a subspace* Y *isomorphic to* X *such that* $T_{|Y}$ *is an isomorphism and* TY *is complemented in* X .

Proof: By 9.10 there exist a block basis $\{g_{n,i}\}_{i=1,n=0}^{2^n\ \ \infty}$ of the Haar system in X and a sequence $\{h'_{n,i}\}_{i=1,n=0}^{2^n\ \ \infty}$, equivalent to the Haar system such that

(6) $$\Lambda(E) = \lim_{n\to\infty} S^2\Big(\sum_{h_{n,i}\subset E} g_{n,i}\Big)$$

exists a.e. for every $E \in \mathcal{S}$,

(7) $$\inf_n \int_0^1 \max_{1\le i\le 2^n} \Lambda^{1/2}(E_{n,i})d\mu > 0$$

and

(8) $$\sum_{n,i} \|Th'_{n,i} - g_{n,i}\|\ \|Th'_{n,i}\|^{-1} < \infty \ .$$

It was remarked after 9.10 that Λ is an additive vector valued measure on $\mathcal{S}$ taking values in the set of all the non-negative measurable functions. Since the functions $\max_{1\le i\le 2^n} \Lambda^{1/2}(E_{n,i})$, $n = 0,1,\ldots$ are uniformly bounded by the $L_1(0,1)$ function $\Lambda^{1/2}(0,1)$, we have from Egoroff's theorem that there exists a set $G \subset [0,1]$ such that

(9) $\Lambda(0,1)$ is bounded on G, say by C_1,

(10) the convergence in (6) is uniform on G for every $E \in \mathcal{S}$,

and

(11) $$\inf_n \int_G \max_{1\le i\le 2^n} \Lambda^{1/2}(E_{n,i})d\mu > 0 \ .$$

(9) and (11) imply that

(12) $$\inf_n \int_G \max_{1\le i\le 2^n} \Lambda(E_{n,i})d\mu > 0 \ ,$$

while (9) and (10) imply that for every $E \in \mathcal{S}$,

(13)
$$\left\{\begin{aligned}\int_G \Lambda(E)\,d\mu &\le C_1^{1/2}\int_G \Lambda^{1/2}(E)\,d\mu = \\ &\le C_1^{1/2}\lim_{n\to\infty}\int_G S\Big(\sum_{h_{n,i}\subset E}\tilde{T}h_{n,i}\Big)d\mu \quad \text{(cf. 9.10(a))} \\ &\le C_1^{1/2}\limsup_{n\to\infty}\Big\|S\Big(\sum_{h_{n,i}\subset E}\tilde{T}h_{n,i}\Big)\Big\|_X \\ &\le C_1^{1/2}K_4\|\tilde{T}\|\limsup_{n\to\infty}\Big\|\sum_{h_{n,i}\subset E}h_{n,i}\Big\|_X \quad \text{(by 9.7)} \\ &= C_1^{1/2}K_4\|\tilde{T}\|\cdot\|\chi_E\|_X\end{aligned}\right.$$

(where $\tilde{T}$ is defined as in 9.10 by $\tilde{T}h_{n,i} = g_{n,i}$). Since $X \neq L_\infty(0,1)$, the last term tends to zero when $\mu(E)$ tends to zero. We are thus in a position to apply Lemma 9.8 for the measure $\nu = \Lambda \cdot \chi_G$ (notice that, by (13), ν can be extended from $\mathcal{E}$ to the Borel σ field) in order to conclude that there exists a constant C_2, a tree $\{F_{n,i}\}_{i=1,n=0}^{2^n\ \infty}$, with $F_{n,i} \in \mathcal{E}$, and a C_2-tree $\{G_{n,i}\}_{i=1,n=0}^{2^n\ \infty}$, with $G_{n,i} \subset G$, $i = 1,\dots,2^n$, $n = 0,1,\dots,$ such that

(14)
$$\int_{G_{n,i}} \Lambda(F_{n,i})\,d\mu > C_2^{-1}\,2^{-n}, \quad i = 1,\dots,2^n;\ n = 0,1,\dots .$$

It follows now from (6) and (10) that there exists a sequence of integers $\{n(m,j)\}_{j=1,m=0}^{2^m\ \infty}$, with $n(m,j) > \inf\{\ell; F_{m,j} \in \mathcal{E}_\ell\}$, such that

(15)
$$\int_{G_{m,j}} S^2\Big(\sum_{h_{n(m,j),i}\subset F_{m,j}} g_{n(m,j),i}\Big)d\mu > 2^{-1}C_2^{-1}\,2^{-n}$$

and

(16) $$S^2\Big(\sum_{h_{n(m,j),i}\subset F_{m,j}} g_{n(m,j),i}\Big)(t) < 2C_1 \quad \text{for } t \in G_{m,j}$$

$$j = 1,\ldots,2^m;\ m = 0,1,\ldots$$

Define

$$k_{m,j} = \sum_{h_{n(m,j),i}\subset F_{m,j}} g_{n(m,j),i}\ ,\ j = 1,\ldots,2^m;\ m = 0,1,\ldots$$

The sequences $\{k_{m,j}\}_{j=1,m=0}^{2^m\ \infty}$, $\{G_{m,j}\}_{j=1,m=0}^{2^m\ \infty}$ and $C = \max(\sqrt{2C_1}, 2C_2)$ satisfy the assumptions of 9.6 and thus $\{k_{m,j}\}_{j=1,m=0}^{2^m\ \infty}$ is equivalent to the Haar system and $[k_{m,j}]_{j=1,m=0}^{2^m\ \infty}$ is complemented in X . The proof is completed by applying 9.10(a).

□

We are ready now for the proof of the theorem. Since the general proof is based on the proof of theorem 6.1 we prefer to give first a proof for $X = L_p(0,1)$, $p > 2$ which is independent of 6.1 and which illustrates the method of the general proof. In an appendix to this section we give a proof for $L_p(0,1)$, $p < 2$, which is independent of 6.1; this proof, however, is more complicated.

Proof of 9.1 for $X = L_p(0,1)$, $p > 2$. Let $T: L_p(0,1) \to L_p(0,1)$ be an isomorphism. By 9.11, it is enough to show that T has property A . By the reproducibility of the Haar system (cf. [45]) and Lemma 9.10 we can assume, without loss of generality, that $\{Th_{n,i}\}_{i=1,n=0}^{2^n\ \infty}$ is a block basis of the Haar system in $L_p(0,1)$. Define now, for $n = 0,1,\ldots,$ a vector valued measure v_n^2 on $\mathcal{S}_n$ by

$$v_n^2(E) = S^2\Big(\sum_{h_{n,i}\subset E} Th_{n,i}\Big)\ ,\ E \in \mathcal{S}_n$$

where S is the square function with respect to the Haar system. Note first that

$$\|v_n^2(E)\|_{p/2}^{p/2} \leq K^p\|T\|^p \mu(E) \tag{17}$$

where K is the unconditionality constant of the Haar system. Thus, since $\mathcal{E}$ is countable, there exists a subsequence $\{v_{n_k}\}_{k=1}^{\infty}$ of $\{v_n\}_{n=1}^{\infty}$ such that $v_{n_k}^2(E)$ converges weakly in $L_{p/2}(0,1)$ for every $E \in \mathcal{E}$. Denote the limit by $\Lambda(E)$. $\Lambda(E)$ is clearly an additive measure on $\mathcal{E}$. For every $E \in \mathcal{E}$ there exists a sequence $\{\sigma_j\}_{j=1}^{\infty}$ of disjoint finite subsets of the integers and non-negative numbers $\{\alpha_n\}_{n=1}^{\infty}$ such that $\sum_{n\in\sigma_j} \alpha_n^2 = 1$ and

$$\sum_{n\in\sigma_j} \alpha_n^2 v_n^2(E) \to \Lambda(E)\ , \quad \text{as} \quad j \to \infty \tag{18}$$

where the convergence is in $L_{p/2}(0,1)$. We can thus choose disjoint sets of integers $\{\sigma_{m,j}\}_{j=1,m=0}^{2^m\ \ \infty}$ such that $\sum_{n\in\sigma_{m,j}} \alpha_n^2 = 1$ and

$$\|\sum_{n\in\sigma_{m,j}} \alpha_n^2 v_n^2(E_{m,j}) - \Lambda(E_{m,j})\|_{p/2} < 2^{-m} \mu(E_{m,j})\ ; \tag{19}$$

$$j = 1,\dots,2^m\ ,\ m = 0,1,\dots$$

Define a gaussian Haar system by

$$h'_{m,j} = \sum_{n\in\sigma_{m,j}} \alpha_n \sum_{h_{n,i}\subset E_{m,j}} h_{n,i}\,;\ j = 1,\dots,2^m\ ,\ m = 0,1,\dots$$

By 6.2, $\{h'_{m,j}\}_{j=1,m=0}^{2^m\ \ \infty}$ is equivalent to the Haar system. From (19) we conclude easily that if $E \in \mathcal{E}_\ell$ and $m < \ell$ then

$$\|S^2(\sum_{h_{m,j}\subset E} Th'_{m,j}) - \Lambda(E)\|_{p/2} =$$

$$= \|\sum_{n\in\sigma_{m,j}} \alpha_n^2 v_n^2(E) - \Lambda(E)\|_{p/2} \le 2^{-m}\mu(E) .$$

In particular, we have that

$$S^2(\sum_{h_{m,j}\subset E} Th'_{m,j}) \to \Lambda(E) \quad \mu\text{-a.e. as} \quad m\to\infty ,$$

for all $E\in\mathcal{E}$. In order to show that T has property A it is thus enough to prove that

$$\inf_n \int_0^1 \max_{1\le i\le 2^n} \Lambda^{1/2}(E_{n,i})d\mu = \int_0^1 \lim_{n\to\infty} \max_{1\le i\le 2^n} \Lambda^{1/2}(E_{n,i})d\mu > 0$$

(the equality holds since $\{\max_{1\le i\le 2^n} \Lambda(E_{n,i})\}_{n=1}^{\infty}$ is a non-increasing sequence). We claim that, for every $E\in\mathcal{E}$

$$(20)\qquad (K^2\|T^{-1}\|B_p)^{-p}\mu(E) \le \|\Lambda(E)\|_{p/2}^{p/2} \le (K\|T\|)^p \mu(E) .$$

Indeed, by (18)

$$\|\Lambda(E)\|_{p/2}^{p/2} = \lim_{j\to\infty} \int_0^1 (ST(\sum_{n\in\sigma_j} \alpha_n \sum_{h_{n,i}\subset E} h_{n,i}))^p d\mu$$

$$\ge K^{-p}B_p^{-p}\|T^{-1}\|^{-p} \limsup_{j\to\infty} \int_0^1 |\sum_{n\in\sigma_j} \alpha_n \sum_{h_{n,i}\subset E} h_{n,i}|^p d\mu$$

$$\ge (K^2B_p\|T^{-1}\|)^{-p} \limsup_{j\to\infty} \int_0^1 (\sum_{n\in\sigma_j} \alpha_n^2 \sum_{h_{n,i}\subset E} h_{n,i}^2)^{p/2} d\mu$$

$$= (K^2B_p\|T^{-1}\|)^{-p} \mu(E)$$

while the other side inequality follows from (17) and the weak convergence of $v_{n_k}^2(E)$ to $\Lambda(E)$.

By (20) and Holder's inequality we have

$$(K^2\|T^{-1}\|B_p)^{-p} \leq \int_0^1 \sum_{i=1}^{2^n} \Lambda(E_{n,i})^{p/2} d\mu \leq$$

$$\leq \int_0^1 \Big(\sum_{i=1}^{2^n} \Lambda(E_{n,i})\Big) \max_{1\leq i\leq 2^n} \Lambda(E_{n,i})^{p/2-1} d\mu \leq$$

$$\leq \Big(\int_0^1 \Big(\sum_{i=1}^{2^n} \Lambda(E_{n,i})\Big)^{p/2} d\mu\Big)^{2/p} \Big(\int_0 ^1\max_{1\leq i\leq 2^n} \Lambda(E_{n,i})^{p/2} d\mu\Big)^{(p-2)/p}$$

$$\leq (K\|T\|)^2 \Big(\int_0^1 \max_{1\leq i\leq 2^n} \Lambda(E_{n,i})^{p/2} d\mu\Big)^{(p-2)/p}$$

i.e., the sequence $\{\int_0^1 \max_{1\leq i\leq 2^n} \Lambda(E_{n,i})^{p/2} d\mu\}_{n=1}^{\infty}$ is bounded away from zero. Since $\{\max_{1\leq i\leq 2^n} \Lambda(E_{n,i})\}_{i=1}^{2^n}$ is a decreasing sequence this is equivalent to $\lim_{n\to\infty} \max_{1\leq i\leq 2^n} \Lambda(E_{n,i}) \neq 0$, which is what we need. □

Theorem 9.1 is a trivial consequence of the following:

Proposition 9.12: Let X *be a r.i. function space on* [0,1] *such that* X *is s-concave for some* $s < \infty$, $\alpha_X < 1$ *and the Haar system in* X *is not equivalent to a sequence of disjoint vectors in* X . *For an operator* $T: X\to X$, *the following three conditions are equivalent.*

(i) T *has property A*

(ii) There exists a subspace Y *of* X *isomorphic to* X *such that* $T_{|Y}$ *is an isomorphism and* TY *is complemented in* X .

(iii) There exists a subspace Y *of* X *isomorphic to* X *such that* $T_{|Y}$ *is an isomorphism.*

Proof: In view of 9.11, the only thing to prove is (iii) → (i). To prove this we can clearly assume that Y = X ; i.e., that T is an isomorphism.

By the reproducibility of the Haar system in X (cf. [45]) and lemma 9.10 we can also assume that $\{Th_{n,i}\}_{i=1,n=0}^{2^n\ \ \infty}$ is a block basis of $\{h_{n,i}\}_{i=1,n=0}^{2^n\ \ \infty}$. Let S be the square function with respect to the Haar system in X and define, for $n = 0,1,\dots$, an additive measure v_n^2 on $\mathcal{E}_n$ by

$$v_n^2(E) = S^2\Big(\sum_{h_{n,i}\subset E} Th_{n,i}\Big) , \ E \in \mathcal{E}_n .$$

As in the proof of 6.8, we get that there exist an additive measure Λ on $\mathcal{E}$ which takes as values positive measurable functions on $[0,1]$, a sequence $\{\sigma_m\}_{m=1}^\infty$ of disjoint finite sets of integers and positive numbers $\{\alpha_n\}_{n=1}^\infty$ such that $\sum_{n\in\sigma_m} \alpha_n^2 = 1$ for $m = 1,2,\dots$ and

$$(21) \qquad \sum_{n\in\sigma_m} \alpha_n^2 v_n^2(E) \to \Lambda(E) \quad \mu\text{-a.e. as } m \to \infty ,$$

for every $E \in \mathcal{E}$.

Set

$$\Lambda_n = \max_{1\le i\le 2^n} \Lambda(E_{n,i})$$

and let $\widetilde{\Lambda}$ be the pointwise decreasing limit of $\{\Lambda_n\}_{n=1}^\infty$. If $\|\widetilde{\Lambda}^{1/2}\| = 0$, then, noticing that the proof of Case II in 6.8 can be carried over without any change with the above definition of v_n, Λ, Λ_n and $\widetilde{\Lambda}$, we conclude that (ii) in 6.8 does occur. This in turn implies that either (ii) or (iii) in 6.1 occur which is impossible by the assumption of 9.1. Thus $\widetilde{\Lambda} \neq 0$ which means that

$$(22) \qquad \inf_n \int_0^1 \max_{1\le i\le 2^n} \Lambda(E_{n,i})d\mu > 0 .$$

Now, by (21) and Egoroff's theorem, we can choose disjoint sets of integers $\{\sigma_{m,j}\}_{j=1,m=0}^{2^n\ \ \infty}$ with $\sigma_{m,j} > m$ and positive numbers $\{\alpha_n\}_{n=1}^\infty$ such that

$\sum_{n\in\sigma_{m,j}} \alpha_n^2 = 1$; $j = 1,\dots,2^m$, $m = 0,1,\dots$, and

(23) $$\Big| \sum_{n\in\sigma_{m,j}} \alpha_n^2 v_n^2(E_{m,j}) - \Lambda(E_{m,j})\Big| < 2^{-m}\mu(E_{m,j})$$

on a set of measure $1-4^{-m}$.

Set

$$h'_{m,j} = \sum_{n\in\sigma_{m,j}} \alpha_n \sum_{h_{n,i}\subset E_{m,j}} h_{n,i} ;\ j = 1,\dots,2^m,\ m = 0,1,\dots$$

By 6.2, $\{h'_{m,j}\}_{j=1,m=0}^{2^m\ \infty}$ is equivalent to $\{h_{m,j}\}_{j=1,m=0}^{2^m\ \infty}$, while by (23), for each $E \in \mathcal{S}$ and each $m > \inf\{\ell ;\ E \in \mathcal{S}_\ell\}$

$$\Big|S^2\Big(\sum_{h_{m,j}\subset E} Th'_{m,j}\Big) - \Lambda(E)\Big| \le$$

$$\le \sum_{E_{m,j}\subset E} \Big| \sum_{n\in\sigma_{m,j}} \alpha_n^2 v_n^2(E_{m,j}) - \Lambda(E_{m,j})\Big| < 2^{-m}$$

on a set of measure $\ge 1-2^{-m}$. In particular,

$$S^2\Big(\sum_{h_{m,j}\subset E} Th'_{m,j}\Big) \to \Lambda(E) \quad \mu\text{-a.e.} \quad \text{as} \quad m \to \infty$$

which together with (22) implies that T has property A . □

Remark 1. We have tried to prove the assertions in this section in the most general form. If we replace the assumption "$0 < \beta_X$" in 9.3-9.11 by "X is s-concave for some $s < \infty$" the proofs are simplified. In this case we can take the square function with respect to any unconditional basis for X so that 9.6, 9.10 and 9.11 can be generalized to some extent.

2. Example 10.4 below shows that the assumption that $\{h_{n,i}\}_{i=1,n=0}^{2^n\ \ \infty}$ is not equivalent to a sequence of disjoint vectors in X cannot be removed from the statement of 9.1.

3. The proof of 9.1 can be carried over with only minor changes to prove that: *If* X *is a r.i. function space on* $[0,\infty)$ *which is s-concave for some* $s < \infty$ *, and such that* $\alpha_X < 1$ *and the Haar system in* X(0,1) *is not equivalent to a sequence of disjoint vectors in* X *then every subspace of* X *, isomorphic to* X(0,1) *, contains a further subspace isomorphic to* X(0,1) *and complemented in* X .

Appendix. Proof of 9.1 for $X = L_p(0,1)$; $1 < p < 2$.

As usual, S will denote the square function with respect to the Haar system (in view of the first remark after 9.12 above, the proof does not change if we take S to be the square function with respect to any other unconditional basis for $L_p(0,1)$). As in the proof of 9.11, we can assume without loss of generality that $\{Th_{n,i}\}_{i=1,n=0}^{2^n\ \infty}$ is disjointly supported with respect to $\{h_{n,i}\}_{i=1,n=0}^{2^n\ \infty}$. This is useful because $S^2(f+g) = S^2(f) + S^2(g)$, when f and g have disjoint supports relative to the Haar system. Define

$$v_n = S(\sum_{i=1}^{2^n} Th_{n,i}) \quad n = 1,2,\ldots .$$

Notice that the sequence $\{v_n\}_{n=1}^{\infty}$ is equi-integrable in $L_p(0,1)$; i.e., for every $\varepsilon > 0$ there exists $R < \infty$ such that

$$\int_0^1 v_n^p \chi_{\{v_n \geq R\}}\, d\mu < \varepsilon ; \quad n = 1,2,\ldots .$$

Indeed, otherwise there exist an $\varepsilon_0 > 0$, a subsequence $\{v_{n_k}\}_{k=1}^{\infty}$ of $\{v_n\}_{n=1}^{\infty}$ and disjoint sets $\{A_k\}_{k=1}^{\infty}$ which satisfy

$$(\int_{A_k} v_{n_k}^p\, d\mu)^{1/p} \geq \varepsilon_0 .$$

Thus, for every $\{a_k\}_{k=1}^{\infty} \in \ell_2$,

$$(\sum_{k=1}^{\infty} |a_k|^2)^{1/2} = \| \sum_{k=1}^{\infty} a_k \sum_{i=1}^{2^{n_k}} h_{n_k,i} \|_2 \geq \| \sum_{k=1}^{\infty} a_k \sum_{i=1}^{2^{n_k}} h_{n_k,i} \|_p$$

$$\geq \|T\|^{-1} K^{-1} A_p (\int_0^1 \sum_{k=1}^{\infty} |a_k|^2 v_{n_k}^2)^{p/2} d\mu)^{1/p} \geq \|T\|^{-1} K^{-1} A_p (\sum_{k=1}^{\infty} \int_{A_k} |a_k|^p v_{n_k}^p\, d\mu)^{1/p}$$

$$\geq \|T\|^{-1} K^{-1} A_p \varepsilon_0 (\sum_{k=1}^{\infty} |a_k|^p)^{1/p}$$

(K is the unconditionality constant of the Haar system) which is a contradiction since $p < 2$.

Let R and η be such that

$$\int_0^1 v_n^p \chi_{\{v_n < R\}}\, d\mu \geq \eta \quad \text{for } n = 1,2,\ldots \tag{24}$$

and for each n , define an $L_{p/2}(0,1)$ valued measure on $\mathcal{S}_n$ by

$$\nu_n(A) = S^2\Big(\sum_{h_{n,i} \subset A} Th_{n,i}\Big)\, \chi_{\{v_n < R\}} .$$

Since, for each $A \in \mathcal{S}$, $\{\nu_n(A)\}_{n=1}^\infty$ is a uniformly bounded sequence of functions there is a subsequence $\{\nu_{n_k}\}_{k=1}^\infty$ of $\{\nu_n\}_{n=1}^\infty$ such that $\{\nu_{n_k}(A)\}_{k=1}^\infty$ converges weakly in L_2 to a limit which will be denoted by $\nu(A)$. Thus, there exist disjoint sets N_j , $j = 1,2,\ldots,$ of the integers and numbers $\alpha_n \geq 0$, $n = 1,2,\ldots$ with $\sum_{n \in N_j} \alpha_n = 1$, $j = 1,2,\ldots$ such that

$$\sum_{n \in N_j} \alpha_n \nu_n(A) \to \nu(A) \quad \text{as} \quad j \to \infty \tag{25}$$

where the convergence is in $L_2(0,1)$ (and thus also in $L_1(0,1)$ and $L_{p/2}(0,1)$) and almost everywhere.

Since the convergence is also in $L_{p/2}(0,1)$ we get that for every $A \in \mathcal{S}$

$$\tag{26} \left\{ \begin{aligned} \int_0^1 \nu(A)^{p/2} d\mu &= \lim_{j\to\infty} \int_0^1 \Big(\sum_{n \in N_j} \alpha_n \nu_n(A)\Big)^{p/2} d\mu \\ &\leq \limsup_{j \to \infty} \int_0^1 \Big(S\Big(T \sum_{n \in N_j} \alpha_n^{1/2} \sum_{h_{n,i} \subset A} h_{n,i}\Big)\Big)^p d\mu \\ &\leq (K^2 A_p^{-1} \|T\|)^p \mu(A) . \end{aligned} \right.$$

The next thing we are going to show is that $\nu(0,1) \neq 0$. Notice that by (25)

$$\int_0^1 \nu(0,1)d\mu = \lim_{j\to\infty} \sum_{n\in N_j} \alpha_n \int_0^1 \nu_n(0,1)d\mu$$

so that it is enough to show that $\int_0^1 \nu_n(0,1)d\mu$, $n = 0,1,\ldots$ is uniformly bounded away from zero (since $\sum_{n\in N_j} \alpha_n = 1$) . But by (24),

$$\int_0^1 \nu_n(0,1)d\mu \geq (\int_0^1 \nu_n(0,1)^{p/2}d\mu)^{2/p} \geq \eta^{2/p} .$$

It follows that $\int_0^1 \nu(0,1)d\mu \geq \eta^{2/p}$ and

$$\int_0^1 \nu(0,1)^{p/2}d\mu \geq \int_0^1 \nu(0,1)d\mu\cdot R^{p/2-1} \geq \eta^{2/p}\cdot R^{p/2-1} . \tag{27}$$

From (26) and (27) we conclude that there exists $\epsilon > 0$ such that

$$\int_0^1 \max_{1\leq i\leq 2^n} \nu(E_{n,i})^{p/2}d\mu \geq \epsilon^{p/2} , \quad n = 0,1,\ldots \tag{28}$$

Indeed, by Holder's inequality,

$$R^{p/2-1}\eta^{2/p} \leq \int_0^1 \nu(0,1)^{p/2}d\mu = \int_0^1 (\sum_{i=1}^{2^n} \nu(E_{n,i}))^{p/2}d\mu \leq$$

$$\leq \int_0^1 (\sum_{i=1}^{2^n} \nu(E_{n,i})^{p/2})^{p/2} \max_{1\leq i\leq 2^n} \nu(E_{n,i})^{(1-p/2)p/2}d\mu$$

$$\leq (\int_0^1 \sum_{i=1}^{2^n} \nu(E_{n,i})^{p/2}d\mu)^{p/2}(\int_0^1 \max_{1\leq i\leq 2^n} \nu(E_{n,i})^{p/2}d\mu)^{(2-p)/2}$$

$$\leq (K^2A_p^{-1}\|T\|)^{p^2/2}(\int_0^1 \max_{1\leq i\leq 2^n} \nu(E_{n,i})^{p/2}d\mu)^{(2-p)/2} .$$

It follows from (28) that

$$\int_0^1 \max_{1\leq i\leq 2^n} \nu(E_{n,i})d\mu \geq (\int_0^1 \max_{1\leq i\leq 2^n} \nu(E_{n,i})^{p/2}d\mu)^{2/p} \geq \epsilon . \tag{29}$$

We also have (from (26)) that for all $A \in \mathcal{B}$

$$\int_0^1 \nu(A)d\mu \leq R^{1-p/2}\int_0^1 \nu(A)^{p/2}d\mu \leq R^{1-p/2}(K^2A_p^{-1}\|T\|)^p\cdot\mu(A) .$$

It follows that ν extends to an $L_1^+(0,1)$ valued measure (even to a positive $L_1(0,1)$ operator) and we are in a position to apply Lemma 9.8 in order to conclude that there exist a $C > 0$, a tree $\{F_{n,i}\}_{i=1,n=0}^{2^n\ \ \infty}$ with $F_{n,i} \in \mathcal{S}$ and a C-tree $\{G_{n,i}\}_{i=1,n=0}^{2^n\ \ \infty}$ such that

$$\int_{G_{n,i}} \nu(F_{n,i})d\mu \geq C^{-1}2^{-n} , \quad i = 1,\dots,2^n ; \ n = 0,1,\dots$$

From (25) we get now that there exist a sequence of disjoint finite sets $\{N_{m,j}\}_{j=1,m=0}^{2^m\ \ \infty}$ of integers with $N_{m,j} > \inf\{\ell;\ F_{m,j} \in \mathcal{S}_\ell\}$ and a sequence of non-negative numbers $\{\alpha_n\}_{n=1}^{\infty}$ such that $\sum_{n\in N_{m,j}} \alpha_n = 1$, $j = 1,\dots,2^m$, $m = 0,1,\dots$ and

$$\int_{G_{m,j}} \sum_{n\in N_{m,j}} \alpha_n \nu_n(F_{m,j})d\mu > \frac{1}{2}\, C^{-1}2^{-m} . \tag{30}$$

Define

$$k_{m,j} = \sum_{n\in N_{m,j}} \alpha_n^{1/2} \sum_{h_{n,i}\subset F_{m,j}} Th_{n,i} \ ; \ i = 1,\dots,2^m; \ m = 0,1,\dots .$$

Then $\{k_{m,j}\}_{j=1,m=0}^{2^m\ \ \infty}$ is equivalent to the Haar system by 6.2.

We define now a projection onto $[k_{m,j}]_{j=1,m=0}^{2^m\ \ \infty}$ by

$$Px = \sum_{m,j} \int_{G_{m,j}} \sum_{n\in N_{m,j}} \alpha_n^{1/2}(x, \sum_{h_{n,i}\subset F_{m,j}} Th_{n,i})\chi_{\{v_n\leq R\}}d\mu\Big[\int_{G_{m,j}} \sum_{n\in N_{m,j}} \alpha_n v_n(F_{m,j})\Big]^{-1}k_{m,j}$$

(where (x,y) has the same meaning as in 9.6). First note that P is algebraically a projection since if $(m,j) \neq (\ell,r)$ then $(k_{\ell,r}, \sum_{h_{n,i} \subset F_{m,j}} Th_{n,i}) = 0$ for any $n \in N_{m,j}$ (since $\{Th_{n,i}\}$ sit on disjoint supports with respect to the $h_{n,i}$'s) and

$$\sum_{n \in N_{m,j}} \alpha_n^{1/2} (k_{m,j}, \sum_{h_{n,i} \subset F_{m,j}} Th_{n,i}) \chi_{\{v_n \leq R\}} =$$

$$= \sum_{n \in N_{m,j}} \alpha_n^{1/2} \alpha_n^{1/2} (\sum_{h_{n,i} \subset F_{m,j}} Th_{n,i}, \sum_{h_{n,i} \subset F_{m,j}} Th_{n,i}) \chi_{\{v_n \leq R\}}$$

$$= \sum_{n \in N_{m,j}} \alpha_n S^2 (\sum_{h_{n,i} \subset F_{m,j}} Th_{n,i}) \chi_{\{v_n \leq R\}} = \sum_{n \in N_{m,j}} \alpha_n \nu_n (F_{m,j}) .$$

To prove that P is bounded notice that, as in the proof of Proposition 9.6, we get from (29) and the fact that $\{k_{m,j}\}_{j=1,m=0}^{2^m \ \infty}$ is equivalent to the Haar system that there exists a constant D such that

$$\|Px\| \leq D \| (\sum_{m,j} (\int_{G_{m,j}} \sum_{n \in N_{m,j}} \alpha_n^{1/2} (x, \sum_{h_{n,i} \subset F_{m,j}} Th_{n,i}) \chi_{\{v_n \leq R\}} d\mu)^2 \mu(G_{m,j})^{-2} \chi_{G_{m,j}})^{1/2} \|$$

for $x \in L_p(0,1)$.

Now, for $n \in N_{m,j}$, define Q_n to be the natural basis projection onto the support (with respect to $\{h_{n,i}\}_{i=1,n=0}^{2^n \ \infty}$ of $\sum_{h_{n,i} \subset F_{m,j}} Th_{n,i}$, and let P_m be the natural projection onto the support of $\sum_{j=1}^{2^m} \sum_{n \in N_{m,j}} \sum_{h_{n,i} \subset F_{m,j}} h_{n,i}$.

Then, by the Cauchy-Schwartz inequality, we get that for every m and j ,

$$\left|\int_{G_{m,j}} \sum_{n\in N_{m,j}} \alpha_n^{1/2}\Big(x, \sum_{h_{n,i}\subset F_{m,j}} Th_{n,i}\Big)\chi_{\{v_n\le R\}}\, d\mu\right| =$$

$$= \left|\int_{G_{m,j}} \sum_{n\in N_{m,j}} \alpha_n^{1/2}\Big(Q_n x, \sum_{h_{n,i}\subset F_{m,j}} Th_{n,i}\Big)\chi_{\{v_n\le R\}}\, d\mu\right|$$

$$\le \int_{G_{m,j}} \sum_{n\in N_{m,j}} \alpha_n^{1/2}\, S(Q_n x) S\Big(\sum_{h_{n,i}\subset F_{m,j}} Th_{n,i}\Big)\chi_{\{v_n\le R\}}\, d\mu$$

$$\le R \int_{G_{m,j}} \sum_{n\in N_{m,j}} \alpha_n^{1/2}\, S(Q_n x)\, d\mu \le R \int_{G_{m,j}} \Big(\sum_{n\in N_{m,j}} S^2(Q_n X)\Big)^{1/2} d\mu$$

$$\le R \int_{G_{m,j}} S(P_m x)\, d\mu \quad .$$

We thus get that

$$\|Px\| \le D\cdot R \Big\| \sum_{m,j} \Big(\int_{G_{m,j}} S(P_m x)^2 \mu(G_{m,j})^{-2}\, \chi_{G_{m,j}}\, d\mu\Big)^{1/2} \Big\| \quad .$$

The rest of the proof is an exact repetition of the last steps in the proof of Proposition 9.6. □

10. EXAMPLES

Examples 10.1 and 10.4 show that the pathological cases of the embedding theorems for r.i. spaces (case (ii) in 5.1 and case (ii) in 6.1) really do occur, and that 5.1, the uniqueness of r.i. structure result for p-concave $(p < 2)$ spaces, cannot be extended to 2-concave spaces.

Example 10.1: Let $1 < p < 2$. *There is a Banach space* $U = U(p)$ *which has the following properties.*

(1) U *has a normalized unconditional basis* $\{e_n\}_{n=1}^{\infty}$ *which is p-convex and 2-concave.*

(2) Every p-convex, 2-concave normalized unconditional basis sequence is equivalent to a subsequence of $\{e_n\}_{n=1}^{\infty}$.

(3) U *is unique in the sense that if* U' *satisfies (1) and* U *is isomorphic to a complemented subspace of* U' *, then* U *and* U' *are isomorphic.*

(4) U *is isomorphic to uncountably many different r.i. function spaces on* $[0,1]$.

(5) U *is isomorphic to uncountably many different r.i. function spaces on* $[0,\infty)$. *Uncountably many of these have their restriction to* $[0,1]$ *equal to* $L_2(0,1)$ *up to equivalent renormings.*

Before constructing U , we recall (cf. [25]) a procedure for building a q-concave norm, $1 \leq q < \infty$. Let $\{e_n\}_{n=1}^{\infty}$ be a basis for its linear span $X = \text{span}\,\{e_n\}_{n=1}^{\infty}$, which is endowed with a norm $\|\cdot\|$. For $x = \sum_{i=1}^{n} x(i)e_i$ in X , define

$$|x| = \sum_{i=1}^{n} |x(i)|e_i \quad ,$$

and for a sequence $\{x_k\}$ in X, let

$$(\sum_k |x_k|^q)^{1/q} = \sum_i (\sum_k |x_k(i)|^q)^{1/q} e_i .$$

Define a new function $|||\cdot|||$ on X by

$$|||x||| = \sup\{(\sum_{k=1}^n \|x_k\|^q)^{1/q} ; |x| = (\sum_{k=1}^n |x_k|^q)^{1/q}\} .$$

Lemma 10.2: (a) The function $|||\cdot|||$ *defines a norm on* X *for which the basis* $\{e_n\}_{n=1}^\infty$ *is 1-unconditional and q-concave with constant* 1 .

(b) If $\{e_{n_i}\}_{i=1}^\infty$ *is a subsequence of* $\{e_n\}_{n=1}^\infty$ *which is unconditional and q-concave under the norm* $\|\cdot\|$ *, then* $\|\cdot\|$ *and* $|||\cdot|||$ *are equivalent norms on* $\operatorname{span}\{e_{n_i}\}_{i=1}^\infty$.

(c) If $\{e_n\}_{n=1}^\infty$ *is 1-unconditional and p-convex with constant* C *for the norm* $\|\cdot\|$ *,* $1 \le p \le q$ *, then* $\{e_n\}_{n=1}^\infty$ *is p-convex with constant* C *for the norm* $|||\cdot|||$.

Proof: Let $\mathcal{a}$ be the set of all matrices $a = \{a_{k,n}\}_{k=1,n=1}^{\infty\ \ \infty}$ which satisfy, for all n, $\sum_{k=1}^\infty |a_{k,n}|^q \le 1$. For each $a = \{a_{k,n}\}_{k=1,n=1}^{\infty\ \ \infty}$ in $\mathcal{a}$, define

$$|||x|||_a = (\sum_{k=1}^\infty \|\sum_{n=1}^\infty a_{k,n} x(n) e_n\|^q)^{1/q} .$$

It is clear that $|||\cdot|||_a$ is a semi-norm on X, and, for $x \in X$,

$$|||x||| = \sup_{a \in \mathcal{a}} |||x|||_a ,$$

so $|||\cdot|||$ is a norm on X. The other parts of (a) and (b) are obvious from the definition of $|||\cdot|||$.

To prove (c) we repeat the computation in [25]: Given any $a = \{a_{k,n}\}_{k=1,n=1}^{\infty\ \ \infty}$ in $\mathcal{a}$ and $\{x_j\}_{j=1}^m$ in X, we have

$$|||(\sum_{j=1}^{m} |x_j|^p)^{1/p}|||_a = (\sum_{k=1}^{\infty} \| \sum_{n=1}^{\infty} a_{k,n}(\sum_{j=1}^{m} |x_j(n)|^p)^{1/p} e_n\|^q)^{1/q}$$

$$= (\sum_{k=1}^{\infty} \| \sum_{n=1}^{\infty} (\sum_{j=1}^{m} |a_{k,n}x_j(n)|^p)^{1/p} e_n\|^q)^{1/q}$$

$$\leq C(\sum_{k=1}^{\infty} (\sum_{j=1}^{m} \| \sum_{n=1}^{\infty} a_{k,n}x_j(n)e_n\|^p)^{q/p})^{1/q}$$

$$\leq C(\sum_{j=1}^{m} (\sum_{k=1}^{\infty} \| \sum_{n=1}^{\infty} a_{k,n}x_j(n)e_n\|^q)^{p/q})^{1/p}$$

$$= C(\sum_{j=1}^{m} |||x_j|||_a^p)^{1/p} .$$

(c) follows by taking sup's. □

We are now ready to construct the space U which satisfies (1) and (2) of 10.1. Let $\{e_n^*\}_{n=1}^{\infty}$ be a normalized 1-unconditional basis such that every normalized unconditionally basic sequence is equivalent to a subsequence of $\{e_n^*\}_{n=1}^{\infty}$ and set $X = \text{span}\,\{e_n^*\}_{n=1}^{\infty}$. (We recall the simplified construction in [74] of this version of Pelczynski's universal basis space [64]: let $\{f_n\}_{n=1}^{\infty}$ be a dense sequence in the unit sphere of $C(0,1)$ and define

$$\|\sum_{i=1}^{n} a_i e_i^*\| = \sup_{\pm} \{\|\sum_{i=1}^{n} \pm a_i f_i\|_{C(0,1)}\} .)$$

Now apply 10.2 with $q = p^*$ to get a 1-unconditional norm $|||\cdot|||$ on X which is p^*-concave with constant 1, and such that every p^*-concave normalized unconditional basic sequence is equivalent to a subsequence of $\{e_n^*\}_{n=1}^{\infty}$ in the completion of $(X, ||| \; |||)$. By duality it follows that the biorthogonal functionals $\{e_n\}_{n=1}^{\infty}$ to $\{e_n^*\}_{n=1}^{\infty}$ are 1-unconditional and p-convex with constant 1 under the dual norm to $||| \; |||$ (which we also denote by $||| \; |||$), and that every normalized unconditionally basic, p-convex sequence is equivalent to a subsequence of $\{e_n\}_{n=1}^{\infty}$ in the completion of $(\text{span}\{e_n\}_{n=1}^{\infty}, ||| \; |||)$.

Now apply 10.2 with $q = 2$ to $(\{e_n\}_{n=1}^{\infty}, |||\ |||)$ to get another norm $|||\ |||_o$ on span $\{e_n\}_{n=1}^{\infty}$; it is clear that the completion U of $(\mathrm{span}\{e_n\}_{n=1}^{\infty}, |||\ |||_o)$ has properties (1) and (2).

To see (3), note that the natural unconditional basis for $(U+U+\dots)_{\ell_2}$ is p-convex and 2-concave, hence by (2), $(U+U+\dots)_{\ell_2}$ embeds into U as a complemented subspace, whence by the decomposition method [63], U and $(U+U+\dots)_{\ell_2}$ are isomorphic. Again by the decomposition method, any complemented subspace of U into which U embeds as a complemented subspace is isomorphic to U, so (3) follows.

The derivation of (4) and (5) from (1) - (3) is purely formal, but requires some machinery and construction. To prove (4), we build many different r.i. function spaces on $[0,1]$ into which U embeds as a complemented subspace; in fact, the basis $\{e_n\}_{n=1}^{\infty}$ for U will be equivalent to a sequence of the form $\{a_n \chi_{A_n}\}_{n=1}^{\infty}$ in each of these r.i. function spaces (where $\{A_n\}_{n=1}^{\infty}$ is an appropriate sequence of disjoint intervals), and so U will be isomorphic to the complemented subspace $[\chi_{A_n}]_{n=1}^{\infty}$ in each such r.i. function space. Since the Haar system for each constructed r.i. function space will be unconditional, p-convex, and 2-concave, (4) will follow from the construction and (3). For $p \leq q < 2$, we follow a modification by Davis [17] of the construction in [18] to build an interpolation space between $L_q(0,1)$ and $L_2(0,1)$ which will be isomorphic to U. To this end, choose an increasing sequence $\{n_k\}_{k=1}^{\infty}$ of positive integer so that $\sum_{k=1}^{\infty} n_k^{4p/(2-p)} < 1$ (not essential) and, for $k = 1,2,\dots,$ satisfies

$$(+) \qquad n_k^{-1} \sum_{i=1}^{k-1} n_i + n_k \sum_{i=k+1}^{\infty} n_1^{-1} < 2^{-k-1} .$$

Define a Banach space Y_k (which is just $L_q(0,1)$ with an equivalent norm) whose norm $\|\cdot\|_k$ is the gauge of the convex set

$$\operatorname{conv}(n_k^{-1} B_{L_q(0,1)} \cup n_k B_{L_2(0,1)}) ;$$

i.e., for $f \in L_q(0,1)$,

$$\|f\|_k = \inf\{n_k\|g\|_{L_q} + n_k^{-1}\|h\|_{L_2} : f = g + h\} .$$

The space $Y = Y(q,\{n_k\}_{k=1}^{\infty})$ is the set of all f in $L_q(0,1)$ for which the norm $\| \ \|_Y$ defined by

$$\|f\|_Y = \| \sum_{k=1}^{\infty} \|f\|_k e_k\|_U$$

is finite. Note that Y is complete under $\| \ \|_Y$, since the mapping $j: Y \to (\Sigma Y_k)_U$ defined by $jf = (f,f,\dots)$ is an isometry from Y onto a closed subspace of $(\Sigma Y_k)_U$.

The space Y is actually an abstract generalization of the Lions-Peetre interpolation spaces [52], and Y does indeed have the interpolation property: if T is a linear operator with norm ≤ 1 from $L_q(0,1)$ to $L_q(0,1)$ and also from $L_2(0,1)$ to $L_2(0,1)$, then T defines an operator from Y into itself with norm ≤ 1 . Thus the common isometries of $L_q(0,1)$ and $L_2(0,1)$ are also isometries of Y . It is easily checked that the Haar system forms an unconditional basis for Y (cf., e.g., part (x) of lemma 1 of [18]). In particular, $L_\infty(0,1)$ is dense in Y , so that Y is a r.i. function space on $[0,1]$ (except that $\|\chi_{(0,1)}\|_Y$ is not 1).

We need one simple computation: For any measurable subset A of $[0,1]$,

$$(\neq) \qquad \|\chi_A\|_k = \min(n_k\|\chi_A\|_{L_q(0,1)}, n_k^{-1}\|\chi_A\|_{L_2(0,1)}) = \min(n_k\mu(A)^{1/q}, n_k^{-1}\mu(A)^{1/2}).$$

Of course, "$\leq$" is obvious. To prove "$\geq$" , let E^A be the conditional expectation onto χ_A (defined by $E^A h = \mu(A)^{-1} \int_A h(t)dt\, \chi_A$) , which is a norm one projection in $L_q(0,1)$ and $L_2(0,1)$. If $\chi_A = f + g$, then

$\chi_A = E^A f + E^A g \equiv a\,\chi_A + b\,\chi_A$ (where $a+b=1$; $a,b \geq 0$), so that

$$n_k\|f\|_{L_q(0,1)} + n_k^{-1}\|g\|_{L_2(0,1)} \geq a n_k\|\chi_A\|_{L_q(0,1)} + b n_k^{-1}\|\chi_A\|_{L_2(0,1)}$$

$$\geq \min(n_k\mu(A)^{1/q}, n_k^{-1}\mu(A)^{1/2}) .$$

This gives "$\geq$" in ($\neq$).

Let $\{A_k\}_{k=1}^{\infty}$ be a sequence of disjoint subintervals of $(0,1)$ such that for $k = 1,2,\ldots$, $n_k\mu(A_k)^{1/q} = n_k^{-1}\mu(A_k)^{1/2}$; that is,

$$\mu(A_k) = n_k^{4q/(q-2)} .$$

In view of the lacunarity condition (+), $\|\chi_{A_k}\|_Y$ is essentially equal to $\|\chi_{A_k}\|_k = n_k^{(q+2)/(q-2)}$, and $j\chi_{A_k}$ is essentially supported in the kth coordinate, so that $\{n_k^{(q+2)/(2-q)}\chi_{A_k}\}_{k=1}^{\infty}$ in Y is equivalent to the basis $\{e_n\}_{n=1}^{\infty}$ of U. Indeed, for each $k = 1,2,\ldots$, we have by ($\neq$) and (+) that

$$\sum_{\substack{i=1\\ i\neq k}}^{\infty} n_k^{(q+2)/(2-q)}\|\chi_{A_k}\|_i = \sum_{\substack{i=1\\ i\neq k}}^{\infty} n_k^{(q+2)/(2-q)} \min\{n_i\mu(A_k)^{1/q}, n_i^{-1}\mu(A_k)^{1/2}\}$$

$$= \sum_{i=1}^{k-1} n_i n_k^{(q+2)/(2-q)-4/(2-q)} + \sum_{i=k+1}^{\infty} n_i^{-1} n_k^{(q+2)/(2-q)-2q/(2-q)}$$

$$= n_k^{-1}\sum_{i=1}^{k-1} n_i + n_k \sum_{i=k+1}^{\infty} n_i^{-1} < 2^{-k-1} .$$

A standard perturbation argument now yields that $\{n_k^{(q+2)/(2-q)}\chi_{A_k}\}_{k=1}^{\infty}$ in Y is equivalent to $\{P_k j n_k^{(q+2)/(2-q)}\chi_{A_k}\}_{k=1}^{\infty}$ in $(Y+Y+\ldots)_U$ (here P_k is the projection from $(Y+Y+\ldots)_U$ onto the k-th component), which in turn is isometrically equivalent to the basis $\{e_k\}_{k=1}^{\infty}$ of U.

Finally, the closed span of $\{\chi_{A_k}\}_{k=1}^{\infty}$ in Y is complemented by means of a conditional expectation projection since U is reflexive. More formally, if $\mathcal{A}_n$ is the algebra generated by $\{A_k\}_{k=1}^{n}$, the conditional expectation (restricted to $\bigcup_{k=1}^{n} A_n$) projection $E^{\mathcal{A}_n}$ onto span $\{\chi_{A_k}\}_{k=1}^{n}$ is a norm one operator on Y since Y has the interpolation property, and since U is reflexive, we can pass to the limit to conclude that the closed span of $\{\chi_{A_k}\}_{k=1}^{\infty}$ in Y is norm one complemented.

We already have mentioned that the Haar system forms an unconditional basis for Y, but in order to apply (3) and conclude that U is isomorphic to Y, we need to check that the Haar system is p-convex and 2-concave in Y. To this end, we need the following lemma.

Lemma 10.3: *Assume that* $1 < r \le 2 \le s < \infty$ *and that* Z *is an r-convex and s-concave lattice. Every C-unconditional basic sequence* $\{g_n\}_{n=1}^{\infty}$ *in* Z *is r-convex and s-concave, with constants depending only on* r, s, C *and* Z.

Proof: Assume, without loss of generality, that the r-convexity and s-concavity constant of Z are one. Since $\{g_n\}_{n=1}^{\infty}$ is C-unconditional, for all choices $\{a_n\}_{n=1}^{\infty}$ of scalars

$$\|\sum_{n=1}^{\infty} a_n g_n\| \sim \|(\sum_{n=1}^{\infty} |a_n g_n|^2)^{1/2}\| ,$$

where the equivalence constant for "$\sim$" depends only on s and C. Now if $x_k = \sum_{n=1}^{\infty} x_k(n) g_n$, then

$$\|\sum_{n=1}^{\infty} (\sum_{k=1}^{\infty} |x_k(n)|^r)^{1/r} g_n\| \sim \|(\sum_{n=1}^{\infty} (\sum_{k=1}^{\infty} |x_k(n))^r)^{2/r} |g_n|^2)^{1/2}\|$$

$$\le \|(\sum_{k=1}^{\infty} (\sum_{n=1}^{\infty} |x_k(n) g_n|^2)^{r/2})^{1/r}\|$$

$$\leq (\sum_{k=1}^{\infty} \|(\sum_{n=1}^{\infty} |x_k(n)g_n|^2)^{1/2}\|^r)^{1/r} \sim (\sum_{k=1}^{\infty} \|x_k\|^r)^{1/r} ,$$

which gives the r-convexity of $\{g_n\}_{n=1}^{\infty}$. The s-concavity is checked in a similar way (and does not use the hypothesis that Z is r-convex):

$$\|\sum_{n=1}^{\infty} (\sum_{k=1}^{\infty} |x_k(n)|^s)^{1/s} g_n\| \sim \|\sum_{n=1}^{\infty} (\sum_{k=1}^{\infty} |x_k(n)|^s)^{2/s} |g_n|^2)^{1/2}\|$$

$$\geq \|(\sum_{k=1}^{\infty} (\sum_{n=1}^{\infty} |x_k(n)g_n|^2)^{s/2})^{1/s}\|$$

$$\geq (\sum_{k=1}^{\infty} \|(\sum_{n=1}^{\infty} |x_k(n)g_n|^2)^{1/2}\|^s)^{1/s} \sim (\sum_{k=1}^{\infty} \|x_k\|^s)^{1/s} .$$

□

The dual norm to $\|\cdot\|_k$ is given by

$$\|f\|_k^* = \max\{n_k^{-1}\|f\|_{L_q*(0,1)} , n_k\|f\|_{L_2(0,1)}\} .$$

By 10.3, the Haar basis $\{h_n\}_{n=1}^{\infty}$ is 2-convex and p*-concave under the norm $\|\cdot\|_k^*$ with constants independent of k, so that $\{h_n\}_{n=1}^{\infty}$ is 2-concave and p-convex in Y_k with uniform constants (all less than, say $K < \infty$), which in view of the 2-concavity and p-convexity of the basis $\{e_n\}_{n=1}^{\infty}$ of U yields that $\{h_n\}_{n=1}^{\infty}$ is 2-concave and p-convex in Y. For example, for the 2-concavity, if $x_j = \sum_{n=1}^{\infty} x_j(n)h_n$, then

$$\|\sum_{n=1}^{\infty} (\sum_{j=1}^{\infty} |x_j(n)|^2)^{1/2} h_n\|_Y = \|\sum_{k=1}^{\infty} \|\sum_{n=1}^{\infty} (\sum_{j=1}^{\infty} |x_j(n)|^2)^{1/2} h_n\|_k e_k\|_U$$

$$\geq K^{-1}\|\sum_{k=1}^{\infty} (\sum_{j=1}^{\infty} \|\sum_{n=1}^{\infty} x_j(n)h_n\|_k^2)^{1/2} e_k\|_U$$

$$\geq K^{-1}(\sum_{j=1}^{\infty} \|\sum_{k=1}^{\infty} \|x_j\|_k e_k\|_U^2)^{1/2} = K^{-1}(\sum_{j=1}^{\infty} \|x_j\|_Y^2)^{1/2} ,$$

and the p-convexity of $\{h_n\}_{n=1}^{\infty}$ in Y is checked similarly. Thus Y is isomorphic to U.

We complete the proof of 10.1(4) by observing that the above construction yields different r.i. function spaces for different values of q. To see this, let $p \leq r < q < 2$, and perform the above construction with parameters q and r, respectively, to get spaces Y and Y' respectively. The k-th norm for q is denoted by $\|\cdot\|_k$ as above, and the k-th norm for r by $\|\cdot\|_k'$; i.e.,

$$\|f\|_k' = \inf\{n_k\|g\|_{L_r} + n_k^{-1}\|h\|_{L_2} : f = g + h\} ,$$

for all $f \in L_r(0,1)$. As before, we choose for $k = 1,2,\ldots$ subsets A_k of $[0,1]$ with $\mu(A_k) = n_k^{4q/(q-2)}$, so that

$$\|\chi_{A_k}\|_k = n_k\mu(A_k)^{1/q} = n_k^{-1}\mu(A_k)^{1/2} = n_k^{(q+2)/(q-2)} .$$

We saw earlier that (+) yields

$$\sum_{\substack{i=1 \\ i\neq k}}^{\infty} \|\chi_{A_k}\|_i \leq 2^{-k-1}n_k^{(q+2)/(2-q)} = 2^{-k-1}\|\chi_{A_k}\|_k$$

and hence for $k = 1,2,\ldots$, we have

$$\|\chi_{A_k}\|_{Y'} \leq \sum_{i=1}^{\infty} \|\chi_{A_k}\|_i' \leq \|\chi_{A_k}\|_k' + \sum_{\substack{i=1 \\ i\neq k}}^{\infty} \|\chi_{A_k}\|_i$$

$$\leq n_k\mu(A_k)^{1/r} + 2^{-k-1}\|\chi_{A_k}\|_k .$$

Therefore

$$\|\chi_{A_k}\|_{Y'} \Big/ \|\chi_{A_k}\|_Y \leq \|\chi_{A_k}\|_{Y'} \Big/ \|\chi_{A_k}\|_K \leq$$

$$\leq n_k\mu(A_k)^{1/r} \Big/ n_k\mu(A_k)^{1/q} + 2^{-k-1}$$

$$\leq \mu(A_k)^{1/r-1/q} + 2^{-k-1} \to 0 \text{ as } k \to \infty .$$

This means that Y and Y' are different r.i. function spaces, which completes the proof of 10.1(4).

The first statement in (5) follows from (4) and the results of Section 8. To prove the second statement in (5), one builds for each $p \le q < 2$ an interpolation space between $L_2(0,\infty)$ and $L_2(0,\infty) \cap L_q(0,\infty)$. For each $k = 1,2,\ldots,$ define a norm $|\cdot|_k$ on $L_2(0,\infty)$ by setting

$$|f|_k = \inf\{n_k\|g\|_{L_2(0,\infty)} + n_k^{-1}\max[\|h\|_{L_2(0,\infty)},\|h\|_{L_q(0,\infty)}]; \; f = g + h\}$$

and let Y be the space of all f in $L_2(0,\infty)$ for which the norm

$$|||f||| = \|\sum_{k=1}^{\infty} |f|_k \, e_k\|_U$$

is finite. Y is an interpolation space between $L_2(0,\infty)$ and $L_2(0,\infty) \cap L_q(0,\infty)$, so the union over $n = 0,1,\ldots$ of the Haar systems on $[n,n+1)$ is an unconditional basis for Y which is p-convex and 2-concave, hence Y is isomorphic to a complemented subspace of U. Also, Y is a r.i. function space on $[0,\infty)$, and it is clear that the restriction of Y to $[0,1]$ is $L_2(0,1)$, up to an equivalent renorming. One can check that a version of the argument for (4) yields that U embeds into Y as a complemented subspace, which yields that U and Y are isomorphic. Alternatively, one can observe that the norm one complemented subspace $[\chi_{[n,n+1)}]_{n=0}^{\infty}$ of Y is an interpolation space between ℓ_2 and ℓ_p which Davis [17] showed contains a complemented isomorph of U. Finally, by varying the parameter q which defines Y, it is readily seen (as in (4) above) that the described construction yields uncountably many different representations of U. □

Remark: It is possible to do all the constructions in 10.1 using a slightly different universal space $\tilde{U}$. There exists a space $\tilde{U}$ with normalized unconditional basis $\{e_n\}_{n=1}^{\infty}$ which is 2-concave and has an upper

ℓ_p estimate for disjoint blocks, and such that every normalized unconditional basic sequence which is 2-concave and has an upper ℓ_p estimate for disjoint blocks is equivalent to a subsequence of $\{e_k\}_{k=1}^{\infty}$. The proof of 10.1 yields that $\tilde{U}$ has uncountably many representations as a r.i. function space. According to Figiel [24], the universal property of $\tilde{U}$ has the following formulation: every Banach space X with an unconditional basis which has modulus of smoothness $\rho_X(\tau) \leq c\tau^p$ and modulus of convexity $\delta_X(\epsilon) \geq \frac{1}{c}\epsilon^2$ is isomorphic to a complemented subspace of U (and, of course, the modulus of convexity and smoothness for U admit such estimates).

Example 10.4: *Let* $1 < p < 2$. *There is a subspace* V *of* $L_p(0,1)$ *which has the following properties:*

(1) V *has a normalized unconditional basis* $\{e_n\}_{n=1}^{\infty}$ *such that every normalized unconditional basic sequence in* $L_p(0,1)$ *is equivalent to a subsequence of* $\{e_n\}_{n=1}^{\infty}$.

(2) V *has uncountably many mutually non-equivalent symmetric bases.*

(3) V *is isomorphic to uncountably many different r.i. function spaces on* $[0,1]$.

The space V was constructed in [74]: Let $\{f_n\}_{n=1}^{\infty}$ be a sequence of unit vectors which is dense in the unit sphere of $L_p(0,1)$ and let $\{e_n\}_{n=1}^{\infty}$ be represented as a sequence $\{\tilde{e}_n\}_{n=1}^{\infty}$ in $L_p(0,1)^2$ by setting $\tilde{e}_n(s,t) = f_n(s)r_n(t)$, where as usual $\{r_n\}_{n=1}^{\infty}$ is the sequence of Rademacher functions. It is clear that $\{e_n\}_{n=1}^{\infty}$ satisfies (1). In [56] it was shown that V has a symmetric basis, and the construction there in fact yields (2) (for example, in Lemma 2 of [56] vary the parameter "p" between our p and 2).

To prove (3), we do a continuous version of the construction of [56]. As in 10.1, for fixed q , $p \leq q < 2$, let Y_k be $L_2(0,1)$ under the norm

$$\|f\|_k = \inf\{n_k\|g\|_{L_q} + n_k^{-1}\|h\|_{L_2} ;\ f = g + h\} ,$$

where $\{n_k\}_{k=1}^{\infty}$ satisfies the conditions of 10.1(4); in particular, (+). Observe that each Y_k embeds isometrically into the space $Y_q(0,\infty) = L_q(0,\infty) + L_2(0,\infty)$ discussed in Section 8 (cf. 8.5) when $Y_q(0,\infty)$ is given its natural norm, defined by

$$\|f\|_{Y_q} = \inf\{\|g\|_{L_q(0,\infty)} + \|h\|_{L_2(0,\infty)}: f = g + h\} .$$

To see this, fix $0 < s \leq 1$ and let D_s be the dilation operator defined in Section 8: for $f \in L_q(0,1)$,

$$(D_s f)(t) = \begin{cases} f(st) & , \text{ if } t \in [0,1/s) \\ 0 & , \text{ if } t > 1/s \end{cases}$$

Thus for $f \in L_q(0,1)$,

$$\|D_s f\|_{Y_q} = \inf\{(\int_0^{1/s} |g(st)|^q dt)^{1/q} + (\int_0^{1/s} |h(st)|^2)^{1/2} : D_s f = g + h\}$$

$$= \inf\{s^{-1/q}\|g\|_{L_q(0,1)} + s^{-1/2}\|h\|_{L_2(0,1)}: f = g + h\} .$$

Setting, for $k = 1,2,\ldots,$

$$s_k^{1/q-1/2} = n_k^{-2} , \quad c_k = n_k s_k^{1/q}$$

we have that $c_k D_{s_k}$ maps $(Y_k, \|\cdot\|)_k$ isometrically onto the ideal $Y_q(0,s_k^{-1})$ of $Y_q(0,\infty)$. For further reference, note that

$$s_k = n_k^{4q/(q-2)} \text{ and } c_k = n_k^{(q+2)/(q-2)} .$$

Letting $\widetilde{D}_k$ be D_{s_k} followed by the translation $g(t) \to g(t + \sum_{i=1}^{k-1} s_i^{-1})$, we have that $c_k \widetilde{D}_k$ is an isometry from $(Y_k, \|\cdot\|_k)$ onto $Y_q(\sum_{i=1}^{k-1} s_i^{-1}, \sum_{i=1}^{k} s_i^{-1})$.

Let T be the symmetrized Poisson embedding from $Y_q(0,\infty)$ into $L_q(0,1)$ discussed in Section 8, and define a mapping S from the simple functions $\mathfrak{F}(0,1)$ on $[0,1]$ into the measurable functions on the square $[0,1]^2$ by

$$Sf(r,t) = \sum_{k=1}^{\infty} T\, c_k \widetilde{D}_k\, f(r)\, e_k(t) .$$

For $f \in \mathfrak{F}(0,1)$, set

$$|||f||| = \|Sf\|_{L_p(L_q)} \equiv \left(\int_0^1 \left(\int_0^1 |Sf(r,t)|^q dr\right)^{p/q} dt\right)^{1/p} .$$

Since T is an isomorphism and $\|f\|_k = \|c_k \widetilde{D}_k f\|_{Y_q(0,\infty)}$ for each $k = 1,2,\ldots,$ the triangle inequality yields that $|||f||| < \infty$ for each $f \in \mathfrak{F}(0,1)$.

Let $Z = Z(q)$ be the completion of $(\mathfrak{F}(0,1), |||\cdot|||)$. The fact that Z is a r.i. function space on $[0,1]$ (except that $|||\chi_{(0,1)}|||$ is not one) will follow once we observe that for each $n = 1,2,\ldots,$ $\{Sx_{n,i}(r,t)\}_{i=1}^{2^n}$ forms, for fixed t, an independent, identically distributed sequence of symmetric random variables in the "r" variable since this property clearly implies that $\{Sx_{n,i}\}_{i=1}^{2^n}$ is 1-symmetric in $L_p(L_q)$. But by the definition of the $\widetilde{D}_k$'s, for each fixed k $\{\widetilde{D}_k x_{n,i}\}_{i=1}^{2^n}$ are disjoint functions in $Y_q(0,\infty)$ which have the same distribution, and these functions are disjointly supported as k varies. Since T takes disjoint functions in $Y_q(0,\infty)$ into independent symmetric random variables in $L_q(0,1)$, and dist Tg = dist Th whenever dist g = dist h, it follows that $\{Tx_{n,i}(r,t)\}_{i=1}^{2^n}$ has the desired property.

Next we want to show that V embeds into Z as a complemented subspace. Since Z embeds into $L_p(L_q)$ which, in turn, embeds into L_p, this will yield in view of the universal property of V (cf. the argument for 10.1(3)) that Z and V are isomorphic.

As in 10.1(4), let $\{A_k\}_{k=1}^{\infty}$ be a sequence of disjoint subintervals of $[0,1)$ so that for $k = 1,2,\ldots,$ $\mu(A_k) = n_k^{4q/(q-2)} = s_k^{-1}$. In 10.1(4) it was computed for $k = 1,2,\dots,$ that

$$\|\chi_{A_k}\|_k = n_k^{(q+2)/(q-2)} = c_k \quad \text{and}$$

$$\sum_{\substack{i=1\\ i\neq k}}^{\infty} c_k^{-1} \|\chi_{A_k}\|_i < 2^{-k-1} ,$$

which yields that $\{c_k^{-1} S\chi_{A_k}\}_{k=1}^{\infty}$ is a small perturbation in $L_p(L_q)$ of $\{(T\tilde{D}_k f)(r)e_k(t)\}_{k=1}^{\infty}$ and, hence, that $\{c_k^{-1}\chi_{A_k}\}_{k=1}^{\infty}$ in Z is equivalent to $\{(T\tilde{D}_k\chi_{A_k})(r)e_k(t)\}_{k=1}^{\infty}$ in $L_p(L_q)$. But $\{\tilde{D}_k\chi_{A_k}\}_{k=1}^{\infty}$ in $Y_q(0,\infty)$ is a sequence of characteristic functions of disjoint intervals of length one, and hence is equivalent to the unit vector basis of ℓ_2, whence also $\{(T\tilde{D}_k\chi_{A_k})\}_{k=1}^{\infty}$ in $L_q(0,1)$ is equivalent to the usual ℓ_2-basis. This easily yields that the sequence $\{(T\tilde{D}_k\chi_{A_k})(r)e_k(t)\}_{k=1}^{\infty}$ in $L_p(L_q)$ is equivalent to the basis $\{e_k\}_{k=1}^{\infty}$ of $V \subset L_p(0,1)$. Indeed, $\{e_k\}_{k=1}^{\infty}$ is unconditional, so we compute for arbitrary scalars $\{a_k\}_{k=1}^{\infty}$, by using (*) from the Introduction that

$$\Big\|\sum_{k=1}^{\infty} a_k (T\tilde{D}_k\chi_{A_k})(r)e_k(t)\Big\|_{L_p(L_q)}$$

$$= \int_0^1\Big(\int_0^1 \Big|\sum_{k=1}^{\infty} a_k e_k(t)(T\tilde{D}_k\chi_{A_k})(r)\Big|^q dr\Big)^{p/q} dt\Big)^{1/p}$$

$$\sim \Big(\int_0^1 \Big(\sum_{k=1}^{\infty} |a_k e_k(t)|^2\Big)^{p/2} dt\Big)^{1/p} \sim \Big\|\sum_{k=1}^{\infty} a_k e_k\Big\|_{L_p} .$$

Thus we have seen that $\{c_k^{-1}\chi_{A_k}\}_{k=1}^{\infty}$ in Z is equivalent to $\{e_k\}_{k=1}^{\infty}$ in V, and, of course, the closed span in Z of $\{\chi_{A_k}\}_{k=1}^{\infty}$ is complemented since Z is a reflexive r.i. space (see the argument of 10.1(4)). Finally, as in 10.1(4), by varying q in the interval $(p,2)$, one produces uncountably many different r.i. spaces which are isomorphic to V. □

Remark: Notice that if Y is a subspace of V which is isomorphic to L_p then V embeds into Y but no subspace of Y isomorphic to V is complemented in V. This shows that the assumption in 9.1 that the Haar system for X is not equivalent to a disjoint sequence in X is necessary.

For our next two examples we need a localized version of the Kadec-Pelczynski [38] disjointification procedure for lattices (cf. [26]).

Lemma 10.5: *Suppose that* Y *is a Banach lattice which does not contain* ℓ_∞^n *uniformly for all* n, X *is a space with a 2-convex normalized unconditional basis* $\{e_n\}_{n=1}^\infty$, *and* X *is not* ℓ_2, *even up to an equivalent renorming.*

(a) If, for each n, $T_n\colon [e_i]_{i=1}^n \to Y$ *is an embedding such that* $\|T_n\| = 1$, *and* $\sup_n \|T_n^{-1}\| \le K < \infty$, *then for each* $\varepsilon > 0$ *and each integer* k *there is an integer* $N = N(k,\varepsilon)$ *and* $1 \le n_1 < n_2 < \ldots < n_k \le N$ *so that*

$$\|\, |T_n x_{n_i}| \wedge |T_n x_{n_j}| \,\| < \varepsilon \quad \text{for} \quad 1 \le i < j \le k .$$

(b) If $\{e_n\}_{n=1}^\infty$ *has subsymmetry constant* $< C$, *and* X *is finitely crudely representable in* Y *up to a constant* K, *then* X *is lattice finitely crudely representable in* Y *up to constant* CK.

Proof: (b) is an obvious consequence of (a). We prove (a) by taking ultraproducts and applying the infinite dimensional disjointification technique. Let $\mathcal{U}$ be a free ultrafilter on the positive integers and let $\tilde{Y} = \prod Y/\mathcal{U}$ be the corresponding ultrapower of Y; i.e., $\tilde{Y}$ is the quotient of $\ell_\infty(Y) \equiv (Y + Y + \ldots)_{\ell_\infty}$ by the closed subspace $Z = \{(y_1, y_2, \ldots)\colon \lim_{n \in \mathcal{U}} \|y_n\|_Y = 0\}$. The quotient mapping $\ell_\infty(Y) \to \tilde{Y}$ is denoted by Q. Now Z is an order ideal in $\ell_\infty(Y)$ (where $\ell_\infty(Y)$ has the natural order $(y_1, y_2, \ldots) \ge 0$ iff $y_n \ge 0$ in Y for all $n = 1,2,\ldots$), so Q induces on $\tilde{Y}$ an order which makes $\tilde{Y}$ into a Banach lattice, and $\tilde{Y}$ is lattice finitely representable in Y. Note that if $y = (y_1, y_2, \ldots)$ and $z = (z_1, z_2, \ldots)$ are in $\ell_\infty(Y)$, the infimum $Qy \wedge Qz$ in $\tilde{Y}$ is $Q(y_1 \wedge z_1, y_2 \wedge z_2, \ldots)$; in particular,

$$\| |Qy| \wedge |Qz| \|_{\tilde{Y}} = \lim_{n \in \mathcal{U}} \| |Qy_n| \wedge |Qz_n| \|_Y .$$

Define a mapping $\tilde{T}: X \to \ell_\infty(Y)$ by

$$\tilde{T}\Big(\sum_{i=1}^{\infty} \alpha_i e_i\Big) = \Big(T_1\alpha_1 e_1, T_2(\alpha_1 e_1 + \alpha_2 e_2), T_3\Big(\sum_{i=1}^{3} \alpha_i e_i\Big), \ldots\Big)$$

and let $T: X \to \tilde{Y}$ be the composition $Q\tilde{T}$. It is well-known and easy to see that $\|T\| = 1$ and $\|T^{-1}\| \le K$ since each T_n satisfies the same condition.

We are now in a position to use infinite dimensional reasoning on the sequence $\{Te_n\}_{n=1}^{\infty}$. Let Y' be a separable sublattice of $\tilde{Y}$ which contains TX and let F be a strictly positive functional of norm one on Y' (i.e., $F(y) > 0$ whenever $0 \ne y \in Y'$ and $y \ge 0$). Define a norm $\|\cdot\|_F$ on Y' by $\|y\|_F = F(|y|)$. The completion of $(Y', \|\cdot\|_F)$ is an abstract L-space so that by Kakutani's theorem [40], it is isometric to $L_1(\nu)$ for some measure ν. In particular, $(Y', \|\cdot\|_F)$ has cotype 2.

We claim that $\liminf_n \|Te_n\|_F = 0$. Indeed, if $\|Te_n\|_F \ge \delta > 0$ for $n = 1, 2, \ldots,$ then we have for all scalars $\{a_i\}_{i=1}^k$ that

$$\int_0^1 \Big\| \sum_{i=1}^{k} a_i\, Te_i r_i(t) \Big\|_F dt \ge \delta\, 2^{-1/2} \Big(\sum_{i=1}^{k} a_i^2\Big)^{1/2}$$

since the cotype 2 constant of an L_1 space is $A_1^{-1} = \sqrt{2}$. We thus have for all scalars $\{a_i\}_{i=1}^k$ that

$$\delta 2^{-1/2}\Big(\sum_{i=1}^{k} a_i^2\Big)^{1/2} \le \max_{\pm} \Big\| \sum_{i=1}^{k} \pm a_i Te_i \Big\|_F \le \max \Big\| \sum_{i=1}^{k} \pm a_i Te_i \Big\|_{\tilde{Y}}$$

$$\le \max_{\pm} \Big\| \sum_{i=1}^{k} \pm a_i e_i \Big\|_X \le C \Big\| \sum_{i=1}^{k} a_i e_i \Big\|_X .$$

On the other hand, $\widetilde{Y}$ is 2-convex and does not contain ℓ_∞^n uniformly for all n, since $\widetilde{Y}$ is lattice finitely representable in Y, and T is an isomorphism, so $\{e_n\}_{n=1}^\infty$ is a 2-convex basis for X by 10.3. Therefore, there is a constant $K < \infty$ so that for all choices $\{a_i\}_{i=1}^k$ of scalars,

$$\|\sum_{i=1}^k a_i e_i\|_X \le K(\sum_{i=1}^k a_i^2)^{1/2} ,$$

which contradicts the assumption that X is not isomorphic to ℓ_2 and yields the claim.

The argument for Theorem 4.1 in [26] yields that some subsequence of $\{Te_n\}_{n=1}^\infty$ is essentially disjointly supported in $\widetilde{Y}$. In particular, given $\varepsilon > 0$ and k, there are $n_1 < n_2 < \ldots < n_k$ so that

$$\| \, |Te_{n_i}| \wedge |Te_{n_j}| \, \|_{\widetilde{Y}} < \varepsilon \quad \text{for} \quad 1 \le i < j \le k .$$

This gives (a), because

$$\| \, |Te_{n_i}| \wedge |Te_{n_j}| \, \|_{\widetilde{Y}} = \lim_{\ell \in U} \| \, |T_\ell e_{n_i}| \wedge |T_\ell e_{n_j}| \, \|_Y .$$

□

If Y is a space which has a specified symmetric basis $\{e_n\}_{n=1}^\infty$, define $p_n = p_n(Y)$ by

$$n^{1/p_n} = \|\sum_{i=1}^n e_i\|_Y ,$$

and set

$$p(Y) = \liminf_{n\to\infty} p_n(Y) , \quad q(Y) = \limsup_{n\to\infty} p_n(Y) .$$

An extension by Rosenthal [72] of a result of Krivine's [42] yields that if $r \in [p(Y),q(Y)]$, then ℓ_r is lattice finitely representable in Y. The main point of Krivine's paper [42] was that every isomorph of a Banach lattice has some ℓ_r lattice finitely representable in it; the mentioned result from

[72] gave some hope that the set of all such r is an interval when the Banach lattice structure is given by a symmetric basis, or, perhaps, when the Banach lattice is a r.i. function space.

There are, however, counterexamples to these conjectures. Given a Banach lattice Y, let

$$\mathrm{LFR}(Y) = \{r: \ell_r \text{ is lattice finitely representable in } Y\} .$$

(By Krivine's theorem [42], $\emptyset \neq \mathrm{LFR}(Y)$ for every Banach lattice Y, and the definition of $\mathrm{LFR}(Y)$ remains the same if "crudely" is inserted between "finitely" and "representable".) It is easy to see that there are r.i. function spaces Y on $[0,\infty)$ for which $\mathrm{LFR}(Y)$ consists of two points; for example, if $2 < p < \infty$, then

$$\mathrm{LFR}(L_p(0,\infty) \cap L_2(0,\infty)) = \{2,p\} .$$

("$\supset$" is obvious, while "$\subset$" follows from the fact that $L_p(0,\infty) \cap L_2(0,\infty)$ is isomorphic to $L_p(0,1)$.) Our next example shows that there are also counterexamples to the conjecture that $\mathrm{LFR}(Y)$ is an interval when Y is a r.i. function space on $[0,1]$ or when Y has a symmetric basis.

Example 10.6: A r.i. function space Y *on* $[0,1]$ *(or a space* Y *with a 1-symmetric basis) for which* $\mathrm{LFR}(Y)$ *is not an interval.*

We first construct such a r.i. function space. Let $1<p<2<q<\infty$ and let $\{n_k\}_{k=1}^{\infty}$ satisfy the lacunarity condition (+) in 10.1(4). As in 10.1(4), for $k = 1,2,\ldots$ let Y_k be $L_p(0,1)$ under the norm defined by

$$\|f\|_k = \inf\{n_k\|g\|_{L_p} + n_k^{-1}\|h\|_{L_2} : f = g + h\}$$

and let Y be the set of all $f \in L_p(0,1)$ for which

$$\|f\|_Y = (\sum_{k=1}^{\infty} \|f\|_k^q)^{1/q} < \infty .$$

The proof of 10.1(4) shows that $(Y,\|\cdot\|_Y)$ is a r.i. function space on $[0,1]$ (except that $\|\chi_{(0,1)}\|_Y$ is not one) which is even an interpolation space between $L_p(0,1)$ and $L_2(0,1)$, and that the unit vector basis of ℓ_q is equivalent to a sequence of disjoint vectors in Y, so that $q \in \mathrm{LFR}(Y)$.

Claim: For some $p \leq r \leq 2$, $r \in \mathrm{LFR}(Y)$. To see this, let $\mathcal{U}$ be a free ultrafilter on the positive integers and let Z be the completion of the set of finitely non-zero sequences of scalars under the norm

$$\|\sum_{i=1}^{k} a_i e_i\|_Z = \lim_{n\in\mathcal{U}} \|x_{n,i}\|_Y^{-1} \|\sum_{i=1}^{k\wedge 2^n} a_i x_{n,i}\|_Y .$$

It is clear that $\{e_n\}_{n=1}^{\infty}$ forms a 1-symmetric basis for Z and Z is lattice finitely representable in Y.

Let D_t be the dilation operator defined in section 8. Then, since Y is an interpolation space between $L_p(0,1)$ and $L_2(0,1)$, we have for $0 < s \leq st \leq 1$ that

$$\|\chi_{(0,st)}\|_Y = \|D_{t^{-1}}\chi_{(0,s)}\|_Y \leq \|D_{t^{-1}}\|_Y \|\chi_{(0,s)}\|_Y$$

$$\leq \max\{\|D_{t^{-1}}\|_{L_p(0,1)}, \|D_{t^{-1}}\|_{L_2(0,1)}\}\|\chi_{(0,s)}\|_Y$$

$$= t^{1/p}\|\chi_{(0,s)}\|_Y .$$

Similarly,

$$\|\chi_{(0,s)}\|_Y = \|D_t\chi_{(0,st)}\|_Y \leq t^{-1/2}\|\chi_{(0,st)}\|_Y ,$$

so that

$$t^{1/2} \leq \|\chi_{(0,st)}\|_Y \Big/ \|\chi_{(0,s)}\|_Y \leq t^{1/p} .$$

Hence, for each $n = 1,2,\ldots,$ and $k = 1,2,\ldots,2^n$,

$$k^{1/2} \leq \|\sum_{i=1}^{k} x_{n,i}\|_Y \Big/ \|x_{n,1}\|_Y \leq k^{1/p}$$

whence for each $k = 1,2,\ldots,$

$$k^{1/2} \leq \|\sum_{i=1}^{k} e_i\|_Z \leq k^{1/p} .$$

Therefore, $p \leq p(Z) \leq q(Z) \leq 2$, and the claim follows from Krivine's theorem [42].

We finish the proof by showing that for $2 < s < q$, $s \notin \mathrm{LFR}(Y)$. Since the spaces Y_k for $k = 1,2,\ldots$ embed uniformly into L_p , it is enough to show that ℓ_s is not finitely representable in $(L_p + L_p + \ldots)_{\ell_q}$ when $2 < s < q$. However, if ℓ_s is finitely representable in $(L_p + L_p + \ldots)_{\ell_q}$ then by 10.5, ℓ_s is lattice finitely representable in $(L_p + L_p + \ldots)_{\ell_q}$, so by taking the $2/p$ convexification, we see that $\ell_{2s/p}$ is lattice representable in $(L_2 + L_2 + \ldots)_{\ell_{2q/p}}$, which embeds isometrically into $L_{2q/p}$. This means that $\ell_{2s/p}$ embeds into $L_{2q/p}$ (cf. [44] or [15]), which is false [38].

To do the symmetric basis example, define for $k = 1,2,\ldots$ a norm on ℓ_2 by

$$|x|_k = \inf\{n_k^{-1}\|y\|_{\ell_p} + n_k\|z\|_{\ell_2} : x = y + z\}$$

and let Y' be the set of all $x \in \ell_2$ for which

$$\|x\|_{Y'} = (\sum_{k=1}^{\infty} |x|_k^q)^{1/q} < \infty .$$

The unit vectors form a symmetric basis for Y' , and a simpler version of the argument given in the r.i. function space case shows that $\mathrm{LFR}(Y')$ is not an interval.

□

Our last example shows that, in contrast to the situation in Orlicz function spaces, finite symmetric basic sequences in general r.i. function spaces cannot be represented as exchangeable sequences.

Example 10.7: A uniformly convex r.i. function space $Y(0,\infty)$ *such that* ℓ_4 *embeds into* $Y(0,1)$ *as a sublattice but*

$$C_n \equiv \inf\{C: \text{there exists a sequence of functions in } Y(0,\infty) \text{ having the same distribution which is } C\text{-equivalent to the unit vector basis for } \ell_4^n\} \to \infty$$

as $n \to \infty$.

$Y(0,\infty)$ is the Lorentz function space of exponent four and weight function $t^{-1/2}$; that is,

$$\|f\|_{Y(0,\infty)} = \left(\int_0^\infty (t^{1/8} f^*(t))^4 \, t^{-1} dt\right)^{1/4} .$$

It is clear that $Y(0,\infty)$ is 4-convex with constant one, and it is known [2] that $Y(0,\infty)$ is uniformly convex. It is easy to prove (see, e.g., [26] for a stronger statement) that for each $\epsilon > 0$, ℓ_4 is $1+\epsilon$-isometric to a sublattice of $Y(0,1)$. We have written the norm of $Y(0,\infty)$ in such a way to make it clear that the dilation operators D_t act on $Y(0,\infty)$ just as they act on $L_8(0,\infty)$; namely, for $0 < t < \infty$, $t^{1/8}D_t$ is an isometry from $Y(0,\infty)$ onto itself. Now if $\{f_i\}_{i=1}^n$ are disjointly supported functions in $Y(0,\infty)$ which have the same distribution, then the distribution of $\sum_{i=1}^n f_i$ is the same as the distribution of $D_{n^{-1}} f_1$, so that

$$\left\|\sum_{i=1}^n f_i\right\|_{Y(0,\infty)} = n^{1/8}\|f_1\|_{Y(0,\infty)} .$$

But by 10.5(a),

$$\lim_{n\to\infty} C_n = \liminf_{n\to\infty} \{C: \text{There exists a sequence of disjoint functions in } Y(0,\infty) \text{ having the same distribution which is } C\text{-equivalent to the unit vector basis for } \ell_4^n\}$$

and we have just seen that this last limit is ∞. □

REFERENCES

1. D. Alspach, P. Enflo and E. Odell, On the structure of separable $\mathcal{L}_p$ spaces $(1 < p < \infty)$, Studia Math., 60(1977), 79-90.

2. Z. Altshuler, Uniform convexity in Lorentz sequence spaces, Israel J. Math., 20(1975), 260-274.

3. B. Beauzamy, Proprietes geometriques des espaces d'interpolation, Seminaire Maurey-Schwartz 1974-1975, Ecole Polytechnique.

4. __________, Operaturs de type Rademacher entre espaces de Banach, Seminaire Maurey-Schwartz 1975-1976, Ecole Polytechnique.

5. A. Benedeck and R. Panzone, The spaces L^p with mixed norm, Duke Math. J., 28(1961), 301-324.

6. G. Bennett, L. E. Dor, V. Goodman, W. B. Johnson and C. M. Newman, On uncomplemented subspaces of L^p, $1 < p < 2$, Israel J. Math., 26(1977), 178-187.

7. S. J. Bernau and H. E. Lacey, Bicontractive projections and reordering of L_p-spaces, Pacific J. Math., 69(1977), 291-302.

8. C. Bessaga and A. Pelczynski, On bases and unconditional convergence of series in Banach spaces, Studia Math., 17(1958), 151-164.

9. D. B. Boyd, Indices of function spaces and their relationship to interpolation, Canadian J. of Math., 21(1969), 1245-1254.

10. __________, The spectral radius of averaging operators, Pacific J. Math., 24(1968), 19-28.

11. J. Bretagnolle, D. Dacunha-Castelle et J. L. Krivine, Lois stables et espaces L^p, Ann. Inst. Henri Poincare, serie B, Vol. II, 3(1966), 231-263.

12. J. Bretagnolle et D. Dacunha-Castelle, Application de l'etude de certaines formes lineaires aleatoires au plongement d'espaces de Banach dans des espaces L^p, Ann. Ecole Normale Superieure, 2(1969), 437-480.

13. D. L. Burkholder, Distribution function inequalities for martingales, Ann. of Prob., 1(1973), 19-43.

14. D. L. Burkholder, B. J. Davis and R. F. Gundy, Integral inequalities for convex functions of operators on martingales, Proc. of the 6-th Berkeley Symp. on Math. Stat. and Prob., 223-240.

15. D. Dacunha-Castelle et J. L. Krivine, Application des ultraproducts a l'etude des espace et des algebras de Banach, Studia Math., 41(1972), 315-334.

16. D. Dacunha-Castelle et M. Schreiber, Techniques probabilistes pour l'etude de problemes d'isomorphismes entre espaces de Banach, Ann. Inst. Henri Poincare, 10(1974), 229-277.

17. W. J. Davis, Embedding spaces with unconditional bases, Israel J. Math., 20(1975), 189-191.

18. W. J. Davis, T. Figiel, W. B. Johnson and A. Pelczynski, Factoring weakly compact operators, J. Func. Anal., 17(1974), 311-327.

19. L. E. Dor, On projections in L_1 , Ann. of Math., 102(1975), 463-474.

20. L. E. Dor and T. Starbird, Projections of L_p onto subspaces spanned by independent random variables, to appear in Compositio Math.

21. E. Dubinsky, A. Pelczynski and H. P. Rosenthal, On Banach spaces X for which $\pi_2(\mathcal{L}_\infty,X) = B(\mathcal{L}_\infty,X)$, Studia Math., 44(1972), 617-648.

22. P. Enflo and T. W. Starbird, Subspaces of L^1 containing L^1 , to appear in Studia Math., 65.

23. W. Feller, An Introduction to Probability Theory and its Applications, Vol. II, Wiley, 1966.

24. T. Figiel, On the moduli of convexity and smoothness, Studia Math., 61(1976), 121-155.

25. T. Figiel and W. B. Johnson, A uniformly convex Banach space which contains no ℓ_p , Compositio Math., 29(1974), 179-190.

26. T. Figiel, W. B. Johnson and L. Tzafriri, On Banach lattices and spaces having local unconditional structure with applications to Lorentz sequence spaces, J. Approx. Theory, 13(1975), 297-312.

27. T. Figiel, J. Lindenstrauss and V. Milman, The dimension of almost spherical sections of convex bodies, Acta Math., 139(1977), 53-94.

28. J. L. Gamlen and R. J. Gaudet, On subsequences of the Haar system in $L_p[0,1]$, $(1 < p < \infty)$, Israel J. Math., 15(1973), 404-413.

29. V. F. Gaposhkin, On the existence of unconditional bases in Orlicz spaces, Funkcional Anal. i Prilozen, 1(1967), 26-32, (Russian).

30. D. J. H. Garling and Y. Gordon, Relations between some constants associated with finite-dimensional Banach spaces, Israel J. Math., 9(1971), 346-361.

31. D. P. Giesy, On a convexity condition in normed linear spaces, Trans. A.M.S., 125(1966), 114-146.

32. Y. Gordon, D. R. Lewis and J. Retherford, Banach ideals of operators with applications to the finite dimensional structure of Banach spaces, Israel J. Math., 13(1972), 348-360.

33. A. Grothendieck, Resume de la theorie metrique des produits tensoriels topologiques, Bol. Soc. Math., Sao Paulo, 8(1956), 1-79.

34. Uffe Haagerup, The best constants in the Khintchine inequality,

35. W. B. Johnson, On finite dimensional subspaces of Banach spaces with local unconditional structure, Studia Math., 51(1974), 225-240.

36. W. B. Johnson and L. Jones, Every L_p operator is an L_2 operator, Proc. A.M.S. (to appear).

37. W. B. Johnson and L. Tzafriri, On the local structure of subspaces of Banach lattices, Israel J. Math., 20(1975), 292-299.

38. M. I. Kadec and A. Pelczynski, Bases, lacunary sequences and complemented subspaces in the spaces L_p, Studia Math., 21(1962), 161-176.

39. J. P. Kahane, Some Random Series of Functions, D.C. Heath, Lexington, Mass., 1968.

40. S. Kakutani, Concrete representation of abstract (L)-spaces and the mean ergodic theorem, Ann. of Math., 42(1941), 523-537.

41. J. L. Krivine, Theoremes de factorization dans les espaces reticules, Seminaire Maurey-Schwartz, 1973-1974.

42. ____________, Sous-espaces de dimension finie des espaces de Banach reticules, Ann. of Math., 104(1976), 1-29.

43. H. E. Lacey and P. Wojtaszczyk, Banach lattice structures on separable L_p spaces, Proc. A.M.S., 54(1976), 83-89.

44. J. Lindenstrauss and A. Pelczynski, Absolutely summing operators in $\mathcal{L}_p$ spaces and their applications, Studia Math., 29(1968), 275-326.

45. J. Lindenstrauss and A. Pelczynski, Contributions to the theory of the classical Banach spaces, J. Func. Anal., 8(1971), 225-249.

46. J. Lindenstrauss and L. Tzafriri, On Orlicz sequence spaces II, Israel J. Math., 11(1972), 355-379.

47. ______________, On Orlicz sequence spaces III, Israel J. Math., 14(1973), 368-389.

48. ______________, Classical Banach Spaces, Lecture Notes in Mathematics, Springer-Verlag, 1973.

49. ______________, Classical Banach Spaces I, Sequence Spaces, Springer-Verlag, Berlin, 1977.

50. ______________, Classical Banach Spaces II, Springer-Verlag, Berlin (to appear).

51. J. Lindenstrauss and M. Zippin, Banach spaces with unique unconditional basis, Journal of Func. Anal., 3(1969), 115-125.

52. J. L. Lions et J. Peetre, Sur une classe d'espaces d'interpolation, I.H.E.S. Publications Mathematiques, 19(1964), 5-68.

53. B. Maurey, Nouveaux theoremes de Nikishin, Seminaire Maurey-Schwartz, 1973-1974, Ecole Polytechnique.

54. ________, Type et cotype dans les espaces munis de structures locales inconitionnelles, Seminaire Maurey-Schwartz, 1973-1974, Ecole Polytechnique.

55. ________, Theoremes de factorisation pour les operateures a valeurs dans un espace L^p, Asterisque. Soc. Math. France, 11(1974).

56. B. Maurey and G. Schechtman, Some remarks on symmetric basic sequences in L_1, to appear in Compositio Math.

57. P. A. Meyer, Probability and Potentials, Blaisdell, 1966.

58. P. Meyer-Nieberg, Charakterisierung einiger topologischer und ordnungstheroretischer Eigenschaften von Banachverbänden mit Hilfe disjunkter Folgen, Arch. Math., 24(1973), 640-647.

59. ______________, Zur schwachen Kompaktheit in Banachverbänden, Math. Z., 134(1973), 303-316.

60. B. S. Mityagin, The homotopy structure of the linear group of a Banach space, Russian Math. Surv., 25(1970), 59-103.

61. E. M. Nikishin, Resonance theorems and superlinear operators, Trans. of Uspekhi. Mat. Nauk., Vol., 25(1970), 125-187.

62. A. M. Olevskii, Fourier Series with Respect to General Orthogonal Systems, Springer-Verlag, 1975.

63. A. Pelczynski, Projections in certain Banach spaces, Studia Math., 19(1960), 209-228.

64. ______, Universal bases, Studia Math., 32(1969), 247-268.

65. A. Pelczynski and H. P. Rosenthal, Localization techniques in L^p-spaces, Studia Math., 52(1975), 263-289.

66. A. Persson and A. Pietsch, p-nukleare und p-integrable Abbildungen in Banachraumen, Studia Math., 33(1969), 19-62.

67. G. Pisier, Sur les espaces qui ne contiennent pas de ℓ_n^1 uniforment, Seminaire Maurey-Schwartz, 1973-1974.

68. ______, Sur les espaces de Banach qui ne contiennent pas uniforment de ℓ_n^1, C.R. Acad. Sc. Paris, t. 277(1973), 991-993.

69. S. Rolewicz, Metric Linear Spaces, Warszawa, 1972.

70. H. P. Rosenthal, On the subspaces of L^p $(p > 2)$ spanned by sequences of independent random variables, Israel J. Math., 8(1970), 273-303.

71. ______, On the span in L^p of sequences of independent random variables (II), Proc. of the 6-th Berkeley Symp. on Prob. and Stat., Berkeley, CA, 1971.

72. ______, On a theorem of J.L. Krivine concerning block finite-representability of ℓ^p in general Banach spaces, J. Func. Anal., 28(1978), 197-225.

73. G. Schechtman, A remark on unconditional basic sequences in L_p $(1 < p < \infty)$, Israel J. Math., 19(1974), 220-224.

74. ______, On Pelczynski's paper "Universal bases", Israel J. Math., 22(1975), 181-184.

75. C. Schütt, The projection constant of finite dimensional spaces whose unconditional basis constant is 1, Israel J Math., 30(1978), 207-212.

76. E. M. Stein, Topics in harmonic analysis related to the Littlewood-Paley theory, Ann. of Math. Studies, No. 63, Princeton, 1970.

77. St. J. Szarek, On the best constant in the Khinchin inequality, Studia Math., 58(1976), 197-208.

78. W. Szlenk, Sur les suites faiblementes convergentes dans l'espaces L , Studia Math., 25(1965), 337-341.

79. A. E. Tong, Diagonal submatrices of matrix maps, Pacific J. Math., 32(1970), 551-559.

80. L. Tzafriri, An isomorphic characterization of L_p and c_o-spaces, II, Michigan Math. J., 18(1971), 21-31.

81. J.Y.T. Woo, On a class of universal modular sequence spaces, Israel J. Math., 20(1975), 193-215.

82. M. Zippin, On perfectly homogeneous bases in Banach spaces, Israel J. Math., 4(1966), 265-272.

83. N. J. Kalton, Embedding L_1 in a Banach lattice.

84. N. J. Nielsen, On the Orlicz function spaces $L_M(0,\infty)$, Israel J. Math., 20(1975), 237-259.

W. B. Johnson: Institute for Advanced Studies of The Hebrew University of Jerusalem and The Ohio State University.

B. Maurey: Institute for Advanced Studies of The Hebrew University of Jerusalem and Ecole Polytechnique, France.

G. Schechtman: The Hebrew University of Jerusalem and The Ohio State University.

L. Tzafriri: Institute for Advanced Studies of The Hebrew University of Jerusalem and The Ohio State University.

BCDEFGHIJ–CM–8987654321